Nonlocal
Diffusion Problems

Mathematical
Surveys
and
Monographs

Volume 165

Nonlocal Diffusion Problems

Fuensanta Andreu-Vaillo
José M. Mazón
Julio D. Rossi
J. Julián Toledo-Melero

American Mathematical Society
Providence, Rhode Island

Real Sociedad Matemática Española
Madrid, Spain

2010 *Mathematics Subject Classification.* Primary 45E10, 45A05, 45G10, 47H20, 45M05, 35K05, 35K55, 35K57, 35K92.

For additional information and updates on this book, visit
www.ams.org/bookpages/surv-165

Library of Congress Cataloging-in-Publication Data

Nonlocal diffusion problems / Fuensanta Andreu-Vaillo... [et al.].
     p. cm. — (Mathematical surveys and monographs ; v. 165)
   Includes bibliographical references and index.
   ISBN 978-0-8218-5230-9 (alk. paper)
   1. Integral equations.   2. Semigroups of operators.   3. Parabolic operators.   I. Andreu-Vaillo, Fuensanta, 1955–.

QA431.N585   2010
515′.45—dc22

2010020473

To

Mayte

Nati, Alba and Hugo

and, of course, especially to

Fuensanta

Contents

Preface

The goal in this monograph is to present recent results concerning nonlocal evolution equations with different boundary conditions. We deal with existence and uniqueness of solutions and their asymptotic behaviour. We also give some results concerning limits of solutions to nonlocal equations when a rescaling parameter goes to zero. We recover in these limits some of the most frequently used diffusion models such as the heat equation, the p-Laplacian evolution equation, the porous medium equation, the total variation flow and a convection-diffusion equation. This book is based mainly on results from the papers [**14**], [**15**], [**16**], [**17**], [**68**], [**78**], [**79**], [**80**], [**120**], [**121**] and [**140**].

First, let us briefly introduce the prototype of nonlocal problems that will be considered in this monograph. Let $J : \mathbb{R}^N \to \mathbb{R}$ be a nonnegative, radial, continuous function with

$$\int_{\mathbb{R}^N} J(z) \, dz = 1.$$

Nonlocal evolution equations of the form

$$(0.1) \qquad u_t(x,t) = (J * u - u)(x,t) = \int_{\mathbb{R}^N} J(x-y)u(y,t) \, dy - u(x,t),$$

and variations of it, have been recently widely used to model diffusion processes. More precisely, as stated in [**106**], if $u(x,t)$ is thought of as a density at a point x at time t and $J(x-y)$ is thought of as the probability distribution of jumping from location y to location x, then $\int_{\mathbb{R}^N} J(y-x)u(y,t) \, dy = (J * u)(x,t)$ is the rate at which individuals are arriving at position x from all other places and $-u(x,t) = -\int_{\mathbb{R}^N} J(y-x)u(x,t) \, dy$ is the rate at which they are leaving location x to travel to all other sites. This consideration, in the absence of external or internal sources, leads immediately to the fact that the density u satisfies equation (0.1).

Equation (0.1) is called nonlocal diffusion equation since the diffusion of the density u at a point x and time t depends not only on $u(x,t)$ and its derivatives, but also on all the values of u in a neighborhood of x through the convolution term $J * u$. This equation shares many properties with the classical heat equation, $u_t = \Delta u$, such as: bounded stationary solutions are constant, a maximum principle holds for both of them and, even if J is compactly supported, perturbations propagate with infinite speed, [**106**]. However, there is no regularizing effect in general.

Let us fix a bounded domain Ω in $\mathbb{R}^N$. For local problems the two most common boundary conditions are Neumann's and Dirichlet's. When looking at boundary conditions for nonlocal problems, one has to modify the usual formulations for local problems. As an analog for nonlocal problems of Neumann boundary conditions we propose

$$(0.2) \quad \begin{cases} u_t(x,t) = \displaystyle\int_\Omega J(x-y)(u(y,t)-u(x,t))\, dy, & x \in \Omega,\ t > 0, \\ u(x,0) = u_0(x), & x \in \Omega. \end{cases}$$

In this model, the integral term takes into account the diffusion inside Ω. In fact, as we have explained, the integral $\int J(x-y)(u(y,t)-u(x,t))\, dy$ takes into account the individuals arriving at or leaving position x from other places. Since we are integrating over Ω, we are assuming that diffusion takes place only in Ω. The individuals may not enter or leave the domain. This is analogous to what is called homogeneous Neumann boundary conditions in the literature.

As the homogeneous Dirichlet boundary conditions for nonlocal problems we consider

$$(0.3) \quad \begin{cases} u_t(x,t) = \displaystyle\int_{\mathbb{R}^N} J(x-y)u(y,t)\, dy - u(x,t), & x \in \Omega,\ t > 0, \\ u(x,t) = 0, & x \notin \Omega,\ t > 0, \\ u(x,0) = u_0(x), & x \in \Omega. \end{cases}$$

In this model, diffusion takes place in the whole $\mathbb{R}^N$, but we assume that u vanishes outside Ω. In the biological interpretation, we have a hostile environment outside Ω, and any individual that jumps outside dies instantaneously. This is the analog of what is called homogeneous Dirichlet boundary conditions for the heat equation. However, the boundary datum is not understood in the usual sense, since we are not imposing that $u|_{\partial\Omega} = 0$.

The nonlocal problems of the type of (0.1), (0.2) and (0.3) have been used to model very different applied situations, for example in biology ([65], [133]), image processing ([110], [129]), particle systems ([51]), coagulation models ([108]), nonlocal anisotropic models for phase transition ([1], [2]), mathematical finances using optimal control theory ([50], [126]), etc.

We have to mention the close relation between this kind of evolution problems and probability theory. In fact, when one looks at a Levy process ([48]), the nonlocal operator that appears naturally is a fractional power of the Laplacian. This approach is out of the scope of this monograph, and we refer to [21] for a reference concerning the interplay between nonlocal partial differential equations and probability. Nevertheless, let us explain briefly why the concrete problem (0.1) has a clear probabilistic interpretation.

Let $(E, \mathcal{E})$ be a measurable space and $P : E \times \mathcal{E} \to [0, 1]$ a transition probability on E. Then we define a Markovian transition function as follows: for any $x \in E$, $\mathcal{A} \in \mathcal{E}$, let

$$P_t(x, \mathcal{A}) = e^{-t} \sum_{n=0}^{+\infty} \frac{t^n}{n!} P^{(n)}(x, \mathcal{A}), \qquad t \in \mathbb{R}_+,$$

where $P^{(n)}$ denotes the n-th iterate of P. The associated family of Markovian operators, $P_t f(x) = \int f(y) P_t(x, dy)$, satisfy

$$\frac{\partial}{\partial t} P_t f(x) = \int P_t f(y) P(x, dy) - P_t f(x).$$

If we consider a Markov process $(Z_t)_{t\geq 0}$ associated to the transition function $(P_t)_{t\geq 0}$, and if we denote by μ_t the distribution of Z_t, then the family $(\mu_t)_{t\geq 0}$ satisfies also a linear equation of the form

$$\frac{\partial}{\partial t}\mu_t = \int P(y,\cdot)\mu_t(dy) - \mu_t.$$

In particular, for $E = \mathbb{R}^N$, if the transition probability $P(x,dy)$ has a density $y \mapsto J(x,y)$, and μ_t has a density $y \mapsto u(y,t)$, then the following equation is satisfied:

$$(0.4) \qquad \frac{\partial}{\partial t}u(x,t) = \int J(x,y)u(y,t)\,d\lambda(y) - u(x,t).$$

With different particular choices of P we recover the equation studied in the Cauchy, Dirichlet and Neumann cases. For example, if $P(x,dy) = J(y-x)dy$ is the transition probability of a random walk, equation (0.4) is just equation (0.1). In this particular case, the asymptotic behaviour, described in the first chapter, can be obtained as a consequence of the so-called Local Limit Theorem for Random Walks, which is a classical result in probability theory ([104], Theorems 1 & 2]).

In the Dirichlet and Neumann cases, the results described here also give interesting information on the asymptotic behaviour of some natural Markov process in the space.

Let us now summarize the contents of this book.

The book contains two main parts. The first, which consists of Chapters 1 to 4, deals mainly with linear problems, and in this case the main tool to get existence and uniqueness of solutions is the Fourier transform for the Cauchy problem and a fixed point argument for the Dirichlet and Neumann problems. The second part, Chapters 5, 6, 7 and 8, is concerned with nonlinear problems, and here the main tool for proving existence and uniqueness is Nonlinear Semigroup Theory.

For several classical partial differential equations the solutions belong to appropriate Sobolev spaces. Hence, Poincaré type inequalities play a key role in the analysis. When considering nonlocal problems, it is natural to look for solutions in L^p spaces; however, we prove nonlocal analogs of Poincaré type inequalities that also play a fundamental role in this monograph.

Chapter 1 is devoted to the study of the Cauchy problem for a linear nonlocal operator. In this chapter we make an extensive use of the Fourier transform. We show existence and uniqueness of solutions and study their asymptotic behaviour. In addition, we prove convergence to solutions of local equations when the kernel of the nonlocal operator is rescaled in a suitable way. We also deal with nonlocal analogs of linear higher order evolution problems.

In Chapters 2 and 3 we study the analogs for linear nonlocal diffusion of the Dirichlet and Neumann problems for both the homogeneous and the nonhomogeneous case. For these nonlocal problems we find, besides existence and uniqueness, the asymptotic behaviour as well as convergence, under rescaling, to the usual boundary value problems for the heat equation.

The next chapter contains the study of a nonlocal analog of a convection-diffusion problem taking into account a nonsymmetric kernel to model the convective part of the equation.

Chapter 5 deals with the nonlocal Neumann problem for a nonlinear diffusion equation. The local counterpart serves as a model for many applications, for instance, diffusion in porous media and changes of phases (the multiphase Stefan problem and the Hele-Shaw problem). Here we use the Crandall-Liggett Theorem to prove existence and uniqueness of solutions.

In Chapter 6 we study a nonlocal analog of the p-Laplacian evolution equation for $1 < p < \infty$. We deal here with the Cauchy problem as well as Dirichlet or Neumann boundary conditions. As in the previous chapter, one of the main tools is Nonlinear Semigroup Theory. The main ingredient for the proof of convergence to the local problem is a precompactness lemma inspired by a result due to Bourgain, Brezis and Mironescu, [52].

Motivated by problems in image processing, in recent years there has been an increasing interest in the study of the Total Variation Flow, [7]. Chapter 7 is devoted to the Dirichlet and Neumann problems for the nonlocal version of this evolution. After proving existence and uniqueness of solutions, we analyze their asymptotic behaviour as well as the convergence to local problems when the kernel is rescaled.

In the last chapter we present two nonlocal versions of models for the evolution of a sandpile. The first model corresponds to a nonlocal version of the Aronsson-Evans-Wu model obtained as limit as $p \to \infty$ in the local Cauchy problem for the p-Laplacian evolution equation, and the second corresponds to the Prigozhin model. The local sandpile models are based on the requirement that the slope of sandpile is at most one. However, a more realistic model would require the slope constraint only on a larger scale, with no slope requirements on a smaller scale. This is exactly the case for the nonlocal model presented in this chapter. The main tools for the analysis here are convex analysis and accretive operators. In this chapter we also present some explicit formulae for solutions of the nonlocal sandpile models that illustrate the results.

The book ends with an appendix, in which we outline some of the main tools from Nonlinear Semigroup Theory used in the above chapters. This theory has shown to be a very useful technique to deal with nonlinear evolution equations, and it is well suited to treat nonlocal evolution problems.

The Bibliography of this monograph does not escape the usual rule of being incomplete. In general, we have listed those papers which are closer to the topics discussed here. But, even for those papers, the list is far from being exhaustive and we apologize for omissions. At the end of each chapter we have included some bibliographical notes concerning the references used in that chapter and related ones.

It is a pleasure to acknowledge here the gratitude to our coauthors, namely C. Cortazar, M. Elgueta, M. Chaves, E. Chasseigne, L. Ignat, C. Schonlieb and

N. Wolanski. This monograph could not have been written without their contribution. We would also like to thank M. Pérez-Llanos for her continuous encouragement. Finally, we want to thank B. L. G. for the spirit imbued in us at the beginning of this story.

F. Andreu, J. M. Mazón and J. Toledo have been partially supported by the Spanish MEC and FEDER, project MTM2008-03176. J. D. Rossi has been partially supported by MEC project MTM2004-02223, UBA X066 and CONICET (Argentina).

F. Andreu, J. M. Mazón, J. D. Rossi and J. Toledo
Valencia, November 2008

Fuensanta Andreu died 26 December of 2008, before this book was published. All of us, Julio, Julián and, especially, Mazón are greatly indebted to her. Thanks for such good friendship and mathematics.

CHAPTER 1

The Cauchy problem
for linear nonlocal diffusion

The aim of this chapter is to begin the study of the nonlocal evolution problems by the analysis of the asymptotic behaviour of solutions of nonlocal linear diffusion problems in the whole $\mathbb{R}^N$. First, we deal with the simplest model,

$$u_t(x,t) = \int_{\mathbb{R}^N} J(x-y)u(y,t)\,dy - u(x,t),$$

and after that we also treat a nonlocal analog of higher order problems,

$$u_t(x,t) = (-1)^{n-1}\left(J * \mathrm{Id} - 1\right)^n (u(x,t)).$$

We focus our attention on existence and uniqueness of solutions, their asymptotic behaviour as $t \to \infty$ and the convergence of solutions of these nonlocal evolution equations to solutions of classical models, such as the heat equation, when the nonlocal equation is rescaled in an appropriate way. As it happens in the study of the Cauchy problem for the heat equation, the Fourier transform will play a fundamental role, allowing us to obtain an explicit formula for the solution to the nonlocal equation in Fourier variables.

1.1. The Cauchy problem

We consider the linear nonlocal diffusion problem presented in the Preface,

(1.1)
$$\begin{cases} u_t(x,t) = J * u(x,t) - u(x,t) = \displaystyle\int_{\mathbb{R}^N} J(x-y)u(y,t)\,dy - u(x,t), \\[2mm] u(x,0) = u_0(x), \end{cases}$$

for $x \in \mathbb{R}^N$ and $t > 0$. Here J satisfies the following hypothesis, which will be assumed throughout this chapter:

(H) $J \in C(\mathbb{R}^N, \mathbb{R})$ is a nonnegative radial function with $J(0) > 0$ and

$$\int_{\mathbb{R}^N} J(x)\,dx = 1.$$

This means that J is a radial probability density.

As we have mentioned in the Preface, this equation has been used to model diffusion processes. More precisely (see [106]), if $u(x,t)$ is thought of as a density at a point x at time t and $J(x-y)$ is thought of as the probability distribution of jumping from location y to location x, then $\int_{\mathbb{R}^N} J(y-x)u(y,t)\,dy = (J * u)(x,t)$ is the rate at which individuals are arriving at position x from all other places, and $-u(x,t) = -\int_{\mathbb{R}^N} J(y-x)u(x,t)\,dy$ is the rate at which they are leaving location x

to travel to all other sites. This consideration, in the absence of external or internal sources, leads immediately to the fact that the density u satisfies equation (1.1).

A solution of (1.1) is understood as a function $u \in C^0([0, +\infty); L^1(\mathbb{R}^N))$ that satisfies (1.1) in the integral sense; that is, u satisfies

$$u(x,t) = u_0(x) + \int_0^t \int_{\mathbb{R}^N} J(x-y)u(y,s)\,dy - u(x,s)\,ds.$$

This definition of solution is quite natural since every term in the equation is well defined. In Theorem 1.4, it is shown that existence and uniqueness hold for this kind of solutions. As we are dealing with nonlocal diffusion problems, searching for a solution in some Lebesgue space seems the appropriate thing to do.

We shall make an extensive use of the Fourier transform in order to obtain explicit solutions in frequency formulation. Moreover, the main result in this section states that the decay rate as t goes to infinity of solutions of this nonlocal problem is determined by the behaviour of the Fourier transform of J near the origin, and the asymptotic decays are the same as the ones that hold for solutions of the evolution problem with right hand side given by a power of the Laplacian. Therefore, we begin with some preliminaries concerning the Fourier transform. We assume that the reader is familiar with them and hence we refer to [**113**] or [**148**] for details.

In the sequel, $\hat{f}$ denotes the Fourier transform of f, which is given by the following definition.

DEFINITION 1.1. For $f \in L^1(\mathbb{R}^N)$, the *Fourier transform* of f is given by

$$\hat{f}(\xi) = \int_{\mathbb{R}^N} e^{-i\langle x,\xi \rangle} f(x)\,dx,$$

where $\langle \cdot, \cdot \rangle$ denotes the scalar product in $\mathbb{R}^N$. And the *inverse Fourier transform* of f is given by

$$\check{f}(x) = \frac{1}{(2\pi)^N} \int_{\mathbb{R}^N} e^{i\langle x,\xi \rangle} f(\xi)\,d\xi.$$

We set $\mathcal{F}: L^1(\mathbb{R}^N) \to L^\infty(\mathbb{R}^N)$ to be the Fourier transformation, i.e., the linear bounded operator defined by $\mathcal{F}(f) := \hat{f}$.

By $\mathcal{S}(\mathbb{R}^N)$ we denote the space of rapidly decreasing functions, that is, the set of all $\phi \in C^\infty(\mathbb{R}^N)$ such that

$$\sup |x^\beta \partial^\alpha \phi(x)| < \infty$$

for all multi-indices $\alpha = (\alpha_1, \ldots, \alpha_N) \in \mathbb{N}^N$ and $\beta = (\beta_1, \ldots, \beta_N) \in \mathbb{N}^N$, with the usual multi-index notation $|\alpha| = \sum_{i=1}^N \alpha_i$, $\alpha! = \alpha_1! \cdots \alpha_N!$, $x^\beta = x_1^{\beta_1} \cdots x_N^{\beta_N}$ and $\partial^\alpha \phi = \frac{\partial^{|\alpha|}\phi}{\partial^{\alpha_1} x_1 \cdots \partial^{\alpha_N} x_N}$.

In the following proposition we list some of the main properties of the Fourier transform.

PROPOSITION 1.2.

(1) *For* $f \in L^1(\mathbb{R}^N)$, $\hat{f}$ *and* $\check{f}$ *are bounded and continuous. Moreover,*

$$\lim_{|\xi| \to \infty} \hat{f}(\xi) = 0 \quad and \quad \lim_{|x| \to \infty} \check{f}(x) = 0.$$

(2) $\|\hat{f}\|_{L^\infty(\mathbb{R}^N)} \le \|f\|_{L^1(\mathbb{R}^N)}$.

(3) *If* $f \in C^k(\mathbb{R}^N)$ *with* $D^\beta f \in L^1(\mathbb{R}^N)$ *for all* β, $|\beta| \le k$, *then* $\xi^\beta \hat{f} \in L^\infty(\mathbb{R}^N)$ *and* $\widehat{D^\beta f} = i^{|\beta|}\xi^\beta \hat{f}$.

(4) *If* $x^\alpha f \in L^1(\mathbb{R}^N)$ *for all* α, $|\alpha| \le k$, *then* $\hat{f} \in C^k(\mathbb{R}^N)$ *and* $D^\alpha \hat{f} = (-i)^{|\alpha|}\widehat{x^\alpha f}$.

(5) $\widehat{f * g} = \hat{f} \cdot \hat{g}$.

(6) *The Fourier transform operator* $\mathcal{F}$ *is an isomorphism from* $\mathcal{S}(\mathbb{R}^N)$ *onto* $\mathcal{S}(\mathbb{R}^N)$, *and its inverse is given by the inversion formula*

(1.2) $$f(x) = \mathcal{F}^{-1}(\hat{f})(x) = \frac{1}{(2\pi)^N} \int_{\mathbb{R}^N} e^{i\langle x,\xi\rangle}\hat{f}(\xi)\,d\xi.$$

(7) *For* $f \in L^1(\mathbb{R}^N)$ *such that* $\hat{f} \in L^1(\mathbb{R}^N)$, *the inversion formula* (1.2) *holds for almost every* $x \in \mathbb{R}^N$.

(8) *(Plancherel) There is an isomorphism* $\mathcal{P} : L^2(\mathbb{R}^N) \to L^2(\mathbb{R}^N)$ *such that* $\mathcal{P}(f) = \hat{f}$ *for all* $f \in \mathcal{S}(\mathbb{R}^N)$. *Moreover, the following identity holds:*

$$\|\mathcal{P}(f)\|_{L^2(\mathbb{R}^N)} = (2\pi)^{\frac{N}{2}}\|f\|_{L^2(\mathbb{R}^N)}, \quad \forall f \in L^2(\mathbb{R}^N).$$

(9) *(Hausdorff-Young) If* $1 \le p \le 2$, *there is a linear bounded operator* $\mathcal{F}_p : L^p(\mathbb{R}^N) \to L^{p'}(\mathbb{R}^N)$, $\frac{1}{p} + \frac{1}{p'} = 1$, *such that* $\mathcal{F}_p(f) = \hat{f}$ *for all* $f \in \mathcal{S}(\mathbb{R}^N)$. *Moreover, the following inequality holds:*

$$\|\mathcal{F}_p(f)\|_{L^{p'}(\mathbb{R}^N)} \le (2\pi)^{\frac{N}{p'}}\|f\|_{L^p(\mathbb{R}^N)}, \quad \forall f \in L^p(\mathbb{R}^N).$$

Throughout this chapter we denote by G_A^s, $A > 0$, the inverse Fourier transform of $e^{-A|\xi|^s}$, that is,

(1.3) $$\widehat{G_A^s}(\xi) = e^{-A|\xi|^s}.$$

The fractional Laplacian of a function $f : \mathbb{R}^N \to \mathbb{R}$ is expressed by the formula

$$(-\Delta)^{\frac{s}{2}}f(x) := C_{N,s}\int_{\mathbb{R}^N}\frac{f(x) - f(y)}{|x - y|^{N+s}}\,dy,$$

where the parameter s is a real number $0 < s \le 2$, and $C_{N,s}$ is a normalization constant given by

$$C_{N,s} = \frac{\Gamma(\frac{N+s}{2})}{2^{-s}\pi^{\frac{N}{2}}|\Gamma(-\frac{s}{2})|}.$$

It can also be defined as a pseudo-differential operator,

$$\mathcal{F}\left((-\Delta)^{\frac{s}{2}}f\right)(\xi) = |\xi|^s\hat{f}(\xi).$$

The fractional Laplacian can be defined in a distributional sense for functions that are not differentiable, as long as f is not too singular at the origin or, in terms of the x variable, as long as

$$\int_{\mathbb{R}^N}\frac{f(x)}{(1 + |x|)^{N+s}}\,dx < \infty.$$

For references concerning the fractional Laplacian see for instance [130] or [147].

We will use the following notation throughout this monograph:

$$g(\xi) = h(\xi) + o(|\xi|^s) \qquad \text{as } \xi \to 0$$

means

$$\lim_{\xi \to 0} \frac{g(\xi) - h(\xi)}{|\xi|^s} = 0.$$

The hypothesis (H) for J immediately implies that

$$|\hat{J}(\xi)| \leq 1 \qquad \text{and} \qquad \hat{J}(0) = 1.$$

The main result of this section reads as follows.

THEOREM 1.3. *Assume there exist $A > 0$ and $0 < s \leq 2$ such that*

(1.4) $$\hat{J}(\xi) = 1 - A|\xi|^s + o(|\xi|^s) \quad \text{as } \xi \to 0.$$

For any nonnegative u_0 such that $u_0, \widehat{u_0} \in L^1(\mathbb{R}^N)$, there exists a unique solution $u(x,t)$ of (1.1).

The asymptotic behaviour of $u(x,t)$ is given by

(1.5) $$\lim_{t \to +\infty} t^{\frac{N}{s}} \max_x |u(x,t) - v(x,t)| = 0,$$

where v is the solution of

$$v_t(x,t) = -A(-\Delta)^{\frac{s}{2}} v(x,t)$$

with initial condition $v(x,0) = u_0(x)$. Moreover,

$$\|u(\cdot,t)\|_{L^\infty(\mathbb{R}^N)} \leq C \, t^{-\frac{N}{s}},$$

and the asymptotic profile is given by

$$\lim_{t \to +\infty} \max_y \left| t^{\frac{N}{s}} u(yt^{\frac{1}{s}}, t) - \|u_0\|_{L^1} G_A^s(y) \right| = 0.$$

The assumption $u_0 \geq 0$ can be easily removed, but it is adopted here for simplicity.

Condition (1.4) can be reduced to the existence of $A, s > 0$ such that

$$\hat{J}(\xi) = 1 - A|\xi|^s + o(|\xi|^s), \qquad \xi \to 0,$$

since (H) implies $s \leq 2$ (see Lemma 1.8 below).

In the special case $s = 2$, the decay rate is $t^{-\frac{N}{2}}$ and the asymptotic profile is a Gaussian

$$G_A^2(y) = (4\pi A)^{-\frac{N}{2}} e^{-\frac{|y|^2}{4A}},$$

where $A \cdot \mathrm{Id} = -\frac{1}{2} D^2 \hat{J}(0)$. Note that in this case (which occurs, for example, when J is compactly supported) the asymptotic behaviour is the same as the one that holds for the solutions of the heat equation and, as it happens for the heat equation, the asymptotic profile is a Gaussian.

The decay of the solutions in L^∞ together with the conservation of mass (that holds trivially for solutions to the nonlocal evolution problem (1.1)) give the decay in the L^p-norms by interpolation. As a consequence of the previous theorem, we find that this decay is analogous to the decay of the evolution given by the fractional Laplacian, that is,

$$\|u(\cdot, t)\|_{L^p(\mathbb{R}^N)} \leq C t^{-\frac{N}{s}\left(1 - \frac{1}{p}\right)};$$

see Corollary 1.11. We refer to [**74**] for the decay in the L^p-norms for the fractional Laplacian; see also [**66**], [**96**] and [**98**] for finer decay estimates in the L^p-norms for solutions of the heat equation.

1.1.1. Existence and uniqueness. Let us begin by proving existence and uniqueness of solutions using Fourier variables.

THEOREM 1.4. *Let* $u_0 \in L^1(\mathbb{R}^N)$ *such that* $\widehat{u_0} \in L^1(\mathbb{R}^N)$. *There exists a unique solution* $u \in C^0([0, \infty); L^1(\mathbb{R}^N))$ *of* (1.1), *and it is given by*

$$\hat{u}(\xi, t) = e^{(\hat{J}(\xi) - 1)t} \widehat{u_0}(\xi).$$

PROOF. We have, formally,

$$u_t(x, t) = \int_{\mathbb{R}^N} J(x - y) u(y, t) \, dy - u(x, t) = J * u - u(x, t).$$

Applying the Fourier transform to this equation we obtain

$$\hat{u}_t(\xi, t) = \hat{u}(\xi, t)(\hat{J}(\xi) - 1).$$

Consequently,

$$\hat{u}(\xi, t) = e^{(\hat{J}(\xi) - 1)t} \widehat{u_0}(\xi).$$

Since $\widehat{u_0} \in L^1(\mathbb{R}^N)$ and $e^{(\hat{J}(\xi) - 1)t}$ is continuous and bounded, the result follows by taking the inverse Fourier transform. $\square$

REMARK 1.5. One can also understand solutions of (1.1) directly in Fourier variables. This concept of solution is equivalent to the integral solution in the original variables under our hypotheses on the initial condition.

Now we prove a lemma concerning the fundamental solution of (1.1), that is, the solution of (1.1) with initial condition $u_0 = \delta_0$, the Dirac measure at zero.

LEMMA 1.6. *Let* $J \in \mathcal{S}(\mathbb{R}^N)$. *The fundamental solution of* (1.1) *can be decomposed as*

(1.6) $$w(x, t) = e^{-t} \delta_0(x) + K_t(x),$$

where $K_t(x) = K(x, t)$ *is a smooth function defined in Fourier variables by*

$$\widehat{K_t}(\xi) = e^{-t}(e^{\hat{J}(\xi)t} - 1).$$

Moreover, if u *is a solution of* (1.1) *with* $u_0 \in L^1(\mathbb{R}^N)$, *it can be written as*

$$u(x, t) = (w * u_0)(x, t) = \int_{\mathbb{R}^N} w(x - z, t) u_0(z) \, dz.$$

PROOF. As in the previous result we have

$$\hat{w}_t(\xi, t) = \hat{w}(\xi, t)(\hat{J}(\xi) - 1).$$

Hence, as the initial datum satisfies $\hat{\delta}_0 = 1$,

$$\hat{w}(\xi, t) = e^{(\hat{J}(\xi) - 1)t} = e^{-t} + e^{-t}(e^{\hat{J}(\xi)t} - 1).$$

The first part of the lemma is proved by applying the inverse Fourier transform in $\mathcal{S}(\mathbb{R}^N)$.

To finish the proof we observe that $w * u_0$ is a solution of (1.1) (just use Fubini's theorem) with $(w * u_0)(x, 0) = u_0(x)$.   □

REMARK 1.7. The above proof together with the fact that $\hat{J}(\xi) \to 0$ as $|\xi| \to +\infty$ (since $J \in L^1(\mathbb{R}^N)$) shows that if $\hat{J} \in L^1(\mathbb{R}^N)$, then the same decomposition (1.6) holds and the result also applies.

1.1.2. Asymptotic behaviour. We begin by collecting some properties of the function J.

LEMMA 1.8. *Under the hypothesis* (H) *for* J, *we have*

i) $|\hat{J}(\xi)| \leq 1$, $\hat{J}(0) = 1$.

ii) *If* $\int_{\mathbb{R}^N} J(x)|x|\, dx < +\infty$, *then*

$$\frac{\partial \hat{J}}{\partial \xi_i}(0) = \left(\nabla_\xi \hat{J}\right)_i(0) = -i \int_{\mathbb{R}^N} x_i J(x)\, dx = 0,$$

and if $\int_{\mathbb{R}^N} J(x)|x|^2\, dx < +\infty$,

$$\left(D^2 \hat{J}\right)_{ij}(0) = -\int_{\mathbb{R}^N} x_i x_j J(x)\, dx;$$

therefore $\left(D^2 \hat{J}\right)_{ij}(0) = 0$ *when* $i \neq j$ *and* $\left(D^2 \hat{J}\right)_{ii}(0) \neq 0$. *Hence the Hessian matrix of* $\hat{J}$ *at the origin is given by*

$$D^2 \hat{J}(0) = -\left(\frac{1}{N} \int_{\mathbb{R}^N} |x|^2 J(x)\, dx\right) \cdot \mathrm{Id}.$$

iii) *If* $\hat{J}(\xi) = 1 - A|\xi|^s + o(|\xi|^s)$ *as* $\xi \to 0$, *then necessarily* $s \in (0, 2]$, *and if* J *has a first momentum, then* $s \neq 1$. *Finally, if* $s = 2$,

$$A \cdot \mathrm{Id} = -\frac{1}{2} D^2 \hat{J}(0).$$

iv) *If* $\hat{J}(\xi) = 1 - A|\xi|^s + o(|\xi|^s)$ *as* $\xi \to 0$, *then*

(1.7) $$|\hat{J}(\xi) - 1 + A|\xi|^s| \leq |\xi|^s h(\xi),$$

where $h \geq 0$ *is bounded and* $h(\xi) \to 0$ *as* $\xi \to 0$. *Moreover, given* $D > 0$ *there exist* $a > 0$ *and* $0 < \delta < 1$ *such that*

(1.8) $$|\hat{J}(\xi) - 1 + A|\xi|^s| \leq D|\xi|^s \qquad \text{for } |\xi| \leq a,$$

and

(1.9) $$|\hat{J}(\xi)| \leq 1 - \delta \qquad \text{for } |\xi| \geq a.$$

PROOF. Points i) and ii) are rather straightforward (recall that J is a radial probability density). Now we turn to iii). Observe that if $\hat{J}$ has an expansion of the form

$$\hat{J}(\xi) = 1 + i\langle \mathbf{a}, \xi \rangle - \frac{1}{2}\langle \xi, B\xi \rangle + o(|\xi|^2),$$

where $\mathbf{a} = (a_1, \ldots, a_N)$ and $B = (B_{ij})_{i,j=1,\ldots,N}$, then J has a second momentum and

$$a_i = \int x_i J(x) dx, \qquad B_{ij} = \int x_i x_j J(x) dx < \infty.$$

Thus if iii) held for some $s > 2$, it would turn out that the second moment of J is null, which would imply that $J \equiv 0$, a contradiction. When $s = 2$, since by (ii) $B_{ij} = -\left(D^2 \hat{J}\right)_{ij}(0)$, the Hessian is diagonal. Finally, (1.7) is evident and implies the existence of $a > 0$ satisfying (1.8). Once a is fixed, on account that J is radial and $|\hat{J}(\xi)| \leq 1$, there exists $\delta > 0$ such that (1.9) holds. □

Observe that $\hat{J}(\xi)$ is real for $\xi \in \mathbb{R}^N$ due to the symmetry of J.

Next, we prove Theorem 1.3. Throughout this chapter we denote by C any constant independent of the relevant quantities that may vary from line to line.

PROOF OF THEOREM 1.3. By Theorem 1.4 we have

$$\hat{u}(\xi, t) = e^{(\hat{J}(\xi)-1)t} \widehat{u_0}(\xi)$$

and

$$\hat{u}_t(\xi, t) = \hat{u}(\xi, t)(\hat{J}(\xi) - 1).$$

On the other hand, let $v(x, t)$ be a solution of

$$v_t(x, t) = -A(-\Delta)^{s/2} v(x, t),$$

with the same initial datum $v(x, 0) = u_0(x)$. Solutions of this equation are understood in the sense that

(1.10) $$\hat{v}(\xi, t) = e^{-A|\xi|^s t} \widehat{u_0}(\xi);$$

see [21]. Hence in Fourier variables,

$$\int_{\mathbb{R}^N} |\hat{u} - \hat{v}|(\xi, t)\, d\xi = \int_{\mathbb{R}^N} \left| \left(e^{t(\hat{J}(\xi)-1)} - e^{-A|\xi|^s t} \right) \widehat{u_0}(\xi) \right| d\xi$$

$$\leq \int_{|\xi| \geq r(t)} \left| \left(e^{t(\hat{J}(\xi)-1)} - e^{-A|\xi|^s t} \right) \widehat{u_0}(\xi) \right| d\xi$$

$$+ \int_{|\xi| < r(t)} \left| \left(e^{t(\hat{J}(\xi)-1)} - e^{-A|\xi|^s t} \right) \widehat{u_0}(\xi) \right| d\xi = [I] + [II],$$

where $[I]$ and $[II]$ denote the first and second integral, respectively, in the left hand side of the last identity, and $r(t)$, with $r(t) \to 0$, will be determined later.

To get a bound for $[I]$ we proceed as follows. We decompose it into two parts,

$$t^{\frac{N}{s}} [I] \leq t^{\frac{N}{s}} \int_{|\xi| \geq r(t)} \left| e^{-A|\xi|^s t} \widehat{u_0}(\xi) \right| d\xi + t^{\frac{N}{s}} \int_{|\xi| \geq r(t)} \left| e^{t(\hat{J}(\xi)-1)} \widehat{u_0}(\xi) \right| d\xi.$$

First, we deal with the first term on the right hand side in the above expression. Changing variables, $\eta = \xi t^{1/s}$, we have

$$t^{\frac{N}{s}} \int_{|\xi| > r(t)} e^{-A|\xi|^s t} |\widehat{u_0}(\xi)| d\xi \leq \|\widehat{u_0}\|_{L^\infty(\mathbb{R}^N)} \int_{|\eta| > r(t) t^{1/s}} e^{-A|\eta|^s} \to 0$$

as $t \to \infty$ if we impose that

(1.11) $r(t)t^{\frac{1}{s}} \to \infty$ as $t \to \infty$.

Let us now deal with the second term. Let a and δ be as in Lemma 1.8 iv) for $D = \frac{A}{2}$. By (1.9),

$$t^{\frac{N}{s}} \int_{|\xi| \geq r(t)} \left| e^{t(\hat{J}(\xi)-1)} \widehat{u_0}(\xi) \right| d\xi$$

$$\leq t^{\frac{N}{s}} \int_{a \geq |\xi| \geq r(t)} \left| e^{t(\hat{J}(\xi)-1)} \widehat{u_0}(\xi) \right| d\xi + t^{\frac{N}{s}} \int_{|\xi| \geq a} \left| e^{t(\hat{J}(\xi)-1)} \widehat{u_0}(\xi) \right| d\xi$$

$$= t^{\frac{N}{s}} \int_{a \geq |\xi| \geq r(t)} \left| e^{t(\hat{J}(\xi)-1)} \widehat{u_0}(\xi) \right| d\xi + t^{\frac{N}{s}} e^{-t} \int_{|\xi| \geq a} e^{t|\hat{J}(\xi)|} |\widehat{u_0}(\xi)| d\xi$$

$$\leq t^{\frac{N}{s}} \int_{a \geq |\xi| \geq r(t)} \left| e^{t(\hat{J}(\xi)-1)} \widehat{u_0}(\xi) \right| d\xi + C t^{\frac{N}{s}} e^{-\delta t}.$$

Using the above inequality, (1.8) and changing variables, $\eta = \xi t^{1/s}$,

$$t^{\frac{N}{s}} \int_{|\xi| \geq r(t)} \left| e^{t(\hat{J}(\xi)-1)} \widehat{u_0}(\xi) \right| d\xi$$

$$\leq t^{\frac{N}{s}} \int_{a \geq |\xi| \geq r(t)} e^{-tA|\xi|^s} e^{t|\hat{J}(\xi)-1+A|\xi|^s|} |\widehat{u_0}(\xi)| d\xi + C t^{\frac{N}{s}} e^{-\delta t}$$

$$\leq t^{\frac{N}{s}} \int_{a \geq |\xi| \geq r(t)} e^{-t\frac{A}{2}|\xi|^s} |\widehat{u_0}(\xi)| d\xi + C t^{\frac{N}{s}} e^{-\delta t}$$

$$\leq \int_{at^{1/s} \geq |\eta| \geq t^{1/s} r(t)} e^{-\frac{A}{2}|\eta|^s} \left| \widehat{u_0}(\eta t^{-1/s}) \right| d\eta + C t^{\frac{N}{s}} e^{-\delta t}$$

$$\leq C \int_{|\eta| \geq t^{1/s} r(t)} e^{-\frac{A}{2}|\eta|^s} d\eta + C t^{\frac{N}{s}} e^{-\delta t} \quad \to 0$$

as $t \to \infty$ if (1.11) holds.

Now we estimate $[II]$ as follows: by (1.7) and the elementary inequality

$$|e^y - 1| \leq C|y|$$

for $|y|$ bounded,

$$t^{\frac{N}{s}} [II] = t^{\frac{N}{s}} \int_{|\xi| < r(t)} \left| e^{(\hat{J}(\xi)-1+A|\xi|^s)t} - 1 \right| e^{-A|\xi|^s t} |\widehat{u_0}(\xi)| d\xi$$

$$\leq C t^{\frac{N}{s}} \int_{|\xi| < r(t)} t|\xi|^s h(\xi) e^{-A|\xi|^s t} d\xi,$$

provided we impose

(1.12) $t \, (r(t))^s h(r(t)) \to 0$ as $t \to \infty$.

In this case, changing variables $\eta = \xi t^{1/s}$ again, we have

$$t^{\frac{N}{s}} [II] \leq C \int_{|\eta| < r(t)t^{1/s}} |\eta|^s h(\eta/t^{1/s}) e^{-A|\eta|^s} d\eta.$$

Since the integrand is dominated by $\|h\|_\infty |\eta|^s \exp(-c|\eta|^s)$, a function which belongs to $L^1(\mathbb{R}^N)$, and $h(\eta/t^{1/s}) \to 0$ as $t \to \infty$, using the Dominated Convergence Theorem, we obtain

$$\lim_{t \to +\infty} t^{\frac{N}{s}}[II] = 0.$$

This shows that

(1.13) $$t^{\frac{N}{s}}([I] + [II]) \to 0 \quad \text{as } t \to \infty,$$

provided we can find a $r(t) \to 0$ as $t \to \infty$ which fulfils both conditions (1.11) and (1.12). This is proved in Lemma 1.9, which is postponed to just after the end of the present proof. To conclude (1.5), we only have to observe that from (1.13) we get

$$t^{\frac{N}{s}} \max_x |u(x,t) - v(x,t)| \le (2\pi)^{-N} t^{\frac{N}{s}} \int_{\mathbb{R}^N} |\hat{u} - \hat{v}|(\xi,t)\, d\xi \to 0 \quad \text{as } t \to \infty.$$

From the above, we obtain that the asymptotic behaviour is the same as the one for solutions of the evolution given by the fractional Laplacian. Now, it is easy to check that this asymptotic behaviour is exactly the one described in the statement. Indeed, in Fourier variables (see (1.10)) we have for $t \to \infty$

$$\hat{v}(t^{-\frac{1}{s}}\eta, t) = e^{-A|\eta|^s} \widehat{u_0}(\eta t^{-\frac{1}{s}}) \longrightarrow e^{-A|\eta|^s} \widehat{u_0}(0) = \widehat{G_A^s}(\eta) \|u_0\|_{L^1(\mathbb{R}^N)}.$$

Therefore

$$\lim_{t \to +\infty} \max_y \left| t^{\frac{N}{s}} v(yt^{\frac{1}{s}}, t) - \|u_0\|_{L^1} G_A^s(y) \right| = 0. \qquad \square$$

The following lemma shows that there exists a function $r(t)$ satisfying (1.11) and (1.12), as required in the proof of the previous theorem.

LEMMA 1.9. *Given a function $h \in C(\mathbb{R}, \mathbb{R})$ such that $h(\rho) \to 0$ as $\rho \to 0$ with $h(\rho) > 0$ for small ρ, there exists a function r with $r(t) \to 0$ as $t \to \infty$ which satisfies*

$$\lim_{t \to \infty} r(t) t^{\frac{1}{s}} = \infty$$

and

$$\lim_{t \to \infty} t(r(t))^s h(r(t)) = 0.$$

PROOF. For fixed t large enough, we choose $r(t)$ as a small solution of

(1.14) $$r(h(r))^{\frac{1}{2s}} = t^{-\frac{1}{s}}.$$

This equation defines a function $r = r(t)$ which, by continuity arguments, goes to zero as t goes to infinity. Indeed, if there exists $t_n \to \infty$ with no solution of (1.14) for $r \in (0, \delta)$, then $h(r) \equiv 0$ in $(0, \delta)$, a contradiction. $\qquad \square$

REMARK 1.10. In the case $h(t) = t^\alpha$ with $\alpha > 0$, we can look for a function r of power type, $r(t) = t^\beta$ with $\beta < 0$, and the two conditions read as follows:

(1.15) $$\beta + 1/s > 0, \quad 1 + \beta s + \alpha\beta < 0.$$

This implies that $\beta \in (-1/s, -1/(\alpha + s))$, which is, of course, always possible.

Now we find the decay rate in L^p of solutions of (1.1).

COROLLARY 1.11. *Let* $1 < p < \infty$. *Under the hypotheses in Theorem 1.3, the decay in the* L^p-*norm of the solution of* (1.1) *is given by*

$$\|u(\cdot, t)\|_{L^p(\mathbb{R}^N)} \leq C t^{-\frac{N}{s}\left(1 - \frac{1}{p}\right)}.$$

PROOF. By interpolation (see [56]) we have

$$\|u\|_{L^p(\mathbb{R}^N)} \leq \|u\|_{L^1(\mathbb{R}^N)}^{\frac{1}{p}} \|u\|_{L^\infty(\mathbb{R}^N)}^{1 - \frac{1}{p}}.$$

Integrating and using Fubini's Theorem it is easy to see that solutions of (1.1) preserve the total mass and consequently the L^1-norm. Hence, the result follows from the above inequalities and the previous results that give the decay in L^∞ of the solutions. $\qquad\square$

1.2. Refined asymptotics

The goal now is to get refined asymptotic expansions for the solution u of the nonlocal evolution problem (1.1).

For the heat equation a precise asymptotic expansion in terms of the fundamental solution and its derivatives was found in [96]. In fact, for the fundamental solution v of the heat equation, under adequate assumptions on the initial condition, we have

$$(1.16) \qquad \left\| v(x, t) - \sum_{|\alpha| \leq k+1} \frac{(-1)^{|\alpha|}}{\alpha!} \left(\int_{\mathbb{R}^N} u_0(x) x^\alpha \right) \partial^\alpha G_t^2 \right\|_{L^q(\mathbb{R}^N)} \leq C t^{-\theta},$$

where

$$\theta = \frac{N}{2}\left(\frac{k+1}{N} + 1 - \frac{1}{q}\right) = \frac{k+1}{2} + \frac{N}{2}\left(1 - \frac{1}{q}\right),$$

and G_t^2 is given in (1.3).

As pointed out by the authors in [96], the same asymptotic expansion can be done in a more general setting, dealing with the equations $v_t = -(-\Delta)^{\frac{s}{2}} v$, $s > 0$.

The main objective here is to study if an expansion analogous to (1.16) holds for the nonlocal problem (1.1). We find a complete expansion for $u(x, t)$ in terms of the derivatives of the regular part of the fundamental solution K_t given in Lemma 1.6.

Concerning the first term, it has been shown that (see Theorem 1.3) if J satisfies $\hat{J}(\xi) = 1 - |\xi|^s + o(|\xi|^s)$ as $\xi \to 0$, then the asymptotic behaviour of the solution $u(x, t)$ of (1.1) is given by

$$\lim_{t \to +\infty} t^{\frac{N}{s}} \max_x |u(x, t) - v(x, t)| = 0,$$

where v is the solution of $v_t(x, t) = -(-\Delta)^{\frac{s}{2}} v(x, t)$ with initial condition $v(x, 0) = u_0(x)$. As a consequence, the decay rate is given by $\|u(\cdot, t)\|_{L^\infty(\mathbb{R}^N)} \leq C t^{-\frac{N}{s}}$ and the asymptotic profile is as follows:

$$\lim_{t \to +\infty} \left\| t^{\frac{N}{s}} u(y t^{\frac{1}{s}}, t) - \left(\int_{\mathbb{R}^N} u_0 \right) G_1^s(y) \right\|_{L^\infty(\mathbb{R}^N)} = 0.$$

In contrast with the analysis done in the previous section, where the long time behaviour is studied in the $L^\infty(\mathbb{R}^N)$-norm, here we also consider $L^q(\mathbb{R}^N)$-norms for $q \geq 1$. In the sequel we denote by $L^1(\mathbb{R}^N, a(x))$ the following weighted space:

$$L^1(\mathbb{R}^N, a(x)) = \left\{ \varphi : \mathbb{R}^N \to \mathbb{R} \text{ measurable } : \int_{\mathbb{R}^N} a(x)|\varphi(x)|dx < \infty \right\}.$$

THEOREM 1.12. *Let $0 < s \leq 2$, $m > 0$ and $c > 0$ satisfy*

$$(1.17) \qquad \widehat{J}(\xi) = 1 - |\xi|^s + o(|\xi|^s) \quad as\ \xi \to 0$$

and

$$(1.18) \qquad |\widehat{J}(\xi)| \leq \frac{c}{|\xi|^m} \quad as\ |\xi| \to \infty.$$

Then, for any $2 \leq q \leq \infty$ and $k + 1 < m - N$, there exists a constant

$$C(q, k, u_0) = C(q, k)\||x|^{k+1}u_0\|_{L^1(\mathbb{R}^N)}$$

such that

$$(1.19) \qquad \left\| u(x,t) - \sum_{|\alpha| \leq k+1} \frac{(-1)^{|\alpha|}}{\alpha!} \left(\int_{\mathbb{R}^N} u_0(x)x^\alpha \right) \partial^\alpha K_t \right\|_{L^q(\mathbb{R}^N)} \leq C(q, k, u_0)t^{-\theta}$$

for all $u_0 \in L^1(\mathbb{R}^N, 1 + |x|^{k+1})$, where

$$\theta = \frac{k+1}{s} + \frac{N}{s}\left(1 - \frac{1}{q}\right).$$

REMARK 1.13. The condition $k + 1 < m - N$ guarantees that all the partial derivatives $\partial^\alpha K_t$ of order $|\alpha| = k+1$ make sense. In addition if $\widehat{J}$ decays at infinity faster than any polynomial,

$$(1.20) \qquad \forall m \in \mathbb{N} \setminus \{0\}, \ \exists\, c(m) \text{ such that } |\widehat{J}(\xi)| \leq \frac{c(m)}{|\xi|^m}, \quad |\xi| \to \infty,$$

then the expansion (1.19) holds for all k.

To deal with L^q-norms for $1 \leq q < 2$ more restrictive assumptions have to be imposed. In the sequel $[s]$ stands for the floor function of s.

THEOREM 1.14. *Let $N \leq 3$. Assume J satisfies (1.17) with $[s] > N/2$ and that for any $m \geq 0$ and any index α there exists $C(m, \alpha)$ such that*

$$(1.21) \qquad |\partial^\alpha \widehat{J}(\xi)| \leq \frac{C(m, \alpha)}{|\xi|^m}, \qquad |\xi| \to \infty.$$

Then for any $1 \leq q < 2$, the asymptotic expansion (1.19) holds for any k.

REMARK 1.15. Recall that (1.17) implies $0 < s \leq 2$, hence the hypothesis $[s] > N/2$ implies that if $N = 1$ then $1 \leq s \leq 2$, and if $N = 2$ or $N = 3$ then $s = 2$ in the previous theorem.

1.2.1. Estimates on the regular part of the fundamental solution.
To prove the previous results we need some estimates on the kernel K_t and its
derivatives for large t. The strategy is as follows: for $2 \leq q \leq \infty$ the behaviour in
the $L^q(\mathbb{R}^N)$-norms follows from the estimates in the L^∞-norm and the L^2-norm,
for which we use Plancherel's identity. The case $1 \leq q < 2$ is more tricky. In order
to evaluate the $L^1(\mathbb{R}^N)$-norm of K_t we use the Carlson type inequality (see for
instance [31], [54])

$$(1.22) \qquad \|f\|_{L^1(\mathbb{R}^N)} \leq \|f\|_{L^2(\mathbb{R}^N)}^{1-\frac{N}{2n}} \||x|^n f\|_{L^2(\mathbb{R}^N)}^{\frac{N}{2n}},$$

which holds for $n > N/2$. The use of the above inequality with $f = \partial^\alpha K_t$ imposes
that $|x|^n \partial^\alpha K_t$ belongs to $L^2(\mathbb{R}^N)$. To guarantee that property and to obtain the
decay rate in the $L^2(\mathbb{R}^N)$-norm of $|x|^n \partial^\alpha K_t$, the additional hypothesis of Theo-
rem 1.14 will be imposed in Lemma 1.17.

The following lemma gives us the decay rate in the $L^q(\mathbb{R}^N)$-norms of the kernel
K_t and its derivatives for $2 \leq q \leq \infty$.

LEMMA 1.16. *Let $2 \leq q \leq \infty$ and J satisfy (1.17) and (1.18). Then for all
multi-indices α such that $|\alpha| < m - N$ there exists a constant $C(q, \alpha)$ such that*

$$\|\partial^\alpha K_t\|_{L^q(\mathbb{R}^N)} \leq C(q, \alpha)\, t^{-\frac{N}{s}(1-\frac{1}{q})-\frac{|\alpha|}{s}}$$

holds for sufficiently large t.

*Moreover, if J satisfies (1.20), then the same result holds with no restriction
on α.*

PROOF. We consider the cases $q = 2$ and $q = \infty$. The other cases follow by
interpolation. We denote by *e.s.* the exponentially small terms.

First, let us consider the case $q = \infty$. Using the definition of K_t,

$$\widehat{K_t}(\xi) = e^{-t}(e^{t\widehat{J}(\xi)} - 1),$$

we get, for any $x \in \mathbb{R}^N$,

$$|\partial^\alpha K_t(x)| \leq Ce^{-t} \int_{\mathbb{R}^N} |\xi|^{|\alpha|} \left| e^{t\widehat{J}(\xi)} - 1 \right| d\xi.$$

Since $|e^y - 1| \leq 2|y|$ for $|y|$ small, say $|y| \leq c_0$, by (1.18), we obtain

$$\left| e^{t\widehat{J}(\xi)} - 1 \right| \leq 2t|\widehat{J}(\xi)| \leq \frac{2ct}{|\xi|^m}$$

for all $|\xi| \geq c(t) := \left(\frac{ct}{c_0}\right)^{\frac{1}{m}}$ with t large enough. Then

$$e^{-t} \int_{|\xi|\geq c(t)} |\xi|^{|\alpha|} \left| e^{t\widehat{J}(\xi)} - 1 \right| d\xi \leq 2cte^{-t} \int_{|\xi|\geq c(t)} \frac{|\xi|^{|\alpha|}}{|\xi|^m} d\xi = Ct^{\frac{N}{m}+\frac{|\alpha|}{m}} e^{-t}$$

provided that $|\alpha| < m - N$.

It is easy to see that if (1.20) holds, no restriction on the multi-index α has to
be assumed.

It remains to estimate

$$e^{-t} \int_{|\xi| \leq c(t)} |\xi|^{|\alpha|} |e^{t\widehat{J}(\xi)} - 1| \, d\xi.$$

We observe that the term

$$e^{-t} \int_{|\xi| \leq c(t)} |\xi|^{|\alpha|} d\xi$$

is exponentially small, so we concentrate on

$$I(t) = e^{-t} \int_{|\xi| \leq c(t)} \left| e^{t\widehat{J}(\xi)} \right| |\xi|^{|\alpha|} \, d\xi.$$

Now, let us choose a and δ as in Lemma 1.8 iv) for $D = \frac{1}{2}$ (recall that now $A = 1$). Taking into account (1.8) and (1.9), and using the change of variables $\eta = \xi t^{1/s}$,

$$|I(t)| = e^{-t} \int_{|\xi| \leq a} \left| e^{t\widehat{J}(\xi)} \right| |\xi|^{|\alpha|} d\xi + e^{-t} \int_{a \leq |\xi| \leq c(t)} \left| e^{t\widehat{J}(\xi)} \right| |\xi|^{|\alpha|} d\xi$$

$$\leq \int_{|\xi| \leq a} e^{-\frac{t|\xi|^s}{2}} |\xi|^{|\alpha|} + e^{-t\delta} \int_{a \leq |\xi| \leq c(t)} |\xi|^{|\alpha|} d\xi$$

$$\leq \int_{|\xi| \leq a} e^{-\frac{t|\xi|^s}{2}} |\xi|^{|\alpha|} + e.s.$$

$$= t^{-\frac{|\alpha|}{s} - \frac{N}{s}} \int_{|\eta| \leq a t^{\frac{1}{s}}} e^{-\frac{|\eta|^s}{2}} |\eta|^{|\alpha|} + e.s. \leq C t^{-\frac{|\alpha|}{s} - \frac{N}{s}}.$$

Now, for $q = 2$, by Plancherel's identity we have

$$\|\partial^\alpha K_t\|^2_{L^2(\mathbb{R}^N)} \leq C e^{-2t} \int_{\mathbb{R}^N} |e^{t\widehat{J}(\xi)} - 1|^2 |\xi|^{2|\alpha|} d\xi.$$

Putting out the exponentially small terms, it remains to estimate

$$\int_{|\xi| \leq a} \left| e^{t(\widehat{J}(\xi)-1)} \right|^2 |\xi|^{2|\alpha|} d\xi,$$

where a is as above. Then, by (1.8), and working as before, we get

$$\int_{|\xi| \leq a} \left| e^{t(\widehat{J}(\xi)-1)} \right|^2 |\xi|^{2|\alpha|} d\xi \leq \int_{|\xi| \leq a} e^{-t|\xi|^s} |\xi|^{2|\alpha|} d\xi \leq C t^{-\frac{N}{s} - \frac{2|\alpha|}{s}},$$

which finishes the proof. $\qquad\qquad\qquad\qquad\qquad\qquad\qquad\qquad\qquad\qquad\qquad\square$

Once the case $2 \leq q \leq \infty$ has been analyzed the next step is to obtain similar decay rates for the L^q-norms with $1 \leq q < 2$. These estimates follow from an L^1-estimate and interpolation.

LEMMA 1.17. *Let $N \leq 3$. Assume that J satisfies (1.17) and (1.21) with $[s] > N/2$. Then, for any multi-index α and any $1 \leq q < 2$,*

$$(1.23) \qquad \|\partial^\alpha K_t\|_{L^q(\mathbb{R}^N)} \leq C t^{-\frac{N}{s}(1-\frac{1}{q}) - \frac{|\alpha|}{s}} \qquad \textit{for large } t.$$

REMARK 1.18. There is no restriction on s if J is such that

$$|\partial^\alpha \widehat{J}(\xi)| \leq \min\{|\xi|^{s-|\alpha|}, 1\}, \quad |\xi| \leq 1.$$

This happens if s is a positive integer and $\widehat{J}(\xi) = 1 - |\xi|^s$ in a neighborhood of the origin.

REMARK 1.19. The case $\alpha = (0, \ldots, 0)$ can be easily treated. Let w be the fundamental solution of (1.1) given in Lemma 1.6. As a consequence of the mass conservation,

$$\int_{\mathbb{R}^N} w(x, t) \, dx = 1,$$

we obtain

$$\int_{\mathbb{R}^N} |K_t| \leq 1,$$

and therefore (1.23) follows with $\alpha = (0, \ldots, 0)$.

REMARK 1.20. The condition (1.21) is satisfied, for example, for any smooth, compactly supported function J.

PROOF OF LEMMA 1.17. Fix α. The estimates for $1 < q < 2$ follow from the cases $q = 1$ and $q = 2$ (Lemma 1.16) using interpolation. Let us deal with $q = 1$. We use inequality (1.22) with $f = \partial^\alpha K_t$ and n such that $[s] \geq n > N/2$. We take $n = 1$ if $N = 1$ (in this case, $1 \leq s \leq 2$) and $n = 2$ if $N = 2$ or 3 (in this case, $s = 2$). We have

$$\|\partial^\alpha K_t\|_{L^1(\mathbb{R}^N)} \leq \|\partial^\alpha K_t\|_{L^2(\mathbb{R}^N)}^{1-\frac{N}{2n}} \||x|^n \partial^\alpha K_t\|_{L^2(\mathbb{R}^N)}^{\frac{N}{2n}}.$$

The condition $n \leq [s]$ guarantees that, for $j = 1, \ldots, N$, $\partial_{\xi_j}^n \widehat{J}$ makes sense near $\xi = 0$ and thus the derivatives $\partial_{\xi_j}^n \widehat{K_t}$ exist. Observe that the moment of order n of K_t imposes the existence of the partial derivatives $\partial_{\xi_j}^n \widehat{K_t}$, $j = 1, \ldots, N$.

In view of Lemma 1.16 and the above inequality, we obtain

$$\|\partial^\alpha K_t\|_{L^1(\mathbb{R}^N)} \leq Ct^{-\left(\frac{N}{2s} + \frac{|\alpha|}{s}\right)\left(1 - \frac{N}{2n}\right)} \||x|^n \partial^\alpha K_t\|_{L^2(\mathbb{R}^N)}^{\frac{N}{2n}}.$$

Thus it is sufficient to prove that

$$\||x|^n \partial^\alpha K_t\|_{L^2(\mathbb{R}^N)} \leq Ct^{\frac{n}{s} - \frac{N}{2s} - \frac{|\alpha|}{s}}$$

for all sufficiently large t. Observe that by Plancherel's Theorem

$$\int_{\mathbb{R}^N} |x|^{2n} |\partial^\alpha K_t(x)|^2 \, dx \leq C \int_{\mathbb{R}^N} (x_1^{2n} + \cdots + x_N^{2n}) |\partial^\alpha K_t(x)|^2 \, dx$$

$$\leq C \sum_{j=1}^N \int_{\mathbb{R}^N} |\partial_{\xi_j}^n (\xi^\alpha \widehat{K_t}(\xi))|^2 \, d\xi.$$

Therefore, it remains to show that, for any $j = 1, \ldots, N$,

$$(1.24) \qquad \int_{\mathbb{R}^N} |\partial_{\xi_j}^n (\xi^\alpha \widehat{K_t}(\xi))|^2 \, d\xi \leq Ct^{\frac{2n}{s} - \frac{N}{s} - \frac{2|\alpha|}{s}} \qquad \text{for } t \text{ large.}$$

We analyze the case $j = 1$; the others follow by the same arguments. Observe that if $N = 1$, $\xi_1 = \xi$, and in this case we have chosen $n = 1$, so

$$\partial_\xi (\xi^\alpha \widehat{K_t}(\xi)) = \alpha \xi^{\alpha-1} \widehat{K_t}(\xi) + \xi^\alpha \partial_\xi \widehat{K_t}(\xi)$$

$(\alpha \xi^{\alpha-1} \widehat{K_t}(\xi) = 0$ if $\alpha = 0)$, and for $n = 2$ (now, $N = 2$ or 3, $s = 2$)

$$\partial_{\xi_1}^2 \left(\xi^\alpha \widehat{K_t}(\xi) \right)$$

$$= \xi_2^{\alpha_2} \cdots \xi_N^{\alpha_N} \left(\alpha_1(\alpha_1 - 1)\xi_1^{\alpha_1-2} \widehat{K_t}(\xi) + 2\alpha_1 \xi_1^{\alpha_1-1} \partial_{\xi_1}^1 \widehat{K_t}(\xi) + \xi_1^{\alpha_1} \partial_{\xi_1}^2 \widehat{K_t}(\xi) \right)$$

$(\alpha_1(\alpha_1 - 1)\xi_1^{\alpha_1-2} \widehat{K_t}(\xi) = 0$ if $\alpha_1 = 0$ or 1, $\alpha_1 \xi_1^{\alpha_1-1} \partial_{\xi_1}^1 \widehat{K_t}(\xi) = 0$ if $\alpha_1 = 0)$. Therefore, (1.24) is reduced to

$$\int_{\mathbb{R}^N} \xi_1^{2(\alpha_1-k)} \xi_2^{2\alpha_2} \cdots \xi_N^{2\alpha_N} |\partial_{\xi_1}^{n-k} \widehat{K_t}(\xi)|^2 \, d\xi \leq C t^{\frac{2n}{s} - \frac{N}{s} - \frac{2|\alpha|}{s}} \qquad \text{for } t \text{ large}$$

for all $0 \leq k \leq \min\{\alpha_1, n\}$, or equivalently to

(1.25)

$$I(k,t) := \int_{\mathbb{R}^N} \xi_1^{2(\alpha_1+k-n)} \xi_2^{2\alpha_2} \cdots \xi_N^{2\alpha_N} |\partial_{\xi_1}^k \widehat{K_t}(\xi)|^2 \, d\xi$$

$$\leq C t^{\frac{2n}{s} - \frac{N}{s} - \frac{2|\alpha|}{s}} \qquad \text{for } t \text{ large},$$

for all $n - \min\{\alpha_1, n\} \leq k \leq n$; k can be 0 or 1 if $n = 1$, and 0, 1 or 2 if $n = 2$. Let us call

$$\beta(k) = (\alpha_1 + k - n, \alpha_2, \ldots, \alpha_N).$$

First we analyze the case $k = 0$ in (1.25) (observe that this case appears only when $\alpha_1 > 0$ if $N = 1$, and when $\alpha_1 > 1$ if $N = 1$ or 2). In this case

$$I(0,t) = \int_{\mathbb{R}^N} \left| \xi^{\beta(0)} \widehat{K_t}(\xi) \right|^2 d\xi,$$

and in view of Lemma 1.16 we obtain the desired decay property.

Let us now analyze the cases $k = 1$ and $k = 2$ in (1.25). Taking into account that $\widehat{K_t}(\xi) = e^{t(\hat{J}(\xi)-1)} - e^{-t}$, we get

$$\partial_{\xi_1}^1 \widehat{K_t}(\xi) = e^{t(\hat{J}(\xi)-1)} t \partial_{\xi_1}^1 \hat{J}(\xi)$$

and

$$\partial_{\xi_1}^2 \widehat{K_t}(\xi) = e^{t(\hat{J}(\xi)-1)} \left(t^2 \left(\partial_{\xi_1}^1 \hat{J}(\xi) \right)^2 + t \partial_{\xi_1}^2 \hat{J}(\xi) \right).$$

Using that all the partial derivatives of $\hat{J}$ decay, as $|\xi| \to \infty$, faster than any polynomial in $|\xi|$, we have that

$$\int_{|\xi|>a} |\xi|^{2|\beta(k)|} |\partial_{\xi_1}^k \widehat{K_t}(\xi)|^2 d\xi \leq C e^{-\delta t} t^{2k},$$

where a and δ are chosen as in Lemma 1.8 iv) for $D = \frac{1}{2}$. Having in mind that $n \leq [s]$ and $\hat{J}(\xi) - 1 + |\xi|^s = o(|\xi|^s)$ as $|\xi| \to 0$, we obtain

$$|\partial_{\xi_1}^j \hat{J}(\xi)| \leq C |\xi|^{s-j}, \quad 1 \leq j \leq n,$$

for all $|\xi| \leq a$. Then for all $|\xi| \leq a$, by (1.8), the following holds:

$$|\partial_{\xi_1}^1 \widehat{K_t}(\xi)|^2 \leq C e^{-t|\xi|^s} t^2 |\xi|^{2(s-1)}$$

and

$$|\partial_{\xi_1}^2 \widehat{K_t}(\xi)|^2 \leq C e^{-t|\xi|^s} \left(t^4 |\xi|^{4(s-1)} + t^2 |\xi|^{2(s-2)} \right).$$

Observe that the second derivative only appears when $n = 2 = s$, so in this case,

$$|\partial^2_{\xi_1} \widehat{K_t}(\xi)|^2 \leq C\, e^{-t|\xi|^2} \left(t^4 |\xi|^4 + t^2\right).$$

Using that for any $l \geq 0$,

$$\int_{\mathbb{R}^N} e^{-t|\xi|^s} |\xi|^l d\xi \leq Ct^{-\frac{N}{s} - \frac{l}{s}},$$

we get

$$\int_{|\xi| \leq a} |\xi|^{2|\beta(1)|} |\partial^1_{\xi_1} K_t(\xi)|^2 d\xi \leq Ct^{2 - \frac{N}{s} - \frac{2|\beta(1)| + 2(s-1)}{s}},$$

from which (1.25) follows for $k = 1$. For $k = 2$ (so $n = 2 = s$),

$$\int_{|\xi| \leq a} |\xi|^{2|\beta(2)|} |\partial^2_{\xi_1} K_t(\xi)|^2 d\xi \leq C \left(t^{4 - \frac{N}{2} - \frac{2|\beta(2)| + 4}{2}} + t^{2 - \frac{N}{2} - \frac{2|\beta(2)|}{2}}\right)$$

$$= Ct^{2 - \frac{N}{2} - |\beta(2)|} = Ct^{2 - \frac{N}{2} - |\alpha|},$$

and (1.25) also holds. $\qquad \square$

Now we are ready to prove Theorems 1.12 and 1.14.

PROOF OF THEOREMS 1.12 AND 1.14. Following [**96**] we obtain that the initial condition $u_0 \in L^1(\mathbb{R}^N, 1 + |x|^{k+1})$ has the following decomposition:

$$u_0 = \sum_{|\alpha| \leq k} \frac{(-1)^{|\alpha|}}{\alpha!} \left(\int u_0 x^\alpha \, dx\right) D^\alpha \delta_0 + \sum_{|\alpha| = k+1} D^\alpha F_\alpha,$$

where

$$\|F_\alpha\|_{L^1(\mathbb{R}^N)} \leq C\|u_0\|_{L^1(\mathbb{R}^N, |x|^{k+1})}$$

for all multi-indices α with $|\alpha| = k + 1$.

In view of (1.1) the solution u satisfies

$$u(x, t) = e^{-t} u_0(x) + (K_t * u_0)(x).$$

Since the first term is exponentially small, it suffices to analyze the long time behaviour of $K_t * u_0$. Using the above decomposition, Lemma 1.16 and Lemma 1.17 we get

$$\left\| K_t * u_0 - \sum_{|\alpha| \leq k} \frac{(-1)^{|\alpha|}}{\alpha!} \left(\int u_0(x) x^\alpha dx\right) \partial^\alpha K_t \right\|_{L^q(\mathbb{R}^N)}$$

$$\leq \sum_{|\alpha| = k+1} \|\partial^\alpha K_t * F_\alpha\|_{L^q(\mathbb{R}^N)}$$

$$\leq \sum_{|\alpha| = k+1} \|\partial^\alpha K_t\|_{L^q(\mathbb{R}^N)} \|F_\alpha\|_{L^1(\mathbb{R}^N)}$$

$$\leq Ct^{-\frac{N}{s}(1 - \frac{1}{q})} t^{-\frac{(k+1)}{s}} \|u_0\|_{L^1(\mathbb{R}^N, |x|^{k+1})}.$$

This ends the proof. $\qquad \square$

1.2.2. Asymptotics for the higher order terms. Next it is studied if the higher order terms of the asymptotic expansion found in Theorem 1.12 have some relation with the corresponding ones for the heat equation. The results show that the difference between them is of lower order. Again we have to distinguish between $2 \leq q \leq \infty$ and $1 \leq q < 2$.

THEOREM 1.21. *Let J be as in Theorem 1.12 and assume in addition that there exists $r > 0$ such that*

$$(1.26) \qquad \widehat{J}(\xi) = 1 - |\xi|^s + B|\xi|^{s+r} + o(|\xi|^{s+r}), \qquad \xi \to 0,$$

for some real number B. Then for any $2 \leq q \leq \infty$ and $|\alpha| \leq m - N$ there exists a positive constant $C = C(\alpha, q, N, s, r)$ such that

$$(1.27) \qquad \|\partial^\alpha K_t - \partial^\alpha G_t^s\|_{L^q(\mathbb{R}^N)} \leq Ct^{-\frac{N}{s}(1-\frac{1}{q})-\frac{|\alpha|+r}{s}} \qquad \text{for } t \text{ large.}$$

THEOREM 1.22. *Let $N \leq 3$. Assume J is as in Theorem 1.21 with $[s] > N/2$. Assume also that all the derivatives of $\widehat{J}$ decay at infinity faster than any polynomial, that is,*

$$|\partial^\alpha \widehat{J}(\xi)| \leq \frac{c(m, \alpha)}{|\xi|^m}, \qquad \xi \to \infty.$$

Then, for any $1 \leq q < 2$ and any multi-index α, the inequality (1.27) holds.

Note that these results do not imply that the asymptotic expansion

$$\sum_{|\alpha| \leq k} \frac{(-1)^{|\alpha|}}{\alpha!} \left(\int u_0(x) x^\alpha \right) \partial^\alpha K_t$$

coincides with the expansion that holds for the equation $v_t = -(-\Delta)^{\frac{s}{2}} v$,

$$\sum_{|\alpha| \leq k} \frac{(-1)^{|\alpha|}}{\alpha!} \left(\int u_0(x) x^\alpha \right) \partial^\alpha G_t^s.$$

They only say that the corresponding terms agree up to a better order. When J is compactly supported or rapidly decaying at infinity, then $s = 2$ and we obtain an expansion analogous to the one that holds for the heat equation.

PROOF OF THEOREM 1.21. We consider the case $q = \infty$; the case $q = 2$ can be handled similarly, and the rest of the cases, $2 < q < \infty$, follow again by interpolation.

Writing each of the two terms in Fourier variables we obtain

$$\|\partial^\alpha K_t - \partial^\alpha G_t^s\|_{L^\infty(\mathbb{R}^N)} \leq C \int_{\mathbb{R}^N} |\xi|^{|\alpha|} |e^{-t}(e^{t\widehat{J}(\xi)} - 1) - e^{-t|\xi|^s}| \, d\xi.$$

Let us choose $a > 0$ such that

$$(1.28) \qquad |\widehat{J}(\xi) - 1 + |\xi|^s| \leq C|\xi|^{r+s}, \qquad \text{for } |\xi| \leq a.$$

And let $\delta > 0$ such that

$$|\hat{J}(\xi)| \leq 1 - \delta, \qquad \text{for } |\xi| \geq a.$$

Then, for $|\xi| \geq a$ all the terms are exponentially small as $t \to \infty$. Thus the behaviour of the difference $\partial^\alpha K_t - \partial^\alpha G_t$ is given by the following integral:

$$I(t) = \int_{|\xi| \leq a} |\xi|^{|\alpha|} \left| e^{t(\hat{J}(\xi)-1)} - e^{-t|\xi|^s} \right| d\xi.$$

In view of the elementary inequality $|e^y - 1| \leq C|y|$ for all $|y|$ bounded, we have

$$I(t) = \int_{|\xi| \leq a} |\xi|^{|\alpha|} e^{-t|\xi|^s} \left| e^{t(\hat{J}(\xi)-1+|\xi|^s)} - 1 \right| d\xi$$

$$\leq C \int_{|\xi| \leq a} |\xi|^{|\alpha|} e^{-t|\xi|^s} \left| t(\hat{J}(\xi) - 1 + |\xi|^s) \right| d\xi$$

$$\leq Ct \int_{|\xi| \leq a} |\xi|^{|\alpha|} e^{-t|\xi|^s} |\xi|^{s+r} d\xi$$

$$\leq Ct^{-\frac{N}{s} - \frac{r}{s} - \frac{|\alpha|}{s}}.$$

This finishes the proof. $\square$

PROOF OF THEOREM 1.22. Using the same ideas as in the proof of Lemma 1.17 it remains to prove that for some $N/2 < n \leq [s]$ the following holds:

$$\||x|^n (\partial^\alpha K_t - \partial^\alpha G_t^s)\|_{L^2(\mathbb{R}^N)} \leq Ct^{-\frac{N}{2s} + \frac{n-(|\alpha|+r)}{s}}.$$

Like there, we chose $n = 1$ if $N = 1$ (in this case $1 \leq s \leq 2$), and necessarily $n = 2$ if $N = 2, 3$ (in this case $s = 2$).

Applying Plancherel's identity the proof of the last inequality is reduced to the proof of the following one:

$$\int_{\mathbb{R}^N} |\partial_{\xi_j}^n [\xi^\alpha (\widehat{K_t}(\xi) - \widehat{G_t^s}(\xi))]|^2 \, d\xi \leq Ct^{-\frac{N}{2s} + \frac{n-(|\alpha|+r)}{s}}, \quad j = 1, \ldots, N,$$

provided that all the above terms make sense. This means that all the partial derivatives $\partial_{\xi_j}^k \widehat{K_t}$ and $\partial_{\xi_j}^k \widehat{G_t^s}$, $j = 1, \ldots, N$, $k = 0, \ldots, n$ must be defined. Thus we need $n \leq [s]$.

As in Lemma 1.17 we must see that

$$\int_{\mathbb{R}^N} \xi^{2\beta(k)} |\partial_{\xi_1}^k (\widehat{K_t}(\xi) - \widehat{G_t^s}(\xi))|^2 \, d\xi \leq Ct^{-\frac{N}{s} + \frac{2(k-|\beta(k)|-r)}{s}}$$

for all $n - \min\{n, \alpha_1\} \leq k \leq n$, where

$$\beta(k) = (\alpha_1 + k - n, \alpha_2, \ldots, \alpha_N).$$

Again, k can be 0 or 1 if $n = 1$, and 0, 1 or 2 if $n = 2$.

The case $k = 0$ follows easily from Theorem 1.21. Let us deal with the cases $k = 1$ and $k = 2$. Using that the integral outside of a ball of radius a decays exponentially, it remains to analyze the decay of the following integral

$$I(k,t) := \int_{|\xi| \leq a} |\xi|^{2|\beta(k)|} |\partial_{\xi_1}^k (\widehat{K_t}(\xi) - \widehat{G_t^s}(\xi))|^2 d\xi,$$

where a is as in the proof of Theorem 1.21. Using the definition of $\widehat{K_t}$ and G_t^s we obtain that

$$\partial_{\xi_1}^1 \widehat{K_t}(\xi) = e^{t(\widehat{J}(\xi)-1)} t \partial_{\xi_1}^1 \widehat{J}(\xi)$$

and

$$\partial_{\xi_1}^1 \widehat{G_t^s}(\xi) = e^{-t|\xi|^s} t \partial_{\xi_1}^1 \left(-|\xi|^s\right);$$

and for $k = 2$, which appears only when $n = 2 = s$,

$$\partial_{\xi_1}^2 \widehat{K_t}(\xi) = e^{t(\widehat{J}(\xi)-1)} \left(t^2 \left(\partial_{\xi_1}^1 \widehat{J}(\xi) \right)^2 + t \partial_{\xi_1}^2 \widehat{J}(\xi) \right)$$

and

$$\partial_{\xi_1}^2 \widehat{G_t^2}(\xi) = e^{-t|\xi|^2} \left(t^2 \left(2\xi_1\right)^2 - 2t \right).$$

Let us first analyze $I(1,t)$. We can write

$$|\partial_{\xi_1}^1 \widehat{K_t}(\xi) - \partial_{\xi_1}^1 \widehat{G_t^s}(\xi)|^2 \leq I_1(\xi,t) + I_2(\xi,t),$$

where

$$I_1(\xi,t) = 2 \left| e^{t(\widehat{J}(\xi)-1)} - e^{-t|\xi|^s} \right|^2 t^2 |\partial_{\xi_1}^1 \widehat{J}(\xi)|^2$$

and

$$I_2(\xi,t) = 2 e^{-2t|\xi|^s} t^2 \left| \partial_{\xi_1}^1 \widehat{J}(\xi) - \partial_{\xi_1}^1 \left(-|\xi|^s\right) \right|^2.$$

Let us consider $I_1(\xi,t)$. Taking into account that $n \leq [s]$ and $|\widehat{J}(\xi) - 1 + |\xi|^s| \leq o(|\xi|^s)$ as $|\xi| \to 0$ we obtain

$$|\partial_{\xi_1}^j \widehat{J}(\xi)| \leq C |\xi|^{s-j}, \ 1 \leq j \leq n,$$

for all $|\xi| \leq a$. On the other hand (see (1.28))

$$\left| e^{t(\widehat{J}(\xi)-1)} - e^{-t|\xi|^s} \right|^2 = e^{-2t|\xi|^s} \left| e^{t(\widehat{J}(\xi)-1+|\xi|^s)} - 1 \right|^2$$

$$\leq C e^{-2t|\xi|^s} \left| t(\widehat{J}(\xi) - 1 + |\xi|^s) \right|^2$$

$$\leq C t^2 e^{-2t|\xi|^s} |\xi|^{2(r+s)}.$$

Therefore,

$$I_1(\xi,t) \leq C t^4 e^{-2t|\xi|^s} |\xi|^{2(r+s)}.$$

Let us now deal with $I_2(\xi,t)$. Choosing eventually a smaller a we can guarantee that for $|\xi| \leq a$, since $k = 1 \leq [s]$, the following inequality holds:

$$\left| \partial_{\xi_1}^1 \widehat{J}(\xi) - \partial_{\xi_1}^1 \left(-|\xi|^s\right) \right| \leq C|\xi|^{s+r-1}.$$

Consequently

$$I_2(\xi,t) \leq C t^2 e^{-2t|\xi|^s} |\xi|^{2(r+s-1)}.$$

Using that for any $l \geq 0$,

$$\int_{\mathbb{R}^N} e^{-t|\xi|^s} |\xi|^l d\xi \leq C t^{-\frac{N}{s} - \frac{l}{s}},$$

we get the desired decay result for $I(1,t)$.

Finally, to study the decay of $I(2,t)$ we do the following decomposition:

$$|\partial_{\xi_1}^2 \widehat{K_t}(\xi) - \partial_{\xi_1}^2 \widehat{G_t^2}(\xi)|^2$$

$$\leq 2t^4 \left| e^{t(\widehat{J}(\xi)-1)} \left(\partial_{\xi_1}^1 \widehat{J}(\xi) \right)^2 - e^{-t|\xi|^2} (2\xi_1)^2 \right|^2$$

$$+ 2t^2 \left| e^{t(\widehat{J}(\xi)-1)} \partial_{\xi_1}^2 \widehat{J}(\xi) - e^{-t|\xi|^2}(-2)) \right|^2 .$$

$$\leq 4t^4 \left| e^{t(\widehat{J}(\xi)-1)} - e^{-t|\xi|^2} \right|^2 \left| \partial_{\xi_1}^1 \widehat{J}(\xi) \right|^4$$

$$+ 4t^4 e^{-2t|\xi|^2} \left| \partial_{\xi_1}^1 \widehat{J}(\xi) - (-2\xi_1) \right|^2 \left(\left| \partial_{\xi_1}^1 \widehat{J}(\xi) \right| + 2|\xi_1| \right)$$

$$+ 4t^2 \left| e^{t(\widehat{J}(\xi)-1)} - e^{-t|\xi|^2} \right|^2 \left| \partial_{\xi_1}^2 \widehat{J}(\xi) \right|^2$$

$$+ 4t^2 e^{-2t|\xi|^2} \left| \partial_{\xi_1}^2 \widehat{J}(\xi) - (-2) \right|^2 ,$$

and proceeding similarly to the previous case we finish the proof. $\square$

1.2.3. A different approach.
In this subsection we obtain the first two terms in the asymptotic expansion of the solution under less restrictive hypotheses on J.

THEOREM 1.23. *Let $u_0 \in L^1(\mathbb{R}^N)$ with $\widehat{u_0} \in L^1(\mathbb{R}^N)$ and $s < l$ be two positive numbers such that*

$$\widehat{J}(\xi) = 1 - |\xi|^s + B|\xi|^l + o(|\xi|^l), \quad \xi \to 0,$$

for some real number B.

Then for any $2 \leq q \leq \infty$

$$(1.29) \qquad \lim_{t\to\infty} t^{\frac{N}{s}(1-\frac{1}{q})+\frac{l-s}{s}} \|u(t) - v(t) - Bt[(-\Delta)^{\frac{l}{2}}v](t)\|_{L^q(\mathbb{R}^N)} \to 0,$$

where v is the solution of $v_t = -(-\Delta)^{\frac{s}{2}}v$ with $v(x,0) = u_0(x)$.

Moreover, if we set h_0 such that $\widehat{h_0}(\xi) = e^{-|\xi|^s}|\xi|^l$, then

$$(1.30) \quad \lim_{t\to\infty} \left\| t^{\frac{N}{s}+\frac{l}{s}-1} \left(u(yt^{\frac{1}{s}},t) - v(yt^{\frac{1}{s}},t) \right) - Bh_0(y) \left(\int_{\mathbb{R}^N} u_0 \right) \right\|_{L^\infty(\mathbb{R}^N)} = 0.$$

Let us point out that the asymptotic expansion given by (1.19) involves K_t (and its derivatives) which is not explicit. On the other hand, the two-terms of the asymptotic expansion (1.29) involves G_t^s, a well known explicit kernel (v is just the convolution of G_t^s and u_0). However, the ideas and methods employed here allow us to find only two terms in the latter expansion. The case $1 \leq q < 2$ in (1.29) can be also treated, but additional hypothesis on J must be imposed.

PROOF OF THEOREM 1.23. The method used here is just to estimate the difference $\|u(t) - v(t) - Bt(-\Delta)^{\frac{l}{2}}v(t)\|_{L^q(\mathbb{R}^N)}$ using Fourier variables.

As before, it is enough to consider the cases $q = 2$ and $q = \infty$. We analyze the case $q = \infty$; the case $q = 2$ follows in the same manner by applying Plancherel's identity.

For $q = \infty$ we have

$$\|u(t) - v(t) - tB(-\Delta)^{\frac{l}{2}}v(t)\|_{L^\infty(\mathbb{R}^N)}$$

$$\leq (2\pi)^{-N} \int_{\mathbb{R}^N} \left| \widehat{u}(\xi, t) - \widehat{v}(\xi, t) - tB\widehat{(-\Delta)^{\frac{l}{2}}v}(\xi, t) \right| d\xi$$

$$= (2\pi)^{-N} \int_{\mathbb{R}^N} \left| e^{t(\widehat{J}(\xi)-1)} - e^{-t|\xi|^s}(1 + tB|\xi|^l) \right| |\widehat{u_0}(\xi)| \, d\xi.$$

Let a and δ be as in Lemma 1.8 iv) for $D = \frac{1}{2}$. Hence,

$$\int_{|\xi|\geq a} |e^{t(\widehat{J}(\xi)-1)}||\widehat{u_0}(\xi)|d\xi \leq Ce^{-\delta t}\|\widehat{u_0}\|_{L^1(\mathbb{R}^N)}$$

and

$$\int_{t^{-\frac{1}{l}}\leq|\xi|\leq a} |e^{t(\widehat{J}(\xi)-1)}||\widehat{u_0}(\xi)|d\xi \leq C\|\widehat{u_0}\|_{L^\infty(\mathbb{R}^N)} \int_{t^{-\frac{1}{l}}\leq|\xi|\leq a} e^{-t|\xi|^s/2} \, d\xi$$

$$\leq Ct^{-\frac{N}{s}} \int_{t^{\frac{1}{s}-\frac{1}{l}}\leq|\eta|\leq t^{\frac{1}{s}}a} e^{-|\eta|^s/2} \, d\eta \leq Ct^{-\frac{N}{s}}e^{-\frac{1}{4}t^{1-\frac{s}{l}}}.$$

Also

$$\int_{|\xi|\geq t^{-\frac{1}{l}}} e^{-t|\xi|^s}(1 + tB|\xi|^l)|\widehat{u_0}(\xi)|d\xi \leq C\|\widehat{u_0}\|_{L^\infty(\mathbb{R}^N)} \int_{|\xi|\geq t^{-\frac{1}{l}}} e^{-t|\xi|^s}t|\xi|^l \, d\xi$$

$$\leq Ct^{1-\frac{N}{s}-\frac{l}{s}} \int_{|\eta|\geq t^{\frac{1}{s}-\frac{1}{l}}} e^{-|\eta|^s}|\eta|^l \, d\eta$$

$$\leq Ct^{1-\frac{N}{s}-\frac{l}{s}}e^{-\frac{1}{2}t^{1-\frac{s}{l}}} \int_{|\eta|\geq t^{\frac{1}{s}-\frac{1}{l}}} e^{-|\eta|^s/2}|\eta|^l \, d\eta.$$

Therefore, we have to analyze

$$I(t) = \int_{|\xi|\leq t^{-\frac{1}{l}}} \left| e^{t(\widehat{J}(\xi)-1)} - e^{-t|\xi|^s}(1 + tB|\xi|^l) \right| |\widehat{u_0}(\xi)| \, d\xi.$$

We write $\widehat{J}(\xi) = 1 - |\xi|^s + B|\xi|^l + |\xi|^l f(\xi)$, where $f(\xi) \to 0$ as $|\xi| \to 0$. Thus

$$I(t) \leq I_1(t) + I_2(t),$$

where

$$I_1(t) = \int_{|\xi|\leq t^{-\frac{1}{l}}} e^{-t|\xi|^s} \left| e^{Bt|\xi|^l + t|\xi|^l f(\xi)} - (1 + Bt|\xi|^l + t|\xi|^l f(\xi)) \right| |\widehat{u_0}(\xi)| \, d\xi$$

and

$$I_2(t) = \int_{|\xi|\leq t^{-\frac{1}{l}}} e^{-t|\xi|^s}t|\xi|^l|f(\xi)||\widehat{u_0}(\xi)| \, d\xi.$$

For I_1 we have, for t large, that

$$I_1(t) \leq C\|\widehat{u_0}\|_{L^\infty(\mathbb{R}^N)} \int_{|\xi| \leq t^{-\frac{1}{s}}} e^{-t|\xi|^s}(t|\xi|^l + t|\xi|^l|f(\xi)|)^2 \, d\xi$$
$$\leq C \int_{|\xi| \leq t^{-\frac{1}{s}}} e^{-t|\xi|^s} t^2 |\xi|^{2l} \, d\xi \leq C t^{-\frac{N}{s} + 2 - \frac{2l}{s}}$$

and then

$$t^{\frac{N}{s} + \frac{l}{s} - 1} I_1(t) \leq C t^{1 - \frac{l}{s}} \to 0, \quad t \to \infty.$$

It remains to prove that

$$t^{\frac{N}{s} + \frac{l}{s} - 1} I_2(t) \to 0, \quad t \to \infty.$$

Making a change of variables we obtain

$$t^{\frac{N}{s} - 1 + \frac{l}{s}} I_2(t) \leq C\|\widehat{u_0}\|_{L^\infty(\mathbb{R}^N)} \int_{|\eta| \leq t^{\frac{1}{s} - \frac{1}{t}}} e^{-|\eta|^s} |\eta|^l |f(\eta t^{-\frac{1}{s}})| \, d\eta.$$

Note that the integrand is dominated by $\|f\|_{L^\infty(\mathbb{R}^N)} |\eta|^l e^{-|\eta|^s}$, which belongs to $L^1(\mathbb{R}^N)$. Hence, as $f(\eta t^{-\frac{1}{s}}) \to 0$ when $t \to \infty$,

$$t^{\frac{N}{s} + \frac{l}{s} - 1} I_2(t) \to 0,$$

and the proof of (1.29) is complete.

Thanks to (1.29), the proof of (1.30) is reduced to showing that

$$\lim_{t \to \infty} \left\| t^{\frac{N}{s} + \frac{l}{s}} [(-\Delta)^{\frac{l}{2}} v](y t^{\frac{1}{s}}, t) - h_0(y) \left(\int_{\mathbb{R}^N} u_0 \right) \right\|_{L^\infty(\mathbb{R}^N)} = 0.$$

For any $y \in \mathbb{R}^N$, by making a change of variables,

$$I(y, t) = t^{\frac{N}{s} + \frac{l}{s}} [(-\Delta)^{\frac{l}{2}} v](y t^{\frac{1}{s}}, t) = \int_{\mathbb{R}^N} e^{-|\xi|^s} |\xi|^l e^{iy\xi} \widehat{u_0}(\xi/t^{\frac{1}{s}}).$$

Thus, using the dominated convergence theorem,

$$\left\| I(y, t) - h_0(y) \int_{\mathbb{R}^N} u_0 \right\|_{L^\infty(\mathbb{R}^N)} \leq C \int_{\mathbb{R}^N} e^{-|\xi|^s} |\xi|^l \left| \widehat{u_0}(\xi t^{-\frac{1}{s}}) - \widehat{u_0}(0) \right| \, d\xi \to 0$$

as $t \to \infty$.   □

1.3. Rescaling the kernel. A nonlocal approximation of the heat equation

In this section it is shown that the problem $v_t(x, t) = \Delta v(x, t)$ can be approximated by nonlocal problems like the ones presented here when they are rescaled in an appropriate way. Concretely, we will rescale the problem

$$\begin{cases} u_t(x, t) = (J * u)(x, t) - u(x, t), \\ u(x, 0) = u_0(x), \end{cases}$$

for $x \in \mathbb{R}^N$ and $t > 0$, with J satisfying condition (H) and such that

$$(1.31) \qquad \widehat{J}(\xi) = 1 - |\xi|^2 + o(|\xi|^2) \quad \text{as } \xi \to 0.$$

THEOREM 1.24. *Assume* (1.31). *Let* u_ε *be the unique solution to*

(1.32)
$$
\begin{cases}
(u_\varepsilon)_t(x,t) = \dfrac{J_\varepsilon * u_\varepsilon(x,t) - u_\varepsilon(x,t)}{\varepsilon^2}, & x \in \mathbb{R}^N, \, t > 0, \\[2mm]
u_\varepsilon(x,0) = u_0(x), & x \in \mathbb{R}^N,
\end{cases}
$$

where the kernel J is rescaled according to

$$
J_\varepsilon(x) = \varepsilon^{-N} J\left(\frac{x}{\varepsilon}\right).
$$

Then, for every $T > 0$, we have

$$
\lim_{\varepsilon \to 0} \|u_\varepsilon - v\|_{L^\infty(\mathbb{R}^N \times (0,T))} = 0,
$$

v being the solution of the local problem $v_t(x,t) = \Delta v(x,t)$ with the same initial condition $v(x,0) = u_0(x)$.

REMARK 1.25. Note that J_ε satisfies

$$
\int_{\mathbb{R}^N} J_\varepsilon(x) \, dx = 1.
$$

PROOF OF THEOREM 1.24. The proof uses once more the explicit formula for the solutions in Fourier variables. We have, arguing exactly as before,

$$
\widehat{u_\varepsilon}(\xi,t) = e^{\frac{\widehat{J_\varepsilon}(\xi)-1}{\varepsilon^2}t}\widehat{u_0}(\xi)
$$

and

$$
\hat{v}(\xi,t) = e^{-|\xi|^2 t}\widehat{u_0}(\xi).
$$

Now, we just observe that $\widehat{J_\varepsilon}(\xi) = \hat{J}(\varepsilon\xi)$ and therefore we obtain

$$
\int_{\mathbb{R}^N} |\widehat{u_\varepsilon} - \hat{v}|\,(\xi,t)\,d\xi = \int_{\mathbb{R}^N} \left| \left(e^{\frac{\hat{J}(\varepsilon\xi)-1}{\varepsilon^2}t} - e^{-|\xi|^2 t}\right)\widehat{u_0}(\xi) \right| d\xi
$$

$$
\leq \|\widehat{u_0}\|_{L^\infty(\mathbb{R}^N)} \left(\int_{|\xi| \geq r(\varepsilon)} \left| e^{\frac{\hat{J}(\varepsilon\xi)-1}{\varepsilon^2}t} - e^{-|\xi|^2 t} \right| d\xi \right.
$$

$$
\left. + \int_{|\xi| < r(\varepsilon)} \left| e^{\frac{\hat{J}(\varepsilon\xi)-1}{\varepsilon^2}t} - e^{-|\xi|^2 t} \right| d\xi \right).
$$

For $t \in [0,T]$ we can proceed as in the proof of the asymptotic behaviour (Theorem 1.3) to obtain that

$$
\max_x |u_\varepsilon(x,t) - v(x,t)| \leq (2\pi)^{-N} \int_{\mathbb{R}^N} |\widehat{u_\varepsilon} - \hat{v}|\,(\xi,t)\,d\xi \to 0, \quad \varepsilon \to 0. \qquad \square
$$

1.4. Higher order problems

This section deals with the asymptotic behaviour of solutions of a nonlocal diffusion operator of higher order in the whole $\mathbb{R}^N$.

Let us consider the following nonlocal evolution problem:

$$(1.33) \quad \begin{cases} u_t(x,t) = (-1)^{n-1} \left(J * Id - 1\right)^n (u(x,t)) \\ \qquad = (-1)^{n-1} \left(\displaystyle\sum_{k=0}^n \binom{n}{k}(-1)^{n-k}(J*)^k(u)\right)(x,t), \\ u(x,0) = u_0(x), \end{cases}$$

for $x \in \mathbb{R}^N$ and $t > 0$, where the hypothesis on the convolution kernel J is again (H), and $(J*)^k(u) = J* \overset{k}{\cdots} *J * u$.

Note that in the problem (1.33) we just have the iteration (n times) of the nonlocal operator $J * u - u$ as right hand side of the equation. This can be seen as a nonlocal generalization of higher order equations of the form

$$(1.34) \qquad\qquad v_t(x,t) = -A^n(-\Delta)^{\frac{\alpha n}{2}} v(x,t),$$

where A and α are positive constants specified later in this section. Observe that when $\alpha = 2$, (1.34) is just $v_t(x,t) = -A^n(-\Delta)^n v(x,t)$. Nonlocal higher order problems have been, for instance, proposed as models for periodic phase separation. Here the nonlocal character of the problem is associated with long-range interactions of "particles" in the system. An example is the nonlocal Cahn-Hilliard equation (cf. e.g. [112], [135], [136]).

Here (1.33) is proposed as a model for higher order nonlocal evolution. Existence and uniqueness of solutions of (1.33) are shown first, but the main aim is to study the asymptotic behaviour as $t \to \infty$ of such solutions.

1.4.1. Existence and uniqueness. As it was done for (1.1), we first prove the existence and uniqueness of a solution, which is understood in the integral sense.

THEOREM 1.26. *Let $u_0 \in L^1(\mathbb{R}^N)$ such that $\widehat{u_0} \in L^1(\mathbb{R}^N)$. There exists a unique solution $u \in C^0([0,\infty); L^1(\mathbb{R})^N)$ of (1.33) that, in Fourier variables, is given by the explicit formula,*

$$\hat{u}(\xi,t) = e^{(-1)^{n-1}(\hat{J}(\xi)-1)^n t}\widehat{u_0}(\xi).$$

PROOF. We have formally

$$u_t(x,t) = (-1)^{n-1} \left(\sum_{k=0}^n \binom{n}{k}(-1)^{n-k}(J*)^k(u)\right)(x,t).$$

Applying the Fourier transform to this equation we obtain

$$\hat{u}_t(\xi,t) = (-1)^{n-1}\sum_{k=0}^n \binom{n}{k}(-1)^{n-k}(\hat{J}(\xi))^k\hat{u}(\xi,t)$$

$$= (-1)^{n-1}(\hat{J}(\xi) - 1)^n\hat{u}(\xi,t).$$

Hence

$$\hat{u}(\xi,t) = e^{(-1)^{n-1}(\hat{J}(\xi)-1)^n t}\widehat{u_0}(\xi).$$

Since $\widehat{u_0}(\xi) \in L^1(\mathbb{R}^N)$ and $e^{(-1)^{n-1}(\hat{J}(\xi)-1)^n t}$ is continuous and bounded, $\hat{u}(\cdot,t) \in L^1(\mathbb{R}^N)$ and the result follows by taking the inverse Fourier transform. $\qquad\square$

The next result is a lemma concerning the fundamental solution of (1.33).

LEMMA 1.27. *Let $J \in \mathcal{S}(\mathbb{R}^N)$. The fundamental solution w of (1.33), that is, the solution of the equation with initial condition $u_0 = \delta_0$, can be decomposed as*

$$(1.35) \qquad w(x,t) = e^{-t}\delta_0(x) + V(x,t),$$

with $V(x,t)$ smooth. Moreover, if u is a solution of (1.33), u can be written as

$$u(x,t) = (w * u_0)(x,t) = \int_{\mathbb{R}^N} w(x-z,t)u_0(z)\,dz.$$

PROOF. By the previous result we have

$$\hat{w}_t(\xi,t) = (-1)^{n-1}(\hat{J}(\xi)-1)^n\hat{w}(\xi,t).$$

Hence, as the initial datum satisfies $\widehat{w_0} = \hat{\delta_0} = 1$, we get

$$\hat{w}(\xi,t) = e^{(-1)^{n-1}(\hat{J}(\xi)-1)^n t} = e^{-t} + e^{-t}\left(e^{\left((-1)^{n-1}(\hat{J}(\xi)-1)^n+1\right)t} - 1\right).$$

The first part of the lemma follows by applying the inverse Fourier transform.

To finish the proof we just observe that $w * u_0$ is a solution of (1.33) with $(w * u_0)(x,0) = u_0(x)$. $\qquad\square$

1.4.2. Asymptotic behaviour. Next, we deal with the asymptotic behaviour as $t \to \infty$.

THEOREM 1.28. *Assume*

$$(1.36) \qquad \hat{J}(\xi) = 1 - A|\xi|^s + o(|\xi|^s) \quad as\ \xi \to 0,$$

$A, s > 0$. Let u be a solution of (1.33) with $u_0, \widehat{u_0} \in L^1(\mathbb{R}^N)$, $u_0 \geq 0$. Then the asymptotic behaviour of $u(x,t)$ is given by

$$\lim_{t\to+\infty} t^{\frac{N}{sn}} \max_x |u(x,t) - v(x,t)| = 0,$$

where v is the solution of

$$v_t(x,t) = -A^n(-\Delta)^{\frac{sn}{2}}v(x,t)$$

with initial condition $v(x,0) = u_0(x)$.

Moreover, there exists a constant $C > 0$ such that

$$\|u(\cdot,t)\|_{L^\infty(\mathbb{R}^N)} \leq C\,t^{-\frac{N}{sn}},$$

and the asymptotic profile is given by

$$\lim_{t\to+\infty} \max_y \left| t^{\frac{N}{sn}}u(yt^{\frac{1}{sn}},t) - \|u_0\|_{L^1(\mathbb{R}^N)}\,G_{A^n}^{sn}(y) \right| = 0.$$

PROOF. By Theorem 1.26, $\hat{u}(\xi,t) = e^{(-1)^{n-1}(\hat{J}(\xi)-1)^n t}\widehat{u_0}(\xi)$. On the other hand, $v(x,t)$ is given by

$$\hat{v}(\xi,t) = e^{-A^n|\xi|^{sn}t}\widehat{u_0}(\xi).$$

Hence in Fourier variables

$$\int_{\mathbb{R}^N} |\hat{u} - \hat{v}|\,(\xi, t)\,d\xi$$

$$\leq \int_{|\xi| \geq r(t)} \left| \left(e^{(-1)^{n-1}(\hat{J}(\xi)-1)^n t} - e^{-A^n |\xi|^{sn} t} \right) \widehat{u_0}(\xi) \right|\,d\xi$$

$$+ \int_{|\xi| < r(t)} \left| \left(e^{(-1)^{n-1}(\hat{J}(\xi)-1)^n t} - e^{-A^n |\xi|^{sn} t} \right) \widehat{u_0}(\xi) \right|\,d\xi$$

$$= [I] + [II],$$

where $[I]$ and $[II]$ denote the first and the second integral respectively, and $r(t)$, with $r(t) \to 0$, will be determined later. To deal with $[I]$ we decompose it into two parts,

$$[I] \leq \int_{|\xi| \geq r(t)} \left| e^{-A^n |\xi|^{sn} t} \widehat{u_0}(\xi) \right|\,d\xi + \int_{|\xi| \geq r(t)} \left| e^{(-1)^{n-1}(\hat{J}(\xi)-1)^n t} \widehat{u_0}(\xi) \right|\,d\xi$$

$$= [I_1] + [I_2].$$

Consider $[I_1]$. Setting $\eta = \xi t^{\frac{1}{sn}}$ and writing $[I_1]$ in the new variable η we get,

$$t^{\frac{N}{sn}} [I_1] \leq \|\widehat{u_0}\|_{L^\infty(\mathbb{R}^N)} \int_{|\eta| \geq r(t) t^{\frac{1}{sn}}} e^{-A^n |\eta|^{sn}}\,d\eta \xrightarrow{t \to \infty} 0$$

if we impose that

$$(1.37) \qquad\qquad\qquad r(t) t^{\frac{1}{sn}} \xrightarrow{t \to \infty} \infty.$$

Let a and δ be as in Lemma 1.8 iv) for D such that

$$d := 2 - \left(1 + \frac{D}{A} \right)^n > 0.$$

Therefore

$$[I_2] \leq \int_{a \geq |\xi| \geq r(t)} \left| e^{(-1)^{n-1}(\hat{J}(\xi)-1)^n t} \widehat{u_0}(\xi) \right|\,d\xi$$

$$+ \int_{|\xi| \geq a} \left| e^{(-1)^{n-1}(\hat{J}(\xi)-1)^n t} \widehat{u_0}(\xi) \right|\,d\xi$$

$$\leq \|\widehat{u_0}\|_{L^\infty(\mathbb{R}^N)} \int_{a \geq |\xi| \geq r(t)} e^{-d^n |\xi|^{sn} t}\,d\xi + C e^{-\delta^n t}.$$

Changing variables as before, $\eta = \xi t^{\frac{1}{sn}}$, we get

$$t^{\frac{N}{sn}} [I_2] \leq \|\widehat{u_0}\|_{L^\infty(\mathbb{R}^N)} \int_{at^{\frac{1}{sn}} \geq |\eta| \geq r(t) t^{\frac{1}{sn}}} e^{-d^n |\eta|^{sn}}\,d\eta + C t^{\frac{N}{sn}} e^{-\delta^n t}$$

$$\leq \|\widehat{u_0}\|_{L^\infty(\mathbb{R}^N)} \int_{|\eta| \geq r(t) t^{\frac{1}{sn}}} e^{-d^n |\eta|^{sn}}\,d\eta + C t^{\frac{N}{sn}} e^{-\delta^n t} \to 0$$

as $t \to \infty$ if (1.37) holds.

It remains only to estimate $[II]$. We proceed as follows:

$$[II] = \int_{|\xi|<r(t)} e^{-A^n|\xi|^{sn}t} \left| e^{t\left((-1)^{n-1}(\hat{J}(\xi)-1)^n+A^n|\xi|^{sn}\right)} - 1 \right| |\widehat{u_0}(\xi)|\, d\xi.$$

Applying the binomial formula,

$$(-1)^{n-1}(\hat{J}(\xi)-1)^n + A^n|\xi|^{sn}$$

$$= (-1)^{n-1}\left(\sum_{k=1}^{n} \binom{n}{k} \left(\hat{J}(\xi) - 1 + A|\xi|^s \right)^k (-A|\xi|^s)^{n-k} \right).$$

Consequently, since $|e^y - 1| \le C|y|$ for y small, and using (1.7),

$$t^{\frac{N}{sn}}[II] \le Ct^{\frac{N}{sn}} \int_{|\xi|<r(t)} e^{-A^n|\xi|^{sn}t} t \left(\sum_{k=1}^{n} \binom{n}{k} A^{n-k} h(\xi)^k \right) |\xi|^{sn}\, d\xi$$

$$\le Ct^{\frac{N}{sn}} \int_{|\xi|<r(t)} e^{-A^n|\xi|^{sn}t} t h(\xi) |\xi|^{sn}\, d\xi,$$

provided that we impose

$$(1.38) \qquad\qquad t(r(t))^{sn} h(r(t)) \to 0 \quad \text{as } t \to \infty.$$

In this case, changing variables, $\eta = \xi t^{\frac{1}{sn}}$, we have

$$t^{\frac{N}{sn}}[II] \le C \int_{|\eta|<r(t)t^{\frac{1}{sn}}} e^{-A^n|\eta|^{sn}} |\eta|^{sn} h(\eta/t^{\frac{1}{sn}})\, d\eta.$$

Since $h(\eta/t^{\frac{1}{sn}}) \to 0$ as $t \to \infty$, and in the above inequality the integrand is dominated by

$$\|h\|_{L^\infty(\mathbb{R}^N)} e^{-A^n|\eta|^{sn}} |\eta|^{sn},$$

which belongs to $L^1(\mathbb{R}^N)$, by the dominated convergence theorem,

$$t^{\frac{N}{sn}}[II] \to 0 \quad \text{as } t \to \infty.$$

Combining this with the previous results we have that

$$(1.39) \qquad t^{\frac{N}{sn}} \int_{\mathbb{R}^N} |\hat{u} - \hat{v}|\,(\xi, t)\, d\xi \le t^{\frac{N}{sn}}([I] + [II]) \to 0 \quad \text{as } t \to \infty,$$

provided we can find an $r(t) \to 0$ as $t \to \infty$ that fulfils both conditions (1.37) and (1.38), which is true by Lemma 1.9 changing there s by sn. To conclude we only have to observe that from the convergence of the Fourier transforms

$$\|\hat{u}(\cdot, t) - \hat{v}(\cdot, t)\|_{L^1(\mathbb{R}^N)} \to 0 \quad \text{as } t \to \infty,$$

the convergence of $u - v$ in L^∞ follows. Indeed, from (1.39) we obtain

$$t^{\frac{N}{sn}} \max_x |u(x,t) - v(x,t)| \le (2\pi)^{-N} t^{\frac{N}{sn}} \int_{\mathbb{R}^N} |\hat{u} - \hat{v}|\,(\xi, t)\, d\xi \to 0, \quad t \to \infty,$$

which ends the first part of the proof of the theorem.

As a consequence of this first part we obtain that the asymptotic behaviour is the same as the one for solutions of the evolution given by a power n of the fractional Laplacian. It is now easy to check that the asymptotic behaviour is in fact the one described in the statement of the second part of the theorem. Indeed, in Fourier variables we have

$$\lim_{t \to \infty} \hat{v}(\eta t^{-\frac{1}{sn}}, t) = \lim_{t \to \infty} e^{-A^n |\eta|^{sn}} \widehat{u_0}(\eta t^{-\frac{1}{sn}})$$

$$= e^{-A^n |\eta|^{sn}} \widehat{u_0}(0) = e^{-A^n |\eta|^{sn}} \|u_0\|_{L^1(\mathbb{R}^N)}.$$

Therefore

$$\lim_{t \to +\infty} \max_y \left| t^{\frac{N}{sn}} v(y t^{\frac{1}{sn}}, t) - \|u_0\|_{L^1(\mathbb{R}^N)} G_{A^n}^{ns}(y) \right| = 0. \qquad \square$$

With similar arguments one can prove that also the asymptotic behaviour of the derivatives of solutions u of (1.33) is the same as the one for derivatives of solutions v of the evolution of a power n of the fractional Laplacian, assuming sufficient regularity of the solutions u of (1.33).

THEOREM 1.29. *Let u be a solution of* (1.33) *with $u_0 \in W^{k,1}(\mathbb{R}^N)$, $k \le sn$ and $\widehat{u_0} \in L^1(\mathbb{R}^N)$. Then the asymptotic behaviour of $D^k u(x,t)$ is given by*

$$\lim_{t \to +\infty} t^{\frac{N+k}{sn}} \max_x \left| D^k u(x,t) - D^k v(x,t) \right| = 0,$$

where v is the solution of

$$v_t(x,t) = -A^n (-\Delta)^{\frac{sn}{2}} v(x,t)$$

with initial condition $v(x,0) = u_0(x)$.

PROOF. We begin again by transforming our problem for u and v into a problem for the corresponding Fourier transforms $\hat{u}$ and $\hat{v}$. To this end we consider

$$\max_x \left| D^k u(x,t) - D^k v(x,t) \right|$$

$$\le (2\pi)^{-N} \int_{\mathbb{R}^N} \left| \widehat{D^k u(\xi,t)} - \widehat{D^k v(\xi,t)} \right| d\xi$$

$$= (2\pi)^{-N} \int_{\mathbb{R}^N} |\xi|^k \left| \hat{u}(\xi,t) - \hat{v}(\xi,t) \right| d\xi.$$

The proof of

$$\int_{\mathbb{R}^N} |\xi|^k \left| \hat{u}(\xi,t) - \hat{v}(\xi,t) \right| d\xi \to 0$$

as $t \to \infty$ works in an analogous way to the proof of the first part of the previous theorem since the additional term $|\xi|^k$ is always dominated by the exponential terms. $\qquad \square$

1.4.3. Rescaling the kernel in a higher order problem.
Now we show that the problem $v_t(x,t) = -A^n (-\Delta)^{\frac{sn}{2}} v(x,t)$ can be approximated by nonlocal problems like the ones presented in this section when they are rescaled in an appropriate way.

THEOREM 1.30. *Assume* (1.36). *Let u_ε be the unique solution to*

$$(1.40) \qquad \begin{cases} (u_\varepsilon)_t(x,t) = (-1)^{n-1} \dfrac{(J_\varepsilon * \mathrm{Id} - 1))^n}{\varepsilon^{sn}} (u_\varepsilon(x,t)), \\[2mm] u_\varepsilon(x,0) = u_0(x), \end{cases}$$

where

$$J_\varepsilon(x) = \varepsilon^{-N} J\left(\frac{x}{\varepsilon}\right).$$

Then, for every $T > 0$, we have

$$\lim_{\varepsilon \to 0} \|u_\varepsilon - v\|_{L^\infty(\mathbb{R}^N \times (0,T))} = 0,$$

where v is the solution of the local problem $v_t(x,t) = -A^n(-\Delta)^{\frac{sn}{2}} v(x,t)$ with the same initial condition $v(x,0) = u_0(x)$.

PROOF. As for the case of the heat equation, the proof uses once more the explicit formula for the solutions in Fourier variables. We have, arguing exactly as before,

$$\widehat{u_\varepsilon}(\xi,t) = e^{(-1)^{n-1}\frac{(\widehat{J_\varepsilon}(\xi)-1)^n}{\varepsilon^{sn}}t}\widehat{u_0}(\xi)$$

and

$$\hat{v}(\xi,t) = e^{-A^n|\xi|^{sn}t}\widehat{u_0}(\xi).$$

Now, since $\widehat{J_\varepsilon}(\xi) = \hat{J}(\varepsilon\xi)$,

$$\int_{\mathbb{R}^N} |\widehat{u_\varepsilon} - \hat{v}|\,(\xi,t)\,d\xi = \int_{\mathbb{R}^N} \left|\left(e^{(-1)^{n-1}\frac{(\hat{J}(\varepsilon\xi)-1)^n}{\varepsilon^{sn}}t} - e^{-A^n|\xi|^{sn}t}\right)\widehat{u_0}(\xi)\right|\,d\xi$$

$$\leq \|\widehat{u_0}\|_{L^\infty(\mathbb{R}^N)} \left(\int_{|\xi|\geq r(\varepsilon)} \left|e^{(-1)^{n-1}\frac{(\hat{J}(\varepsilon\xi)-1)^n}{\varepsilon^{sn}}t} - e^{-A^n|\xi|^{sn}t}\right|\,d\xi\right.$$

$$\left.+ \int_{|\xi|<r(\varepsilon)} \left|e^{(-1)^{n-1}\frac{(\hat{J}(\varepsilon\xi)-1)^n}{\varepsilon^{sn}}t} - e^{-A^n|\xi|^{sn}t}\right|\,d\xi\right).$$

For $t \in [0,T]$ we can proceed as in the proof of Theorem 1.28 to obtain that

$$\max_x |u_\varepsilon(x,t) - v(x,t)| \leq (2\pi)^{-N} \int_{\mathbb{R}^N} |\widehat{u_\varepsilon} - \hat{v}|\,(\xi,t)\,d\xi \to 0, \quad \varepsilon \to 0. \qquad \square$$

Bibliographical notes

The results included in this chapter come from [68], [121] and [140].

There is a large amount of literature dealing with related problems such as the ones considered in this chapter. We refer to equations with a linear nonlocal diffusion operator of the form $J * u - u$. See [106] for a general survey. Next, let us describe briefly some of the existing bibliography.

In [37], [38], [39] and [40] nonlocal diffusion is used to describe phase transition, and in [42] and [65] some biological models are studied. Concerning travelling front type solutions to the parabolic problem a great deal of work has been done. We quote [33], [34], [35], [36], [39], [69], [70], [72], [74], [81], [82], [85], [86], [105], [145] and [155]. In those papers the authors study the existence and stability of travelling waves for different source terms. Note that the equation for the profile of the travelling wave gives an integro-differential equation. The stability of such profiles is a delicate issue. See also [71], [73], [83], [107], [114] and [137], which deal with a source term of logistic type, bistable or power-like nonlinearity. With respect to a singular perturbation problem for a nonlocal equation see [131].

For anisotropic problems we refer to [84]. We mention [115], where some logistic equations and systems of Lotka-Volterra type are studied. See [75] and [77] for interesting features in other related nonlocal problems. Concerning large deviations for nonlocal problems (that can be used to gain some intuition on the underlying probabilistic background) we refer to [53].

There is also an increasing interest in free boundary problems and regularity issues for nonlocal problems such as the obstacle problem for the fractional Laplacian (that can be regarded as a nonlocal operator with a singular kernel), including a theory of viscosity solutions for nonlocal problems, but we are not dealing with such issues in the present work. We refer to [24], [28], [29], [62], [63], [64], [67], [123] and [146]. For initial traces and well-posedness of an evolution governed by the fractional Laplacian we refer to [3], and for fractional mean curvature flows, to [124]. These evolution problems governed by the fractional Laplacian are related to Levy processes in probability theory; see [21].

For nonlocal equations on spacial lattices (note that in this case we deal with an infinite system of ODEs) see, for example, [32], [38], [41] and [122].

CHAPTER 2

The Dirichlet problem
for linear nonlocal diffusion

In this chapter we continue the analysis of the linear nonlocal diffusion operator $J * u - u$ but instead of the Cauchy problem we confine the evolution to a bounded domain prescribing u outside the domain, the nonlocal analog of Dirichlet boundary conditions. Since the operator is nonlocal, it is not enough to prescribe the value of $u(x,t)$ on the boundary of the domain $\partial\Omega \times (0,T)$. In fact, we have to impose that $u(x,t)$ agrees with a given datum $g(x,t)$ outside Ω, that is, $u(x,t) = g(x,t)$ for $(x,t) \in (\mathbb{R}^N \setminus \Omega) \times (0,T)$.

As in the first chapter, we deal with existence and uniqueness of solutions, discuss the asymptotic behaviour as $t \to \infty$ (that in this case is given by an exponential decay) and show that the solutions of this nonlocal problem converge to the solutions of the classical Dirichlet problem for the heat equation when we rescale the equation.

2.1. The homogeneous Dirichlet problem

To impose boundary conditions to this nonlocal model, we consider a bounded smooth domain $\Omega \subset \mathbb{R}^N$. First, we present the nonlocal analog of the usual homogeneous Dirichlet boundary conditions for local problems.

Consider the nonlocal problem

$$(2.1) \quad \begin{cases} u_t(x,t) = \displaystyle\int_{\mathbb{R}^N} J(x-y)u(y,t)\,dy - u(x,t), & x \in \Omega, \ t > 0, \\[2mm] u(x,t) = 0, & x \notin \Omega, \ t > 0, \\[2mm] u(x,0) = u_0(x), & x \in \Omega, \end{cases}$$

where J satisfies hypothesis (H) stated in Chapter 1. In this model diffusion takes place in the whole $\mathbb{R}^N$ but we assume that u vanishes outside Ω. This is analogous to what is called Dirichlet boundary conditions for the heat equation. However, the boundary datum is not understood in the usual sense; see Remark 2.4. As for the Cauchy problem, we understand solutions in the integral sense.

DEFINITION 2.1. A *solution* of (2.1) is a function $u \in C([0,\infty); L^1(\mathbb{R}^N))$ such that

$$u(x,t) = u_0(x) + \int_0^t \int_{\mathbb{R}^N} J(x-y)(u(y,s) - u(x,s))\,dy\,ds, \quad \text{for } x \in \Omega, \ t > 0,$$

and

$$u(x,t) = 0, \quad \text{for } x \notin \Omega, \ t > 0.$$

In this case, when looking at the asymptotic behaviour of the solutions, we find, as happens for the heat equation, that an exponential decay is given by the first eigenvalue of an associated problem and the asymptotic behaviour of solutions is described by the unique (up to a constant) associated eigenfunction.

Existence and uniqueness of solutions are consequence of Banach's fixed point theorem and will be shown in Theorem 2.8 in a more general setting.

2.1.1. Asymptotic behaviour. Let us now analyze the asymptotic behaviour of the solutions.

First, we look for steady states of (2.1).

PROPOSITION 2.2. $u \equiv 0$ *is the unique stationary solution of* (2.1).

PROOF. Let u be a stationary solution of (2.1). Then, using that $\int_{\mathbb{R}^N} J = 1$,

$$0 = \int_{\mathbb{R}^N} J(x-y)u(y)\,dy - u(x) = \int_{\mathbb{R}^N} J(x-y)(u(y) - u(x))\,dy, \qquad x \in \Omega,$$

and $u(x) = 0$ for $x \notin \Omega$. Hence, we obtain that for every $x \in \mathbb{R}^N$,

$$u(x) = \int_{\mathbb{R}^N} J(x-y)u(y)\,dy.$$

This equation, together with $u(x) = 0$ for $x \notin \Omega$, implies that $u \equiv 0$. □

For a function g given in a set D, we define

$$\overline{g}(x) = \begin{cases} g(x) & \text{if } x \in D, \\ 0 & \text{otherwise.} \end{cases}$$

Since zero is the unique stationary solution, it is expected that solutions converge to zero as $t \to \infty$. Hence, the main goal here will be to study the rate of convergence.

Let $\lambda_1 = \lambda_1(\Omega)$ be given by

$$(2.2) \qquad \lambda_1 = \inf_{u \in L^2(\Omega)} \frac{\frac{1}{2} \int_{\mathbb{R}^N} \int_{\mathbb{R}^N} J(x-y)(\overline{u}(x) - \overline{u}(y))^2\,dx\,dy}{\int_{\Omega} (u(x))^2\,dx}$$

and let ϕ_1 be an associated eigenfunction (a function where the infimum is attained).

First, let us look at the eigenvalue given by (2.2). If the minimum is attained, we get, differentiating, that it is a solution of

$$(2.3) \qquad u(x) - \int_{\mathbb{R}^N} J(x-y)\overline{u}(y)\,dy = \lambda_1 u(x), \quad x \in \Omega.$$

Conversely, it is easy to check that if u is a solution to (2.3) (with λ_1 the smallest eigenvalue), then it is a minimizer of (2.2). Hence, we look for the first eigenvalue of (2.3), which is equivalent to

$$(2.4) \qquad (1 - \lambda_1)u(x) = \int_{\mathbb{R}^N} J(x-y)\overline{u}(y)\,dy, \quad x \in \Omega.$$

Let $S : L^2(\Omega) \to L^2(\Omega)$ be the operator given by

$$S(u)(x) := \int_{\mathbb{R}^N} J(x - y)\overline{u}(y)\, dy, \quad x \in \Omega.$$

Hence we are looking for the largest eigenvalue, $1 - \lambda_1$, of S. Since S is compact, this eigenvalue is attained at some function $\phi_1(x)$, which turns out to be an eigenfunction for our original problem (2.3).

By taking $|\phi_1|$ instead of ϕ_1 in (2.2) we may assume that $\phi_1 \geq 0$ in Ω. Indeed, one simply has to use the fact that $(a - b)^2 \geq (|a| - |b|)^2$.

Let us analyze some properties of the eigenvalue problem (2.3).

PROPOSITION 2.3. *Let λ_1 be the first eigenvalue of (2.3) and denote by $\phi_1(x)$ a corresponding non-negative eigenfunction. Then $\phi_1(x)$ is strictly positive in Ω and λ_1 is a positive simple eigenvalue with $\lambda_1 < 1$.*

PROOF. Since $J(0) > 0$, we have that $B(0, d) \subset \mathrm{supp}(J)$ for some $d > 0$. Let us assume, for simplicity, that $\mathrm{supp}(J) = \overline{B}(0, 1)$. First, observe that $\lambda_1 = 1$ cannot be an eigenvalue since then

$$\int_{\mathbb{R}^N} J(x - y)\overline{\phi}_1(y)\, dy = 0, \qquad \phi_1(x) \geq 0,$$

which implies $\phi_1 = 0$. Consequently, we have that

$$(1 - \lambda_1)\phi_1(x) = \int_{\mathbb{R}^N} J(x - y)\overline{\phi}_1(y)\, dy, \quad x \in \Omega, \quad \lambda_1 \neq 1,$$

which implies that ϕ_1 is uniformly continuous in Ω. In what follows, we consider bounded continuous functions in Ω extended in the natural way to $\overline{\Omega}$. We begin with the positivity of the eigenfunction ϕ_1. Assume for contradiction that the set $\mathbf{B} = \{x \in \Omega : \phi_1(x) = 0\}$ is nonempty. Then, from the continuity of ϕ_1 in Ω, we have that $\mathbf{B}$ is closed. We next prove that $\mathbf{B}$ is also open, and hence, since Ω is connected, standard topological arguments allow us to conclude that $\Omega \equiv \mathbf{B}$ yielding a contradiction. Consider $x_0 \in \mathbf{B}$. Since $\phi_1 \geq 0$, we obtain from (2.4) that $\Omega \cap B(x_0, 1) \in \mathbf{B}$ (we use here that $\mathrm{supp}(J) = \overline{B}(0, 1)$). Hence $\mathbf{B}$ is open and the result follows. Analogous arguments apply to prove that ϕ_1 is positive in $\overline{\Omega}$.

Assume now by contradiction that $\lambda_1 \leq 0$ and denote by M^* the maximum of ϕ_1 in $\overline{\Omega}$ and by x^* a point where the maximum is attained. Assume for the moment that $x^* \in \Omega$. By Proposition 2.2, one can choose x^* in such a way that $\phi_1(x) \neq M^*$ in $\Omega \cap B(x^*, 1)$. By using (2.4) we get that

$$M^* \leq (1 - \lambda_1)\phi_1(x^*) = \int_{\mathbb{R}^N} J(x^* - y)\overline{\phi}_1(y) < M^*,$$

and a contradiction follows. If $x^* \in \partial\Omega$, we obtain a similar contradiction after substituting and passing to the limit in (2.4) on a sequence $\{x_n\} \in \Omega$, $x_n \to x^*$ as $n \to \infty$. To obtain the upper bound, assume that $\lambda_1 \geq 1$. Then, from (2.4) we have for every $x \in \Omega$ that

$$0 \geq (1 - \lambda_1)\phi_1(x^*) = \int_{\mathbb{R}^N} J(x^* - y)\overline{\phi}_1(y)dy,$$

a contradiction with the positivity of ϕ_1.

Finally, to prove that λ_1 is a simple eigenvalue, let $\phi_1 \neq \phi_2$ be two different eigenfunctions associated to λ_1 and define

$$C^* = \inf\{C > 0 : \phi_2(x) \leq C\phi_1(x), \, x \in \overline{\Omega}\}.$$

The regularity of the eigenfunctions and the previous analysis show that C^* is nontrivial and bounded. Moreover, from its definition, there must exist $x^* \in \overline{\Omega}$ such that $\phi_2(x^*) = C^*\phi_1(x^*)$. Define $\phi(x) = C^*\phi_1(x) - \phi_2(x)$. From the linearity of (2.3), ϕ is a nonnegative eigenfunction associated to λ_1 with $\phi(x^*) = 0$. From the positivity of the eigenfunctions stated above, it must be $\phi \equiv 0$. Therefore, $\phi_2(x) = C^*\phi_1(x)$ and the result follows. This completes the proof. $\qquad \square$

REMARK 2.4. Observe that the first eigenfunction ϕ_1 is strictly positive in Ω (with a positive continuous extension to $\overline{\Omega}$) and vanishes outside Ω. Therefore a discontinuity occurs on $\partial\Omega$ and the boundary value is not taken in the usual "classical" sense.

Let us recall the concept of ω-limit set of a trajectory $u(t)$ that begins with $u(0) = u_0$,

$$\omega(u_0) := \left\{ g \in L^2(\Omega) \; : \; \exists t_n \to \infty \text{ with } u(t_n) \to g \text{ in } L^2(\Omega) \right\}.$$

The following theorem is the main result of this section.

THEOREM 2.5. Let $u_0 \in L^2(\Omega)$. Then the solution u of (2.1) decays to zero as $t \to \infty$ with an exponential rate

$$(2.5) \qquad \qquad \|u(\cdot, t)\|_{L^2(\Omega)} \leq \|u_0\|_{L^2(\Omega)} e^{-\lambda_1 t}.$$

If u_0 is continuous, positive and bounded, then there exist positive constants C and C^* such that

$$(2.6) \qquad \qquad \|u(\cdot, t)\|_{L^\infty(\Omega)} \leq C\, e^{-\lambda_1 t}$$

and

$$(2.7) \qquad \qquad \lim_{t \to \infty} \|e^{\lambda_1 t} u(\cdot, t) - C^*\phi_1\|_{L^\infty(\Omega)} = 0.$$

PROOF. Using the symmetry of J, we have

$$\frac{\partial}{\partial t}\left(\frac{1}{2}\int_\Omega u^2(x, t)\, dx\right) = \int_{\mathbb{R}^N}\int_{\mathbb{R}^N} J(x - y)\, (u(y, t) - u(x, t))\, u(x, t)\, dy\, dx$$

$$= -\frac{1}{2}\int_{\mathbb{R}^N}\int_{\mathbb{R}^N} J(x - y)\, (u(y, t) - u(x, t))^2\, dy\, dx.$$

From the definition of λ_1, (2.2), we get

$$\frac{\partial}{\partial t}\int_\Omega u^2(x, t)\, dx \leq -2\lambda_1 \int_\Omega u^2(x, t)\, dx.$$

Therefore

$$\int_\Omega u^2(x, t)\, dx \leq e^{-2\lambda_1 t}\int_\Omega u_0^2(x)\, dx$$

and (2.5) is obtained.

We now establish the decay rate and the convergence stated in (2.6) and (2.7) respectively. Consider a nontrivial and nonnegative continuous initial datum $u_0(x)$ and let $u(x, t)$ be the corresponding solution to (2.1). We first note that $u(x, t)$ is a continuous function satisfying $u(x,t) > 0$ for every $x \in \Omega$ and $t > 0$, and the same holds in $\overline{\Omega}$. This instantaneous positivity can be obtained by using analogous topological arguments to those in Proposition 2.3.

In order to deal with the asymptotic analysis, it is more convenient to introduce the rescaled function $v(x,t) = e^{\lambda_1 t}u(x,t)$. By substituting in (2.1), we find that the function $v(x,t)$ satisfies

$$(2.8) \qquad v_t(x,t) = \int_{\mathbb{R}^N} J(x-y)v(y,t)\,dy - (1-\lambda_1)v(x,t), \quad x \in \Omega.$$

On the other hand, we have that $C\phi_1(x)$ is a solution of (2.8) for every $C \in \mathbb{R}$ and moreover, it follows from the above eigenfunction analysis that the set of stationary solutions of (2.8) is given by $\mathbf{S}^* = \{C\phi_1, C \in \mathbb{R}\}$.

Define now for every $t > 0$, the function

$$C^*(t) = \inf\{C > 0 : v(x,t) \leq C\phi_1(x), x \in \overline{\Omega}\}.$$

By definition and by using the linearity of equation (2.8), $C^*(t)$ is a nonincreasing function. In fact, this is a consequence of the comparison principle (given in Corollary 2.11) applied to the solutions $C^*(t_1)\phi_1(x)$ and $v(x,t)$ for t larger than any fixed $t_1 > 0$. It implies that $C^*(t_1)\phi_1(x) \geq v(x,t)$ for every $t \geq t_1$, and therefore $C^*(t_1) \geq C^*(t)$ for every $t \geq t_1$. In an analogous way, one can see that the function

$$C_*(t) = \sup\{C > 0 : v(x,t) \geq C\phi_1(x), x \in \overline{\Omega}\}$$

is nondecreasing. These properties imply that the following two limits exist:

$$\lim_{t\to\infty} C^*(t) = K^* \quad \text{and} \quad \lim_{t\to\infty} C_*(t) = K_*,$$

and also provide the compactness of the orbits, which is necessary to pass to the limit (after extracting subsequences if needed) in order to obtain that $v(\cdot, t+t_n) \to w(\cdot, t)$ as $t_n \to \infty$ uniformly on compact subsets in $\overline{\Omega} \times \mathbb{R}_+$ and that $w(x,t)$ is a continuous function which satisfies (2.8). For every $g \in \omega(u_0)$ we also have

$$K_*\phi_1(x) \leq g(x) \leq K^*\phi_1(x).$$

Moreover, $C^*(t)$ plays the role of a Lyapunov function and this fact allows us to conclude that $\omega(u_0) \subset \mathbf{S}^*$, the set of stationary solutions of (2.8), and the uniqueness of the profile. In more detail, assume that $g \in \omega(u_0)$ does not belong to $\mathbf{S}^*$ and consider $w(x,t)$, the solution of (2.8) with initial data $g(x)$, and define

$$C^*(w)(t) = \inf\{C > 0 : w(x,t) \leq C\phi_1(x), x \in \overline{\Omega}\}.$$

It is clear that $W(x,t) = K^*\phi_1(x) - w(x,t)$ is a nonnegative continuous solution of (2.8) and it becomes strictly positive for every $t > 0$. This implies that there exists $t^* > 0$ such that $C^*(w)(t^*) < K^*$ and, by the convergence, the same holds before passing to the limit. Hence, $C^*(t^* + t_j) < K^*$ if j is large enough, which is a contradiction with the properties of $C^*(t)$. The same arguments allow us to establish the uniqueness of the profile. $\qquad \square$

2.2. The nonhomogeneous Dirichlet problem

This section deals with the following nonlocal nonhomogeneous Dirichlet boundary value problem:

(2.9)
$$
\begin{cases}
u_t(x,t) = \displaystyle\int_{\mathbb{R}^N} J(x-y)(u(y,t) - u(x,t))\, dy, & x \in \Omega,\ t > 0, \\[2mm]
u(x,t) = g(x,t), & x \notin \Omega,\ t > 0, \\[2mm]
u(x,0) = u_0(x), & x \in \Omega,
\end{cases}
$$

where $g(x,t)$ is defined for $x \in \mathbb{R}^N \setminus \Omega$, $t > 0$ and $u_0(x)$ for $x \in \Omega$.

In this model the values of u are prescribed outside Ω, which is analogous to prescribing the so-called Dirichlet boundary conditions for the classical heat equation. However, the boundary datum is not understood in the usual sense as we have seen in Remark 2.4.

Existence and uniqueness of solutions of (2.9) is proved by a fixed point argument. Also a comparison principle is obtained.

2.2.1. Existence, uniqueness and a comparison principle. First, let us state what will be understood by a solution of problem (2.9). Let $u_0 \in L^1(\Omega)$ and $g \in C([0,\infty); L^1(\mathbb{R}^N \setminus \Omega))$.

DEFINITION 2.6. A *solution* of (2.9) is a function $u \in C([0,\infty); L^1(\mathbb{R}^N))$ such that

$$
u(x,t) = u_0(x) + \int_0^t \int_{\mathbb{R}^N} J(x-y)\,(u(y,s) - u(x,s))\, dy\, ds, \quad \text{for } x \in \Omega,\ t > 0,
$$

where it is assumed that

$$
u(x,t) = g(x,t), \quad \text{for } x \notin \Omega,\ t > 0.
$$

Existence and uniqueness of solutions is a consequence of Banach's fixed point theorem. Fix $t_0 > 0$ and consider the Banach space $X_{t_0} = \{w \in C([0,t_0]; L^1(\Omega))\}$ with the norm

$$
|||w||| = \max_{0 \le t \le t_0} \|w(\cdot,t)\|_{L^1(\Omega)}.
$$

The solution will be obtained as a fixed point of the operator $\mathcal{T} : X_{t_0} \to X_{t_0}$ defined by

$$
\mathcal{T}_{w_0}(w)(x,t) = w_0(x) + \int_0^t \int_{\mathbb{R}^N} J(x-y)\,(w(y,s) - w(x,s))\, dy\, ds,
$$

for $x \in \Omega$, $t > 0$, where

$$
w(x,t) = g(x,t), \quad \text{for } x \notin \Omega,\ t > 0.
$$

LEMMA 2.7. *Let $w_0, z_0 \in L^1(\Omega)$, then there exists a constant C depending on J and Ω such that*

$$
|||\mathcal{T}_{w_0}(w) - \mathcal{T}_{z_0}(z)||| \le Ct_0|||w - z||| + \|w_0 - z_0\|_{L^1(\Omega)}
$$

for all $w, z \in X_{t_0}$.

PROOF. We have

$$\int_\Omega |\mathcal{T}_{w_0}(w)(x,t) - \mathcal{T}_{z_0}(z)(x,t)|\, dx \leq \int_\Omega |w_0 - z_0|(x)\, dx$$

$$+ \int_\Omega \left| \int_0^t \int_{\mathbb{R}^N} J(x-y) \left[(w(y,s) - z(y,s)) - (w(x,s) - z(x,s)) \right] dy\, ds \right| dx.$$

Hence, taking into account that $w - z$ vanishes outside Ω,

$$|||\mathcal{T}_{w_0}(w) - \mathcal{T}_{z_0}(z)||| \leq ||w_0 - z_0||_{L^1(\Omega)} + Ct_0|||w - z|||,$$

as we wanted to prove. □

Now all is ready to prove existence and uniqueness of solutions.

THEOREM 2.8. *For every $u_0 \in L^1(\Omega)$ there exists a unique solution of problem (2.9).*

PROOF. We check first that $\mathcal{T}_{u_0}$ maps X_{t_0} into X_{t_0}. Taking $z_0 \equiv 0$ and $z \equiv 0$ in Lemma 2.7 we get that $\mathcal{T}_{u_0}(w) \in C([0, t_0]; L^1(\Omega))$ for any $w \in X_{t_0}$.

Choose t_0 such that $Ct_0 < 1$. Taking $z_0 \equiv w_0 \equiv u_0$ in Lemma 2.7 we get that $\mathcal{T}_{u_0}$ is a strict contraction in X_{t_0} and the existence and uniqueness part of the theorem follows from Banach's fixed point theorem in the interval $[0, t_0]$. To extend the solution to $[0, \infty)$ we may take $u(x, t_0) \in L^1(\Omega)$ as initial datum and obtain a solution up to $[0, 2t_0]$. Iterating this procedure we get a solution defined in $[0, \infty)$. □

Let us define what we understand by sub and supersolutions.

DEFINITION 2.9. A function $u \in W^{1,1}(0, T; L^1(\Omega))$ is a *supersolution* of (2.9) if

(2.10)
$$\begin{cases} u_t(x,t) \geq \int_{\mathbb{R}^N} J(x-y)(u(y,t) - u(x,t))dy, & x \in \Omega,\ t > 0, \\ u(x,t) \geq g(x,t), & x \notin \Omega,\ t > 0, \\ u(x,0) \geq u_0(x), & x \in \Omega. \end{cases}$$

As usual, subsolutions are defined analogously by reversing the inequalities.

LEMMA 2.10. *Let $u_0 \in C(\overline{\Omega})$, $u_0 \geq 0$, and $u \in C(\overline{\Omega} \times [0, T])$ a supersolution of (2.9) with $g \geq 0$. Then $u \geq 0$.*

PROOF. Assume by contradiction that $u(x,t)$ is negative somewhere. Let $v(x,t) = u(x,t) + \varepsilon t$ with ε so small that v is still negative somewhere. Then, if (x_0, t_0) is a point where v attains its negative minimum, we have $t_0 > 0$ and

$$v_t(x_0, t_0) = u_t(x_0, t_0) + \varepsilon > \int_{\mathbb{R}^N} J(x-y)(u(y,t_0) - u(x_0, t_0))\, dy$$

$$= \int_{\mathbb{R}^N} J(x-y)(v(y,t_0) - v(x_0, t_0))\, dy \geq 0,$$

which is a contradiction. Thus, $u \geq 0$. □

COROLLARY 2.11. *Let u_0 and v_0 be given in $L^1(\Omega)$ with $u_0 \geq v_0$ and two data g, $h \in C([0,T]; L^1(\mathbb{R}^N \setminus \Omega))$ with $g \geq h$. Let u be a solution of (2.9) with $u(x,0) = u_0$ and Dirichlet boundary condition g, and let v be a solution of (2.9) with $v(x,0) = v_0$ and boundary condition h. Then $u \geq v$ a.e.*

PROOF. Let $w = u - v$. Then w is a supersolution with nonnegative initial datum $u_0 - v_0 \geq 0$ and nonnegative boundary condition $g - h \geq 0$. Using the continuity of solutions with respect to the data and the fact that $J \in L^\infty(\mathbb{R}^N)$, we may assume that u, $v \in C(\overline{\Omega} \times [0,T])$. By Lemma 2.10 we obtain that $w = u - v \geq 0$. So the corollary is proved.   □

COROLLARY 2.12. *Let $u \in C(\overline{\Omega} \times [0,T])$ (resp. v) be a supersolution (resp. subsolution) of (2.9). Then $u \geq v$.*

PROOF. It follows the lines of the proof of the previous corollary.   □

2.2.2. Convergence to the heat equation when rescaling the kernel.
Consider the classical Dirichlet problem for the heat equation,

$$(2.11) \quad \begin{cases} v_t(x,t) - \Delta v(x,t) = 0 & \text{in } \Omega \times (0,\infty), \\ v(x,t) = g(x,t) & \text{in } \partial\Omega \times (0,\infty), \\ v(x,0) = u_0(x) & \text{in } \Omega. \end{cases}$$

The nonlocal Dirichlet model (2.9) and the classical Dirichlet problem (2.11) share many properties; among them the asymptotic behaviour of their solutions as $t \to \infty$ is similar as it was proved in the previous section.

The main goal now is to show that the Dirichlet problem for the heat equation (2.11) can be approximated by suitable nonlocal problems of the form (2.9).

More precisely, for a given compactly supported J, $\text{supp}(J) = \overline{B}(0,1)$ for simplicity, and a given $\varepsilon > 0$, consider the rescaled kernel

$$J_\varepsilon(\xi) = C_1 \frac{1}{\varepsilon^N} J\left(\frac{\xi}{\varepsilon}\right), \quad \text{with} \quad C_1^{-1} = \frac{1}{2} \int_{\mathbb{R}^N} J(z) z_N^2 \, dz.$$

Here C_1 is a normalizing constant in order to obtain the Laplacian in the limit instead of a multiple of it. Let $u_\varepsilon(x,t)$ be the solution of

$$(2.12) \quad \begin{cases} (u_\varepsilon)_t(x,t) = \int_{\mathbb{R}^N} \frac{1}{\varepsilon^2} J_\varepsilon(x-y)(u_\varepsilon(y,t) - u_\varepsilon(x,t)) \, dy, & x \in \Omega, t > 0, \\ u_\varepsilon(x,t) = g(x,t), & x \notin \Omega, t > 0, \\ u_\varepsilon(x,0) = u_0(x), & x \in \Omega. \end{cases}$$

The main result now reads as follows.

THEOREM 2.13. *Let Ω be a bounded $C^{2+\alpha}$ domain for some $0 < \alpha < 1$. Let $v \in C^{2+\alpha, 1+\alpha/2}(\overline{\Omega} \times [0,T])$ be the solution of (2.11) and let u_ε be the solution of (2.12) with J_ε as above. Then there exists $C = C(T)$ such that*

$$\sup_{t \in [0,T]} \|u_\varepsilon(\cdot,t) - v(\cdot,t)\|_{L^\infty(\Omega)} \leq C\varepsilon^\alpha \to 0 \quad \text{as } \varepsilon \to 0.$$

See [109] for the existence of solutions of (2.11) with the desired regularity.

In order to prove Theorem 2.13 let $\tilde{v}$ be a $C^{2+\alpha,1+\alpha/2}$ extension of v to the whole $\mathbb{R}^N \times [0,T]$.

Let us define the operator

$$\tilde{L}_\varepsilon(z) = \frac{1}{\varepsilon^2} \int_{\mathbb{R}^N} J_\varepsilon(x-y)\big(z(y,t) - z(x,t)\big)dy.$$

Then $\tilde{v}$ satisfies

(2.13)
$$\begin{cases} \tilde{v}_t(x,t) = \tilde{L}_\varepsilon(\tilde{v})(x,t) + F_\varepsilon(x,t), & x \in \Omega,\ t \in (0,T], \\ \tilde{v}(x,t) = g(x,t) + G(x,t), & x \notin \Omega,\ t \in (0,T], \\ \tilde{v}(x,0) = u_0(x), & x \in \Omega, \end{cases}$$

where, since $\Delta v = \Delta \tilde{v}$ in Ω,

$$F_\varepsilon(x,t) = -\tilde{L}_\varepsilon(\tilde{v})(x,t) + \Delta \tilde{v}(x,t).$$

Moreover, as $G(x,t) = \tilde{v}(x,t) - g(x,t)$ is smooth and $G(x,t) = 0$ if $x \in \partial\Omega$, there exists $M_1 > 0$ such that

$$|G(x,t)| \leq M_1 \varepsilon, \quad \text{for } x \text{ such that } \operatorname{dist}(x,\partial\Omega) \leq \varepsilon.$$

We set $w_\varepsilon = \tilde{v} - u_\varepsilon$. Note that

$$\begin{cases} (w_\varepsilon)_t(x,t) = \tilde{L}_\varepsilon(w_\varepsilon)(x,t) + F_\varepsilon(x,t), & x \in \Omega,\ t \in (0,T], \\ w_\varepsilon(x,t) = G(x,t), & x \notin \Omega,\ t \in (0,T], \\ w_\varepsilon(x,0) = 0, & x \in \Omega. \end{cases}$$

First, we claim that, by the choice of C_1, the fact that J is radially symmetric and $\tilde{v} \in C^{2+\alpha,1+\alpha/2}(\mathbb{R}^N \times [0,T])$, there exists $M_2 > 0$ such that

(2.14)
$$\sup_{t\in[0,T]} \|F_\varepsilon(\cdot,t)\|_{L^\infty(\Omega)} = \sup_{t\in[0,T]} \|\Delta\tilde{v}(\cdot,t) - \tilde{L}_\varepsilon(\tilde{v})(\cdot,t)\|_{L^\infty(\Omega)} \leq M_2 \varepsilon^\alpha.$$

In fact,

$$\Delta\tilde{v}(x,t) - \frac{C_1}{\varepsilon^{N+2}} \int_{\mathbb{R}^N} J\left(\frac{x-y}{\varepsilon}\right) (\tilde{v}(y,t) - \tilde{v}(x,t))\, dy$$

becomes, under the change of variables $z = (x-y)/\varepsilon$,

$$\Delta\tilde{v}(x,t) - \frac{C_1}{\varepsilon^2} \int_{\mathbb{R}^N} J(z) (\tilde{v}(x-\varepsilon z,t) - \tilde{v}(x,t))\, dz$$

and hence (2.14) follows by a simple Taylor expansion. This proves the claim.

Let us proceed now to proving Theorem 2.13.

PROOF OF THEOREM 2.13. In order to prove the theorem we first look for a supersolution of (2.13). Let $\bar{w}$ be given by

$$\bar{w}(x,t) = K_1 \varepsilon^\alpha t + K_2 \varepsilon.$$

For $x \in \Omega$ we have, if K_1 is large,

(2.15)
$$\bar{w}_t(x,t) - \tilde{L}(\bar{w})(x,t) = K_1 \varepsilon^\alpha \geq F_\varepsilon(x,t) = (w_\varepsilon)_t(x,t) - \tilde{L}_\varepsilon(w_\varepsilon)(x,t).$$

Since
$$|G(x,t)| \leq M\varepsilon \quad \text{for } x \text{ such that } \operatorname{dist}(x, \partial\Omega) \leq \varepsilon,$$
choosing K_2 large, we obtain

(2.16) $$\bar{w}(x,t) \geq w_\varepsilon(x,t)$$

for $x \notin \Omega$ such that $\operatorname{dist}(x, \partial\Omega) \leq \varepsilon$ and $t \in [0,T]$. Moreover it is clear that

(2.17) $$\bar{w}(x,0) = K_2\varepsilon > w_\varepsilon(x,0) = 0.$$

Thanks to (2.15), (2.16) and (2.17) we can apply the comparison result and conclude that
$$w_\varepsilon(x,t) \leq \bar{w}(x,t) = K_1\varepsilon^\alpha t + K_2\varepsilon.$$

In a similar fashion we prove that $\underline{w}(x,t) = -K_1\varepsilon^\alpha t - K_2\varepsilon$ is a subsolution and hence
$$w_\varepsilon(x,t) \geq \underline{w}(x,t) = -K_1\varepsilon^\alpha t - K_2\varepsilon.$$

Therefore
$$\sup_{t\in[0,T]} \|u_\varepsilon(\cdot,t) - v(\cdot,t)\|_{L^\infty(\Omega)} \leq C\varepsilon^\alpha,$$

C depending on T, and the proof is complete. $\qquad\square$

Bibliographical notes

The results of this chapter are based on [**68**] and [**78**].

CHAPTER 3

The Neumann problem
for linear nonlocal diffusion

In the previous chapter Dirichlet boundary conditions have been studied for nonlocal linear diffusion. The main goal of this chapter is to look for Neumann boundary conditions. Here, the integrals over the whole $\mathbb{R}^N$ are replaced by integrals only in Ω. In this way the diffusion is forced to act only in Ω with no interchange of mass between Ω and the exterior $\mathbb{R}^N \setminus \Omega$. This is a motivation to call this a nonlocal Neumann homogeneous boundary condition. We also treat the case in which the amount of individuals entering or leaving the domain is prescribed (nonhomogeneous Neumann boundary conditions).

The topics treated here are existence, uniqueness, the long time behaviour of the solutions of the homogeneous problem, which in this case go exponentially to the mean value of the initial condition, and convergence of the rescaled nonlocal problems to the local ones.

3.1. The homogeneous Neumann problem

The problem to consider is

$$(3.1) \qquad \begin{cases} u_t(x,t) = \displaystyle\int_\Omega J(x-y)(u(y,t) - u(x,t))\,dy, & x \in \Omega,\ t > 0, \\ u(x,0) = u_0(x), & x \in \Omega, \end{cases}$$

where, as in the previous chapters, J satisfies hypothesis (H) and Ω is a bounded domain in $\mathbb{R}^N$. As in the Dirichlet case, solutions are to be understood in an integral sense.

DEFINITION 3.1. A *solution* of (3.1) is a function $u \in C([0,\infty); L^1(\Omega))$ such that

$$u(x,t) = u_0(x) + \int_0^t \int_\Omega J(x-y)\,(u(y,s) - u(x,s))\,dy\,ds, \qquad x \in \Omega,\ t > 0.$$

In this model the integral term takes into account the diffusion inside Ω. In fact, we recall that the integral $\int J(x-y)(u(y,t) - u(x,t))\,dy$ takes into account the individuals arriving at or leaving position x from other places. Since the integration is only in Ω, the diffusion is assumed to take place only in Ω. The individuals may not enter nor leave Ω. This is analogous to what is called homogeneous Neumann boundary conditions in the literature.

Again in this case we find the asymptotic behaviour analogous to the one that holds for the heat equation with Neumann boundary conditions. The solution

$u(x,t)$ of (3.1) converges exponentially to the mean value of the initial datum, and the decay is determined by the eigenvalue

(3.2) $$\beta_1 = \inf_{u \in L^2(\Omega), \int_\Omega u = 0} \frac{\dfrac{1}{2} \displaystyle\int_\Omega \int_\Omega J(x-y)(u(y) - u(x))^2 \, dy \, dx}{\displaystyle\int_\Omega (u(x))^2 \, dx}.$$

As in the previous chapter, the existence and uniqueness of solutions is a consequence of Banach's fixed point theorem. The main arguments are basically the same and are given in Section 3.2.1 in a more general setting to make this chapter self-contained.

The first result in this section shows that the solution u of (3.1) preserves the total mass.

PROPOSITION 3.2. *For every $u_0 \in L^1(\Omega)$ the unique solution u of (3.1) preserves the total mass in Ω, that is,*

$$\int_\Omega u(y,t) \, dy = \int_\Omega u_0(y) \, dy.$$

PROOF. Since

$$u(x,t) - u_0(x) = \int_0^t \int_\Omega J(x-y)(u(y,s) - u(x,s)) \, dy \, ds,$$

integrating in x and applying Fubini's theorem, we obtain

$$\int_\Omega u(x,t) \, dx - \int_\Omega u_0(x) \, dx = 0. \qquad \square$$

3.1.1. Asymptotic behaviour. The corresponding stationary problem to (3.1) is described by the equation

(3.3) $$0 = \int_\Omega J(x-y)(\varphi(y) - \varphi(x)) \, dy.$$

The only solutions to this equation are constants.

PROPOSITION 3.3. *Every stationary solution of (3.1) is constant in Ω, and, since the total mass is preserved, the unique stationary solution with the same mass as u_0 is*

$$\varphi = \frac{1}{|\Omega|} \int_\Omega u_0.$$

PROOF. Observe that (3.3) implies that φ is a continuous function. Set

$$K = \max_{x \in \overline{\Omega}} \varphi(x)$$

and consider the set

$$\mathcal{A} = \left\{ x \in \overline{\Omega} \ : \ \varphi(x) = K \right\}.$$

The set $\mathcal{A}$ is clearly closed and nonempty. We claim that it is also open in $\overline{\Omega}$. Let $x_0 \in \mathcal{A}$; then

$$0 = \int_\Omega J(x_0 - y)(\varphi(y) - \varphi(x_0)) \, dy.$$

Therefore, since $\varphi(y) \le \varphi(x_0)$, this implies $\varphi(y) = \varphi(x_0)$ for all $y \in \Omega \cap B(x_0, d)$, for any $B(0, d) \subset \text{supp}(J)$. Hence $\mathcal{A}$ is open as claimed. Consequently, as Ω is connected, $\mathcal{A} = \overline{\Omega}$ and φ is constant. $\qquad\square$

We now show that β_1 defined in (3.2) is strictly positive.

PROPOSITION 3.4. *Given J and Ω the quantity*

$$\beta_1 := \beta_1(J, \Omega) = \inf_{u \in L^2(\Omega), \int_\Omega u = 0} \frac{\dfrac{1}{2} \displaystyle\int_\Omega \int_\Omega J(x - y)(u(y) - u(x))^2 \, dy \, dx}{\displaystyle\int_\Omega (u(x))^2 \, dx}$$

is strictly positive.

PROOF. It is clear that $\beta_1 \ge 0$. Let us prove that β_1 is in fact strictly positive. To this end, consider the subspace H of $L^2(\Omega)$ given by the orthogonal to the constants, and the symmetric (self-adjoint) operator $T : H \mapsto H$ given by

$$T(u)(x) = \int_\Omega J(x - y)(u(x) - u(y)) \, dy = -\int_\Omega J(x - y)u(y) \, dy + A(x)u(x),$$

where $A(x) = \int_\Omega J(x - y)dy$. Note that T is the sum of an invertible operator and a compact operator. Since T is symmetric, its spectrum satisfies $\sigma(T) \subset [m, M]$, where

$$m = \inf_{u \in H, \|u\|_{L^2(\Omega)} = 1} \langle Tu, u \rangle$$

and

$$M = \sup_{u \in H, \|u\|_{L^2(\Omega)} = 1} \langle Tu, u \rangle;$$

see [57]. Remark that

$$m = \inf_{u \in H, \|u\|_{L^2(\Omega)} = 1} \langle Tu, u \rangle$$

$$= \inf_{u \in H, \|u\|_{L^2(\Omega)} = 1} \int_\Omega \int_\Omega J(x - y)(u(x) - u(y)) \, dy \, u(x) \, dx$$

$$= \beta_1.$$

Then $m \ge 0$. Let us show now that

$$m > 0.$$

If not, then, since $m \in \sigma(T)$ (see [57]), $T : H \mapsto H$ is not invertible. Using Fredholm's alternative, this implies that there exists a nontrivial $u \in H$ such that $T(u) = 0$, but then, by Proposition 3.3, u must be constant in Ω, which is a contradiction. $\qquad\square$

To study the asymptotic behaviour of the solutions, an upper estimate on β_1 is needed. Here and in what follows, $\mathcal{X}_D$ denotes the characteristic function of the set D.

LEMMA 3.5. *Let β_1 be given by (3.2); then*

$$(3.4) \qquad \beta_1 \leq \min_{x \in \overline{\Omega}} \int_{\Omega} J(x - y)\, dy.$$

PROOF. Let

$$A(x) = \int_{\Omega} J(x - y)\, dy.$$

Since $\overline{\Omega}$ is compact and A is continuous, there exists a point $x_0 \in \overline{\Omega}$ such that

$$A(x_0) = \min_{x \in \overline{\Omega}} A(x).$$

For every small ε let us choose two disjoint balls of radius ε contained in Ω, $B(x_{1,\varepsilon}, \varepsilon)$ and $B(x_{2,\varepsilon}, \varepsilon)$, in such a way that $x_{i,\varepsilon} \to x_0$ as $\varepsilon \to 0$. By using

$$u_\varepsilon(x) = \chi_{B(x_{1,\varepsilon}, \varepsilon)}(x) - \chi_{B(x_{2,\varepsilon}, \varepsilon)}(x)$$

as a test function in the definition of β_1, (3.2), for ε small, we have

$$\beta_1 \leq \frac{\dfrac{1}{2} \displaystyle\int_{\Omega} \int_{\Omega} J(x - y)(u_\varepsilon(y) - u_\varepsilon(x))^2 \, dy \, dx}{\displaystyle\int_{\Omega} (u_\varepsilon(x))^2 \, dx}$$

$$= \frac{\displaystyle\int_{\Omega} A(x) u_\varepsilon^2(x) \, dx - \int_{\Omega} \int_{\Omega} J(x - y) u_\varepsilon(y)\, u_\varepsilon(x) \, dy \, dx}{\displaystyle\int_{\Omega} (u_\varepsilon(x))^2 \, dx}$$

$$= \frac{\displaystyle\int_{\Omega} A(x) u_\varepsilon^2(x) \, dx - \int_{\Omega} \int_{\Omega} J(x - y) u_\varepsilon(y)\, u_\varepsilon(x) \, dy \, dx}{2|B(0, \varepsilon)|}.$$

Using the continuity of A and the explicit form of u_ε we obtain

$$\lim_{\varepsilon \to 0} \frac{\displaystyle\int_{\Omega} A(x) u_\varepsilon^2(x) \, dx}{2|B(0, \varepsilon)|} = A(x_0)$$

and

$$\lim_{\varepsilon \to 0} \frac{\displaystyle\int_{\Omega} \int_{\Omega} J(x - y) u_\varepsilon(y)\, u_\varepsilon(x) \, dy \, dx}{2|B(0, \varepsilon)|} = 0.$$

Therefore, (3.4) follows. $\qquad\square$

The main result of this section is the following.

THEOREM 3.6. *For every $u_0 \in L^2(\Omega)$ the solution $u(x, t)$ of (3.1) satisfies*

$$(3.5) \qquad \left\| u(\cdot, t) - \frac{1}{|\Omega|} \int_{\Omega} u_0 \right\|_{L^2(\Omega)} \leq e^{-\beta_1 t} \left\| u_0 - \frac{1}{|\Omega|} \int_{\Omega} u_0 \right\|_{L^2(\Omega)},$$

where β_1 is given by (3.2). Moreover, if u_0 is continuous and bounded, there exists a positive constant $C > 0$ such that

$$(3.6) \qquad \left\| u(\cdot, t) - \frac{1}{|\Omega|} \int_{\Omega} u_0 \right\|_{L^\infty(\Omega)} \leq C e^{-\beta_1 t}.$$

PROOF. Let

$$H(t) = \frac{1}{2} \int_\Omega \left(u(x,t) - \frac{1}{|\Omega|} \int_\Omega u_0 \right)^2 dx.$$

Differentiating, using (3.2) and the conservation of the total mass, we obtain

$$H'(t) = -\frac{1}{2} \int_\Omega \int_\Omega J(x-y)(u(y,t) - u(x,t))^2 \, dy \, dx$$

$$\leq -\beta_1 \int_\Omega \left(u(x,t) - \frac{1}{|\Omega|} \int_\Omega u_0 \right)^2 dx.$$

Hence

$$H'(t) \leq -2\beta_1 H(t).$$

Therefore, integrating,

$$H(t) \leq e^{-2\beta_1 t} H(0),$$

and (3.5) follows.

In order to prove (3.6) let $w(x,t)$ denote the difference

$$w(x,t) = u(x,t) - \frac{1}{|\Omega|} \int_\Omega u_0.$$

We seek an exponential estimate in L^∞ of the decay of $w(x,t)$. The linearity of the equation implies that $w(x,t)$ is a solution of (3.1) and satisfies

$$w(x,t) = e^{-A(x)t} w_0(x) + e^{-A(x)t} \int_0^t e^{A(x)s} \int_\Omega J(x-y)w(y,s) \, dy \, ds,$$

where $A(x) = \int_\Omega J(x-y)dx$. By using (3.5) and Hölder's inequality we obtain that

$$|w(x,t)| \leq e^{-A(x)t}|w_0(x)| + Ce^{-A(x)t} \int_0^t e^{A(x)s - \beta_1 s} \, ds.$$

Therefore, $w(x,t)$ decays to zero exponentially fast and, moreover, (3.6) holds thanks to Lemma 3.5. □

3.2. The nonhomogeneous Neumann problem

The main purpose of this section is to show that the solutions of the usual Neumann boundary value problem for the heat equation can be approximated by the solutions of a sequence of nonlocal Neumann boundary value problems.

Given a bounded smooth domain Ω, one of the most common boundary conditions considered in the literature for the heat equation, $v_t = \Delta v$, is the *Neumann boundary condition* $\frac{\partial v}{\partial \eta}(x,t) = g(x,t)$, $x \in \partial\Omega$, which leads to the following classical problem:

(3.7)
$$\begin{cases} v_t - \Delta v = 0 & \text{in } \Omega \times (0, +\infty), \\ \dfrac{\partial v}{\partial \eta} = g & \text{on } \partial\Omega \times (0, +\infty), \\ v(x,0) = u_0(x) & \text{in } \Omega. \end{cases}$$

In the sequel we treat the nonlocal Neumann boundary value problem

(3.8)
$$\begin{cases} u_t(x,t) = \displaystyle\int_\Omega J(x-y)\big(u(y,t)-u(x,t)\big)\,dy \\ \qquad\qquad\qquad + \displaystyle\int_{\mathbb{R}^N\setminus\Omega} G(x,x-y)g(y,t)\,dy, \\ u(x,0) = u_0(x), \qquad\qquad\qquad\qquad x\in\Omega,\ t>0, \end{cases}$$

where $G(x,\xi)$ is smooth and compactly supported in ξ uniformly in x, and J satisfies condition (H).

Recall that in the previous section we considered homogeneous boundary data, that is, $g \equiv 0$. In this new model the last integral term takes into account the prescribed flux of individuals that enter or leave the domain.

The nonlocal Neumann model (3.8) and the Neumann problem for the heat equation (3.7) share many properties. For example, a comparison principle holds for both equations when G is nonnegative and the asymptotic behaviour of their solutions as $t \to \infty$ is similar; see the previous section.

First of all, we prove existence and uniqueness for solutions of (3.8) with general G by using a fixed point argument. We also give a comparison principle when $G \geq 0$.

3.2.1. Existence and uniqueness. The existence and uniqueness result for solutions of (3.8) is valid in a general L^1 setting. Let $G \in L^\infty(\Omega \times \mathbb{R}^N)$, $g \in L^\infty_{\text{loc}}([0,\infty); L^1(\mathbb{R}^N \setminus \Omega))$ and $u_0 \in L^1(\Omega)$.

DEFINITION 3.7. A *solution* of (3.8) is a function $u \in C([0,\infty); L^1(\Omega))$ such that

$$u(x,t) = u_0(x) + \int_0^t \int_\Omega J(x-y)\,(u(y,s)-u(x,s))\,dy\,ds$$

$$+ \int_0^t \int_{\mathbb{R}^N\setminus\Omega} G(x,x-y)g(y,s)\,dy\,ds.$$

Existence and uniqueness will be a consequence of Banach's fixed point theorem. Fix $t_0 > 0$ and consider the Banach space

$$X_{t_0} = C([0,t_0]; L^1(\Omega))$$

with the norm

$$|||w||| = \max_{0\leq t\leq t_0} \|w(\cdot,t)\|_{L^1(\Omega)}.$$

The solution will be obtained as a fixed point of the operator $\mathcal{T}_{u_0,g} : X_{t_0} \to X_{t_0}$ defined by

(3.9)
$$\mathcal{T}_{u_0,g}(w)(x,t) = u_0(x) + \int_0^t \int_\Omega J(x-y)\,(w(y,s)-w(x,s))\,dy\,ds$$

$$+ \int_0^t \int_{\mathbb{R}^N\setminus\Omega} G(x,x-y)g(y,t)\,dy\,ds,$$

for $x \in \Omega$ and $t \in [0,t_0]$.

The following lemma is the main ingredient in the proof of existence.

LEMMA 3.8. *Let $G \in L^\infty(\Omega \times \mathbb{R}^N)$. Let g, $h \in L^\infty((0, t_0); L^1(\mathbb{R}^N \setminus \Omega))$ and u_0, $v_0 \in L^1(\Omega)$. Then there exists a constant C depending only on Ω, J and G such that, for $w, z \in X_{t_0}$,*

(3.10)
$$\||\mathcal{T}_{u_0, g}(w) - \mathcal{T}_{v_0, h}(z)\||$$

$$\leq \|u_0 - v_0\|_{L^1} + C t_0 \Big(\||w - z\|| + \|g - h\|_{L^\infty((0, t_0); L^1(\mathbb{R}^N \setminus \Omega))} \Big).$$

PROOF. Since

$$\int_\Omega |\mathcal{T}_{u_0, g}(w)(x, t) - \mathcal{T}_{v_0, h}(z)(x, t)| \, dx \leq \int_\Omega |u_0(x) - v_0(x)| \, dx$$

$$+ \int_\Omega \left| \int_0^t \int_\Omega J(x - y) \left((w(y, s) - z(y, s)) - (w(x, s) - z(x, s)) \right) dy \, ds \right| dx$$

$$+ \int_\Omega \int_0^t \int_{\mathbb{R}^N \setminus \Omega} |G(x, x - y)| |g(y, s) - h(y, s)| \, dy \, ds \, dx,$$

we get that (3.10) holds.   □

THEOREM 3.9. *Let $G \in L^\infty(\Omega \times \mathbb{R}^N)$. Then, for every $u_0 \in L^1(\Omega)$ and $g \in L^\infty_{\text{loc}}([0, \infty); L^1(\mathbb{R}^N \setminus \Omega))$, there exists a unique solution of (3.8).*

PROOF. Let $\mathcal{T} = \mathcal{T}_{u_0, g}$. We check first that $\mathcal{T}$ maps X_{t_0} into X_{t_0}. From (3.9) we see that, for $0 \leq t_1 < t_2 \leq t_0$,

$$\|\mathcal{T}(w)(t_2) - \mathcal{T}(w)(t_1)\|_{L^1(\Omega)} \leq A \int_{t_1}^{t_2} \int_\Omega |w(y, s)| \, dy \, ds + B \int_{t_1}^{t_2} \int_{\mathbb{R}^N \setminus \Omega} |g(y, s)| \, dy \, ds.$$

On the other hand, again from (3.9)

$$\|\mathcal{T}(w)(t) - u_0\|_{L^1(\Omega)} \leq C t \{ \||w\|| + \|g\|_{L^\infty((0, t_0); L^1(\mathbb{R}^N \setminus \Omega))} \}.$$

These two estimates give that $\mathcal{T}(w) \in C([0, t_0]; L^1(\Omega))$. Hence $\mathcal{T}$ maps X_{t_0} into X_{t_0}.

Choose t_0 such that $C t_0 < 1$. From Lemma 3.8 we get that $\mathcal{T}$ is a strict contraction in X_{t_0} and the existence and uniqueness part of the theorem follows from Banach's fixed point theorem in the interval $[0, t_0]$. To extend the solution to $[0, \infty)$ we may take $u(x, t_0) \in L^1(\Omega)$ for the initial datum and obtain a solution in $[0, 2 t_0]$. Iterating this procedure we get a solution defined in $[0, \infty)$.   □

The next aim is to prove a comparison principle for solutions of (3.8) when $G \geq 0$. It is also stated for sub and supersolutions.

DEFINITION 3.10. *A function $u \in W^{1,1}(0, T; L^1(\Omega))$ is a supersolution of (3.8) if $u(x, 0) \geq u_0(x)$ and*

$$u_t(x, t) \geq \int_\Omega J(x - y) \big(u(y, t) - u(x, t) \big) \, dy + \int_{\mathbb{R}^N \setminus \Omega} G(x, x - y) g(y, t) \, dy.$$

Subsolutions are defined analogously by reversing the inequalities.

LEMMA 3.11. *Let $G \geq 0$, $u_0 \geq 0$ and $g \geq 0$. If $u \in C(\overline{\Omega} \times [0, T])$ is a supersolution of (3.8), then $u \geq 0$.*

PROOF. Assume that $u(x,t)$ is negative somewhere. Let $v(x,t) = u(x,t) + \varepsilon t$ with ε so small that v is still negative somewhere. Then, if we take (x_0, t_0) to be a point where v attains its negative minimum, then $t_0 > 0$ and

$$v_t(x_0, t_0) = u_t(x_0, t_0) + \varepsilon > \int_\Omega J(x - y)(u(y, t_0) - u(x_0, t_0))\, dy$$

$$= \int_\Omega J(x - y)(v(y, t_0) - v(x_0, t_0))\, dy \geq 0$$

which is a contradiction. Thus, $u \geq 0$.   □

COROLLARY 3.12. *Let G be nonnegative and bounded. Let u_0 and v_0 be in $L^1(\Omega)$ with $u_0 \geq v_0$ and $g, h \in L^\infty((0,T); L^1(\mathbb{R}^N \setminus \Omega))$ with $g \geq h$. Let u be a solution of (3.8) with initial condition u_0 and flux g and let v be a solution of (3.8) with initial condition v_0 and flux h. Then $u \geq v$.*

PROOF. Let $w = u - v$. Then w is a solution with initial datum $u_0 - v_0 \geq 0$ and boundary datum $g - h \geq 0$. Using the continuity of solutions with respect to the initial and Neumann data (Lemma 3.8) and the fact that $J \in L^\infty(\mathbb{R}^N)$ and $G \in L^\infty(\Omega \times \mathbb{R}^N)$, we may assume that $u, v \in C(\overline{\Omega} \times [0,T])$. By Lemma 3.11 we obtain that $w = u - v \geq 0$. So the corollary is proved.   □

COROLLARY 3.13. *Let G be nonnegative and bounded. Let $u \in C(\overline{\Omega} \times [0,T])$ be a supersolution of (3.8) and let $v \in C(\overline{\Omega} \times [0,T])$ be a subsolution of (3.8). Then $u \geq v$.*

PROOF. The argument follows the lines of the proof of the previous corollary.
□

3.2.2. Rescaling the kernels. Convergence to the heat equation.

In this section we show that the Neumann problem for the heat equation (3.7) can be approximated by suitable nonlocal Neumann problems like (3.8) when they are rescaled appropriately. From now on in this chapter, we will assume that $\mathrm{supp}(J) = B(0,1)$.

For given J and G consider the rescaled kernels

$$(3.11) \qquad J_\varepsilon(\xi) = C_1 \frac{1}{\varepsilon^N} J\left(\frac{\xi}{\varepsilon}\right), \qquad G_\varepsilon(x, \xi) = C_1 \frac{1}{\varepsilon^N} G\left(x, \frac{\xi}{\varepsilon}\right)$$

with

$$C_1^{-1} = \frac{1}{2} \int_{\mathbb{R}^N} J(z) z_N^2\, dz,$$

which is a normalizing constant in order to obtain the Laplacian in the limit instead of a multiple of it. Then consider the solution $u_\varepsilon(x,t)$ of

$$(3.12) \quad \begin{cases} (u_\varepsilon)_t(x,t) = \dfrac{1}{\varepsilon^2} \displaystyle\int_\Omega J_\varepsilon(x - y)(u^\varepsilon(y,t) - u_\varepsilon(x,t))\, dy \\[2mm] \qquad\qquad\qquad + \dfrac{1}{\varepsilon} \displaystyle\int_{\mathbb{R}^N \setminus \Omega} G_\varepsilon(x, x - y) g(y,t)\, dy, \\[2mm] u_\varepsilon(x,0) = u_0(x), \qquad\qquad\qquad\qquad x \in \Omega,\ t > 0. \end{cases}$$

It will be shown that

$$u_\varepsilon \to v,$$

where v is the solution of (3.7), in different topologies according to different choices of the kernel G.

Let us give a heuristic idea in one space dimension, with $\Omega = (0,1)$, of why the scaling involved in (3.11) is the correct one. We assume that

$$\int_1^\infty G(1, 1-y)\, dy = -\int_{-\infty}^0 G(0, -y)\, dy = \int_0^1 J(y)\, y\, dy$$

and, as stated above, that $G(x, \cdot)$ has compact support independent of x. In this case (3.12) reads

(3.13)

$$u_t(x,t) = \mathcal{A}_\varepsilon u(x,t) := \frac{1}{\varepsilon^2}\int_0^1 J_\varepsilon(x-y)\left(u(y,t)-u(x,t)\right)dy$$

$$+\frac{1}{\varepsilon}\int_{-\infty}^0 G_\varepsilon(x, x-y)\, g(y,t)\, dy + \frac{1}{\varepsilon}\int_1^{+\infty} G_\varepsilon(x, x-y)\, g(y,t)\, dy.$$

If $x \in (0,1)$, a Taylor expansion gives that, for any fixed smooth u and ε small enough, the right hand side $\mathcal{A}_\varepsilon u$ in (3.13) becomes

$$\mathcal{A}_\varepsilon u(x) = \frac{1}{\varepsilon^2}\int_0^1 J_\varepsilon(x-y)\left(u(y)-u(x)\right)dy \approx u_{xx}(x),$$

and, if $x = 0$ and ε is small,

$$\mathcal{A}_\varepsilon u(0) = \frac{1}{\varepsilon^2}\int_0^1 J_\varepsilon(-y)\left(u(y)-u(0)\right)dy + \frac{1}{\varepsilon}\int_{-\infty}^0 G_\varepsilon(0,-y)\, g(y)\, dy$$

$$\approx \frac{C_2}{\varepsilon}\left(u_x(0)-g(0)\right).$$

Analogously, $\mathcal{A}_\varepsilon u(1) \approx (C_2/\varepsilon)(-u_x(1)+g(1))$.

However, the proofs of the results are much more involved than simple Taylor expansions due to the fact that for each $\varepsilon > 0$ there are points $x \in \Omega$ for which the ball in which integration takes place, $B(x, \varepsilon)$, is not contained in Ω. Moreover, when working in several space dimensions, one has to take into account the geometry of the domain.

The proofs of the convergence results will follow by getting bounds for the difference $u_\varepsilon - v$. Set $w_\varepsilon = u_\varepsilon - v$ and let $\tilde{v}$ be a $C^{2+\alpha, 1+\alpha/2}$ extension of v to $\mathbb{R}^N \times [0,T]$. Consider the following operators:

$$L_\varepsilon(w)(x,t) := \frac{1}{\varepsilon^2}\int_\Omega J_\varepsilon(x-y)\left(w(y,t)-w(x,t)\right)dy$$

and

$$\tilde{L}_\varepsilon(w)(x,t) := \frac{1}{\varepsilon^2}\int_{\mathbb{R}^N} J_\varepsilon(x-y)\left(w(y,t)-w(x,t)\right)dy.$$

Then

$$(w_\varepsilon)_t(x,t) = L_\varepsilon(u_\varepsilon)(x,t) - \Delta v(x,t) + \frac{1}{\varepsilon} \int_{\mathbb{R}^N \setminus \Omega} G_\varepsilon(x, x-y) g(y,t) \, dy$$

$$= L_\varepsilon(w_\varepsilon)(x,t) + \tilde{L}_\varepsilon(\tilde{v})(x,t) - \Delta v(x,t) + \frac{1}{\varepsilon} \int_{\mathbb{R}^N \setminus \Omega} G_\varepsilon(x, x-y) g(y,t) \, dy$$

$$- \frac{1}{\varepsilon^2} \int_{\mathbb{R}^N \setminus \Omega} J_\varepsilon(x-y) \big(\tilde{v}(y,t) - \tilde{v}(x,t)\big) \, dy,$$

that is,

$$(w_\varepsilon)_t(x,t) - L_\varepsilon(w_\varepsilon)(x,t) = F_\varepsilon(x,t),$$

where, since $\Delta v = \Delta \tilde{v}$ in Ω,

(3.14)
$$F_\varepsilon(x,t) = \tilde{L}_\varepsilon(\tilde{v}) - \Delta \tilde{v} + \frac{1}{\varepsilon} \int_{\mathbb{R}^N \setminus \Omega} G_\varepsilon(x, x-y) g(y,t) \, dy$$

$$- \frac{1}{\varepsilon^2} \int_{\mathbb{R}^N \setminus \Omega} J_\varepsilon(x-y) \big(\tilde{v}(y,t) - \tilde{v}(x,t)\big) \, dy.$$

The main task in order to prove the convergence results is to get bounds on F_ε.

First, observe that, by the choice of C_1, the fact that J is radially symmetric and $\tilde{v} \in C^{2+\alpha, 1+\alpha/2}(\mathbb{R}^N \times [0,T])$, we have

(3.15)
$$\sup_{t \in [0,T]} \|\tilde{L}_\varepsilon(\tilde{v}) - \Delta \tilde{v}\|_{L^\infty(\Omega)} = O(\varepsilon^\alpha).$$

In fact,

$$\frac{C_1}{\varepsilon^{N+2}} \int_{\mathbb{R}^N} J\left(\frac{x-y}{\varepsilon}\right) (\tilde{v}(y,t) - \tilde{v}(x,t)) \, dy - \Delta \tilde{v}(x,t)$$

becomes, under the change of variables $z = (x-y)/\varepsilon$,

$$\frac{C_1}{\varepsilon^2} \int_{\mathbb{R}^N} J(z) (\tilde{v}(x - \varepsilon z, t) - \tilde{v}(x,t)) \, dy - \Delta \tilde{v}(x,t),$$

and hence (3.15) follows by a simple Taylor expansion.

The estimate of the last integral in F_ε follows from the next lemma, which is valid for any smooth function and not only for a solution of the heat equation.

The following notation will be used in the sequel. Remember that Ω is a bounded $C^{2+\alpha}$ domain. Let

(3.16)
$$\Omega_\varepsilon := \{x \in \Omega \; : \; \text{dist}(x, \partial\Omega) < \varepsilon\}.$$

For $x \in \Omega_\varepsilon$ and ε small enough, x can be expressed as

$$x = \bar{x} - s\,\eta(\bar{x}),$$

where $\bar{x}$ is the orthogonal projection of x on $\partial\Omega$, $0 < s < \varepsilon$ and $\eta(\bar{x})$ is the unit exterior normal to Ω at $\bar{x}$.

LEMMA 3.14. *If θ is a $C^{2+\alpha,1+\alpha/2}$ function on $\mathbb{R}^N \times [0,T]$ and $\dfrac{\partial\theta}{\partial\eta} = h$ on $\partial\Omega$, then, for $x \in \Omega_\varepsilon$, ε small,*

$$\frac{1}{\varepsilon^2} \int_{\mathbb{R}^N \setminus \Omega} J_\varepsilon(x-y)\left(\theta(y,t) - \theta(x,t)\right) dy$$

$$= \frac{1}{\varepsilon} \int_{\mathbb{R}^N \setminus \Omega} J_\varepsilon(x-y)\eta(\bar{x}) \cdot \frac{y-x}{\varepsilon} h(\bar{x},t)\, dy$$

$$+ \int_{\mathbb{R}^N \setminus \Omega} J_\varepsilon(x-y) \sum_{|\beta|=2} \frac{D^\beta \theta}{2}(\bar{x},t)\left(\left(\frac{y-\bar{x}}{\varepsilon}\right)^\beta - \left(\frac{x-\bar{x}}{\varepsilon}\right)^\beta\right) dy + O(\varepsilon^\alpha),$$

where $\bar{x}$ is the orthogonal projection of x on the boundary of Ω.

Here and in what follows we use the notation $A = O(b)$ to mean that there exists a constant C such that $|A| \le Cb$.

PROOF. Since $\theta \in C^{2+\alpha,1+\alpha/2}(\mathbb{R}^N \times [0,T])$, we have

$$\theta(y,t) - \theta(x,t) = \theta(y,t) - \theta(\bar{x},t) - (\theta(x,t) - \theta(\bar{x},t))$$

$$= \nabla\theta(\bar{x},t) \cdot (y-x) + \sum_{|\beta|=2} \frac{D^\beta \theta}{2}(\bar{x},t)\left((y-\bar{x})^\beta - (x-\bar{x})^\beta\right)$$

$$+ O(\|\bar{x} - x\|^{2+\alpha}) + O(\|\bar{x} - y\|^{2+\alpha}).$$

Therefore,

$$\frac{1}{\varepsilon^2} \int_{\mathbb{R}^N \setminus \Omega} J_\varepsilon(x-y)\left(\theta(y,t) - \theta(x,t)\right) dy = \frac{1}{\varepsilon} \int_{\mathbb{R}^N \setminus \Omega} J_\varepsilon(x-y)\nabla\theta(\bar{x},t) \cdot \frac{y-x}{\varepsilon}\, dy$$

$$+ \int_{\mathbb{R}^N \setminus \Omega} J_\varepsilon(x-y) \sum_{|\beta|=2} \frac{D^\beta \theta}{2}(\bar{x},t)\left(\left(\frac{y-\bar{x}}{\varepsilon}\right)^\beta - \left(\frac{x-\bar{x}}{\varepsilon}\right)^\beta\right) dy + O(\varepsilon^\alpha).$$

Fix $x \in \Omega_\varepsilon$. Let us take a new coordinate system such that $\eta(\bar{x}) = e_N$. Since $\dfrac{\partial\theta}{\partial\eta} = h$ on $\partial\Omega$, we get

$$\int_{\mathbb{R}^N \setminus \Omega} J_\varepsilon(x-y)\nabla\theta(\bar{x},t) \cdot \frac{y-x}{\varepsilon}\, dy = \int_{\mathbb{R}^N \setminus \Omega} J_\varepsilon(x-y)\eta(\bar{x}) \cdot \frac{y-x}{\varepsilon} h(\bar{x},t)\, dy$$

$$+ \int_{\mathbb{R}^N \setminus \Omega} J_\varepsilon(x-y) \sum_{i=1}^{N-1} \theta_{x_i}(\bar{x},t)\frac{y_i - x_i}{\varepsilon}\, dy.$$

We estimate this last integral. Since Ω is a $C^{2+\alpha}$ domain, we can choose vectors $e_1, e_2, \ldots, e_{N-1}$ so that there exist $\kappa > 0$ and constants $f_i(\bar{x})$ such that

$$B(\bar{x}, 2\varepsilon) \cap \left\{ y \in \mathbb{R}^N : y_N - \left(\bar{x}_N + \sum_{i=1}^{N-1} f_i(\bar{x})(y_i - x_i)^2\right) > \kappa\varepsilon^{2+\alpha} \right\} \subset \mathbb{R}^N \setminus \Omega$$

and

$$B(\bar{x}, 2\varepsilon) \cap \left\{ y \in \mathbb{R}^N : y_N - \left(\bar{x}_N + \sum_{i=1}^{N-1} f_i(\bar{x})(y_i - x_i)^2 \right) < -\kappa\varepsilon^{2+\alpha} \right\} \subset \Omega.$$

Therefore

$$\int_{\mathbb{R}^N \backslash \Omega} J_\varepsilon(x-y) \left(\sum_{i=1}^{N-1} \theta_{x_i}(\bar{x}, t) \frac{y_i - x_i}{\varepsilon} \right) dy$$

$$= \int_{(\mathbb{R}^N \backslash \Omega) \cap \left\{ y \in \mathbb{R}^N : \left| y_N - \left(\bar{x}_N + \sum_{i=1}^{N-1} f_i(\bar{x})(y_i - x_i)^2 \right) \right| \leq \kappa\varepsilon^{2+\alpha} \right\}} J_\varepsilon(x-y)$$

$$\times \left(\sum_{i=1}^{N-1} \theta_{x_i}(\bar{x}, t) \frac{y_i - x_i}{\varepsilon} \right) dy$$

$$+ \int_{\left\{ y \in \mathbb{R}^N : y_N - \left(\bar{x}_N + \sum_{i=1}^{N-1} f_i(\bar{x})(y_i - x_i)^2 \right) > \kappa\varepsilon^{2+\alpha} \right\}} J_\varepsilon(x-y)$$

$$\times \left(\sum_{i=1}^{N-1} \theta_{x_i}(\bar{x}, t) \frac{y_i - x_i}{\varepsilon} \right) dy$$

$$= I_1 + I_2.$$

If we take $z = (y-x)/\varepsilon$ as a new variable, recalling that $\bar{x}_N - x_N = \varepsilon s$, we obtain

$$|I_1| \leq C_1 \sum_{i=1}^{N-1} |\theta_{x_i}(\bar{x}, t)| \int_{\left\{ z \in \mathbb{R}^N : \left| z_N - \left(s + \varepsilon \sum_{i=1}^{N-1} f_i(\bar{x})(z_i)^2 \right) \right| \leq \kappa\varepsilon^{1+\alpha} \right\}} J(z)|z_i| \, dz$$

$$\leq C \kappa \varepsilon^{1+\alpha}.$$

On the other hand,

$$I_2 = C_1 \sum_{i=1}^{N-1} \theta_{x_i}(\bar{x}, t) \int_{\left\{ z \in \mathbb{R}^N : z_N - \left(s + \varepsilon \sum_{i=1}^{N-1} f_i(\bar{x})(z_i)^2 \right) > \kappa\varepsilon^{1+\alpha} \right\}} J(z) z_i \, dz.$$

Fix $1 \leq i \leq N-1$. Then, since J is radially symmetric, $J(z)z_i$ is an odd function of the variable z_i and, since the set

$$\left\{ z \in \mathbb{R}^N : z_N - \left(s + \varepsilon \sum_{i=1}^{N-1} f_i(\bar{x})(z_i)^2 \right) > \kappa\varepsilon^{1+\alpha} \right\}$$

is symmetric in that variable, we get

$$I_2 = 0.$$

Collecting the previous estimates we complete the proof. $\square$

The following inequality is also used.

LEMMA 3.15. *There exist $K > 0$ and $\bar{\varepsilon} > 0$ such that, for $\varepsilon < \bar{\varepsilon}$,*

$$(3.17) \quad \int_{\mathbb{R}^N \backslash \Omega} J_\varepsilon(x-y)\eta(\bar{x}) \cdot \frac{y-x}{\varepsilon} \, dy \geq K \int_{\mathbb{R}^N \backslash \Omega} J_\varepsilon(x-y) \, dy \quad \text{for all } x \in \Omega_\varepsilon.$$

PROOF. Let us put the origin at the point $\bar{x}$ and take a coordinate system such that $\eta(\bar{x}) = e_N$. Then $x = (0, -\mu)$ with $0 < \mu < \varepsilon$. Arguing as before,

$$\int_{\mathbb{R}^N \setminus \Omega} J_\varepsilon(x-y)\eta(\bar{x}) \cdot \frac{y-x}{\varepsilon}\, dy = \int_{\mathbb{R}^N \setminus \Omega} J_\varepsilon(x-y)\frac{y_N + \mu}{\varepsilon}\, dy$$

$$= \int_{\{y_N > \kappa\varepsilon^2\}} J_\varepsilon(x-y)\frac{y_N + \mu}{\varepsilon}\, dy + \int_{\mathbb{R}^N \setminus \Omega \cap \{|y_N| < \kappa\varepsilon^2\}} J_\varepsilon(x-y)\frac{y_N + \mu}{\varepsilon}\, dy$$

$$\geq \int_{\{y_N > \kappa\varepsilon^2\}} J_\varepsilon(x-y)\frac{y_N + \mu}{\varepsilon}\, dy - C\varepsilon.$$

Fix a small c_1 such that

$$\frac{1}{2}\int_{\{z_N > 0\}} J(z)\, z_N\, dz \geq 2c_1 \int_{\{0 < z_N < 2c_1\}} J(z)\, dz.$$

We divide our arguments into two cases according to whether $\mu \leq c_1\varepsilon$ or $\mu > c_1\varepsilon$.

Case I. Assume that $\mu \leq c_1\varepsilon$. In this case we have

(3.18)
$$\int_{\{y_N > \kappa\varepsilon^2\}} J_\varepsilon(x-y)\frac{y_N + \mu}{\varepsilon}\, dy = C_1 \int_{\{z_N > \kappa\varepsilon + \frac{\mu}{\varepsilon}\}} J(z)\, z_N\, dz$$

$$= C_1 \left(\int_{\{z_N > 0\}} J(z)\, z_N\, dz - \int_{\{0 < z_N < \kappa\varepsilon + \frac{\mu}{\varepsilon}\}} J(z)\, z_N\, dz \right)$$

$$\geq C_1 \left(\int_{\{z_N > 0\}} J(z)\, z_N\, dz - 2c_1 \int_{\{0 < z_N < 2c_1\}} J(z)\, dz \right)$$

$$\geq \frac{C_1}{2} \int_{\{z_N > 0\}} J(z)\, z_N\, dz.$$

Then

$$\int_{\mathbb{R}^N \setminus \Omega} J_\varepsilon(x-y)\eta(\bar{x}) \cdot \frac{y-x}{\varepsilon}\, dy - K \int_{\mathbb{R}^N \setminus \Omega} J_\varepsilon(x-y)\, dy$$

$$\geq C_1 \left(\frac{1}{2}\int_{\{z_N > 0\}} J(z)\, z_N\, dz - K \right) - C\varepsilon \geq 0$$

if ε is small enough and

$$K < \frac{1}{4}\int_{\{z_N > 0\}} J(z)\, z_N\, dz.$$

Case II. Assume that $\mu > c_1\varepsilon$. For $y \in (\mathbb{R}^N \setminus \Omega) \cap B(\bar{x}, \varepsilon)$ we have

$$\frac{y_N}{\varepsilon} \geq -\kappa\varepsilon.$$

Then

$$\int_{\mathbb{R}^N\setminus\Omega} J_\varepsilon(x-y)\frac{y_N+\mu}{\varepsilon}\,dy - K\int_{\mathbb{R}^N\setminus\Omega} J_\varepsilon(x-y)\,dy$$

$$\geq (c_1-\kappa\varepsilon)\int_{\mathbb{R}^N\setminus\Omega} J_\varepsilon(x-y)\,dy - K\int_{\mathbb{R}^N\setminus\Omega} J_\varepsilon(x-y)\,dy$$

$$= (c_1-\kappa\varepsilon-K)\int_{\mathbb{R}^N\setminus\Omega} J_\varepsilon(x-y)\,dy \geq 0$$

if ε is small and

$$K < \frac{c_1}{2}.$$

This ends the proof of (3.17).   □

3.2.3. Uniform convergence in the homogeneous case. The first result on convergence under rescaling deals with homogeneous boundary conditions, that is, $g \equiv 0$. The proof of this convergence result follows by constructing adequate super and subsolutions and then using comparison arguments.

THEOREM 3.16. *Assume $g \equiv 0$. Let Ω be a bounded $C^{2+\alpha}$ domain for some $0 < \alpha < 1$. Let $v \in C^{2+\alpha,1+\alpha/2}(\overline{\Omega}\times[0,T])$ be the solution of (3.7) and u_ε the solution to (3.12) with J_ε as above. Then*

$$\lim_{\varepsilon\to 0}\|u_\varepsilon - v\|_{L^\infty(\Omega\times[0,T])} = 0.$$

Note that the assumed regularity in v is guaranteed if $u_0 \in C^{2+\alpha}(\overline{\Omega})$ and $\partial u_0/\partial\eta = 0$. See, for instance, [**109**].

PROOF. We use a comparison argument. First, let us look for a supersolution. Let us pick an auxiliary function v_1 as a solution of

$$\begin{cases} (v_1)_t - \Delta v_1 = h & \text{in } \Omega\times(0,T), \\[2mm] \dfrac{\partial v_1}{\partial\eta} = g_1 & \text{on } \partial\Omega\times(0,T), \\[2mm] v_1(.,0) = v_{10} & \text{in } \Omega, \end{cases}$$

for some smooth functions $h(x,t) \geq 1$, $g_1(x,t) \geq 1$ and $v_{10}(x) \geq 0$ such that the resulting v_1 has an extension $\tilde{v}_1$ that belongs to $C^{2+\alpha,1+\alpha/2}(\mathbb{R}^N\times[0,T])$, and let M be an upper bound for v_1 in $\overline{\Omega}\times[0,T]$. Then

$$(v_1)_t(x,t) = L_\varepsilon v_1(x,t) + (\Delta v_1 - \tilde{L}_\varepsilon\tilde{v}_1)(x,t)$$

$$+\frac{1}{\varepsilon^2}\int_{\mathbb{R}^N\setminus\Omega} J_\varepsilon(x-y)(\tilde{v}_1(y,t) - \tilde{v}_1(x,t))\,dy + h(x,t).$$

Since $\Delta v_1 = \Delta\tilde{v}_1$ in Ω, we have that v_1 is a solution of

$$\begin{cases} (v_1)_t(x,t) - L_\varepsilon v_1(x,t) = H(x,t,\varepsilon) & \text{in } \Omega\times(0,T), \\[2mm] v_1(x,0) = v_{10}(x) & \text{in } \Omega, \end{cases}$$

where by (3.15), Lemma 3.14 and the fact that $h \geq 1$,

$$H(x,t,\varepsilon)$$

$$= (\Delta \tilde{v}_1 - \tilde{L}_\varepsilon \tilde{v}_1)(x,t) + \frac{1}{\varepsilon^2} \int_{\mathbb{R}^N \setminus \Omega} J_\varepsilon(x-y)(\tilde{v}_1(y,t) - \tilde{v}_1(x,t)) \, dy + h(x,t)$$

$$\geq \frac{1}{\varepsilon} \int_{\mathbb{R}^N \setminus \Omega} J_\varepsilon(x-y)\eta(\bar{x}) \cdot \frac{y-x}{\varepsilon} g_1(\bar{x},t) \, dy$$

$$+ \int_{\mathbb{R}^N \setminus \Omega} J_\varepsilon(x-y) \sum_{|\beta|=2} \frac{D^\beta \tilde{v}_1}{2}(\bar{x},t) \left(\left(\frac{y-\bar{x}}{\varepsilon}\right)^\beta - \left(\frac{x-\bar{x}}{\varepsilon}\right)^\beta \right) \, dy + 1 - C\varepsilon^\alpha$$

$$\geq \frac{g_1(\bar{x},t)}{\varepsilon} \int_{\mathbb{R}^N \setminus \Omega} J_\varepsilon(x-y)\eta(\bar{x}) \cdot \frac{y-x}{\varepsilon} \, dy - D_1 \int_{\mathbb{R}^N \setminus \Omega} J_\varepsilon(x-y) \, dy + \frac{1}{2}$$

for some constant D_1 if ε is small so that $C\varepsilon^\alpha \leq 1/2$.

Observe that Lemma 3.15 implies that for every constant $C_0 > 0$ there exists ε_0 such that

$$\frac{1}{\varepsilon} \int_{\mathbb{R}^N \setminus \Omega} J_\varepsilon(x-y)\eta(\bar{x}) \cdot \frac{y-x}{\varepsilon} \, dy - C_0 \int_{\mathbb{R}^N \setminus \Omega} J_\varepsilon(x-y) \, dy \geq 0$$

if $\varepsilon \leq \varepsilon_0$.

Now, since $g = 0$, by (3.15) and Lemma 3.14 we obtain

$$|F_\varepsilon| \leq C\varepsilon^\alpha + \int_{\mathbb{R}^N \setminus \Omega} J_\varepsilon(x-y) \sum_{|\beta|=2} \frac{D^\beta \tilde{v}}{2}(\bar{x},t) \left(\left(\frac{y-\bar{x}}{\varepsilon}\right)^\beta - \left(\frac{x-\bar{x}}{\varepsilon}\right)^\beta \right) \, dy$$

$$\leq C\varepsilon^\alpha + C_2 \int_{\mathbb{R}^N \setminus \Omega} J_\varepsilon(x-y) \, dy.$$

Given $\delta > 0$, let $v_\delta = \delta v_1$. Then v_δ satisfies

$$\begin{cases} (v_\delta)_t - L_\varepsilon v_\delta = \delta H(x,t,\varepsilon) & \text{in } \Omega \times (0,T), \\ v_\delta(x,0) = \delta v_1(x) & \text{in } \Omega. \end{cases}$$

By our previous estimates, there exists $\varepsilon_1 = \varepsilon_1(\delta) \leq \varepsilon_0$ such that, for $\varepsilon \leq \varepsilon_1$,

$$|F_\varepsilon| \leq \delta H(x,t,\varepsilon).$$

So, by the comparison principle, for any $\varepsilon \leq \varepsilon_1$ the inequalities

$$-M\delta \leq -v_\delta \leq w_\varepsilon \leq v_\delta \leq M\delta$$

hold. Therefore, for every $\delta > 0$,

$$-M\delta \leq \liminf_{\varepsilon \to 0} w_\varepsilon \leq \limsup_{\varepsilon \to 0} w_\varepsilon \leq M\delta,$$

and the theorem is proved. $\qquad\square$

3.2.4. An L^1-convergence result in the nonhomogeneous case. For the convergence result in this section,

$$(3.19) \qquad\qquad G(x,\xi) := -J(\xi)\,\eta(\bar{x})\cdot\xi \quad \text{for } x\in\Omega_\varepsilon,$$

where Ω_ε is given in (3.16).

Notice that the last integral in (3.12) only involves points $x\in\Omega_\varepsilon$ since when $y\notin\Omega$, $x-y\in\mathrm{supp}(J_\varepsilon)$ implies that $x\in\Omega_\varepsilon$. Hence the above definition makes sense for ε small.

THEOREM 3.17. *Let Ω be a bounded $C^{2+\alpha}$ domain, $0<\alpha<1$, G defined by (3.19), $g\in C^{1+\alpha,(1+\alpha)/2}\big(\overline{(\mathbb{R}^N\setminus\Omega)}\times[0,T]\big)$, and $v\in C^{2+\alpha,1+\alpha/2}(\overline{\Omega}\times[0,T])$ the solution of (3.7). Let u_ε be the solution of (3.12). Then*

$$\sup_{t\in[0,T]} \|u_\varepsilon(\cdot,t) - v(\cdot,t)\|_{L^1(\Omega)} \to 0 \quad \text{as } \varepsilon\to 0.$$

Observe that G may fail to be nonnegative and hence a comparison principle may not hold. However, in this case our proof of convergence to the solution of the heat equation does not rely on comparison arguments for (3.12). If we want a nonnegative kernel G, in order to have a comparison principle, we can modify G_ε by taking

$$\tilde{G}_\varepsilon(x,\xi) = G_\varepsilon(x,\xi) + \kappa\varepsilon J_\varepsilon(\xi) = \frac{1}{\varepsilon}J_\varepsilon(\xi)\left(-\eta(\bar{x})\cdot\xi + \kappa\varepsilon^2\right).$$

Note that for $x\in\overline{\Omega}$ and $y\in\mathbb{R}^N\setminus\Omega$,

$$\tilde{G}_\varepsilon(x,x-y) = \frac{1}{\varepsilon}J_\varepsilon(x-y)\left(-\eta(\bar{x})\cdot(x-y) + \kappa\varepsilon^2\right)$$

is nonnegative for ε small if we choose the constant κ as a bound for the curvature of $\partial\Omega$, since $|x-y|\le\varepsilon$. As will be seen in Remark 3.19, Theorem 3.17 remains valid with G_ε replaced by $\tilde{G}_\varepsilon$.

Using the previous notation, we first prove that F_ε, given in (3.14), goes to zero as ε goes to zero.

LEMMA 3.18. *If G is given by (3.19), then*

$$\lim_{\varepsilon\to 0} F_\varepsilon = 0 \quad \text{in } L^\infty\big(0,T;L^1(\Omega)\big).$$

PROOF. As $G(x,\xi) = -J(\xi)\,\eta(\bar{x})\cdot\xi$, for $x\in\Omega_\varepsilon$, by (3.15) and Lemma 3.14,

$$F_\varepsilon(x,t) = \frac{1}{\varepsilon}\int_{\mathbb{R}^N\setminus\Omega} J_\varepsilon(x-y)\eta(\bar{x})\cdot\frac{y-x}{\varepsilon}\big(g(y,t)-g(\bar{x},t)\big)\,dy$$

$$-\int_{\mathbb{R}^N\setminus\Omega} J_\varepsilon(x-y)\sum_{|\beta|=2}\frac{D^\beta\tilde{v}}{2}(\bar{x},t)\left(\left(\frac{(y-\bar{x})}{\varepsilon}\right)^\beta - \left(\frac{(x-\bar{x})}{\varepsilon}\right)^\beta\right)dy$$

$$+ O(\varepsilon^\alpha).$$

Since g is smooth, we have that F_ε is bounded in Ω_ε. Recalling the fact that $|\Omega_\varepsilon| = O(\varepsilon)$ and $F_\varepsilon(x,t) = O(\varepsilon^\alpha)$ on $\Omega\setminus\Omega_\varepsilon$ we get the convergence result. $\qquad\square$

We are now ready to prove Theorem 3.17.

PROOF OF THEOREM 3.17. We have that $w_\varepsilon = u_\varepsilon - v$ is a solution of

$$w_t - L_\varepsilon(w) = F_\varepsilon,$$

$$w(x, 0) = 0.$$

Let z_ε be a solution of

$$z_t - L_\varepsilon(z) = |F_\varepsilon|,$$

$$z(x, 0) = 0.$$

Then $-z_\varepsilon$ is a solution of

$$z_t - L_\varepsilon(z) = -|F_\varepsilon|,$$

$$z(x, 0) = 0.$$

By comparison,

$$-z_\varepsilon \leq w_\varepsilon \leq z_\varepsilon \quad \text{and} \quad z_\varepsilon \geq 0.$$

Integrating the equation for z_ε we get

$$\|z_\varepsilon(.,t)\|_{L^1(\Omega)} = \int_\Omega z_\varepsilon(x, t)\, dx = \int_\Omega \int_0^t |F_\varepsilon(x, s)|\, ds\, dx.$$

Applying Lemma 3.18 we obtain

$$\sup_{t \in [0,T]} \|z_\varepsilon(\cdot, t)\|_{L^1(\Omega)} \to 0$$

as $\varepsilon \to 0$. So the theorem is proved. $\qquad\square$

REMARK 3.19. Notice that if we consider a kernel which is a modification of G of the form

$$G_\varepsilon(x, \xi) + A(x, \xi, \varepsilon)$$

with

$$\int_{\mathbb{R}^N \setminus \Omega} |A(x, x - y, \varepsilon)|\, dy \to 0$$

in $L^1(\Omega)$ as $\varepsilon \to 0$, then the conclusion of Theorem 3.17 is still valid. In particular, we can take $A(x, \xi, \varepsilon) = \kappa\varepsilon J_\varepsilon(\xi)$.

3.2.5. A weak convergence result in the nonhomogeneous case. Finally, the other Neumann kernel we propose is just a scalar multiple of J, that is,

$$(3.20) \qquad\qquad G(x, \xi) = C_2 J(\xi),$$

where C_2 is such that

$$(3.21) \qquad\qquad \int_0^1 \int_{\{z_N > s\}} J(z)(C_2 - z_N)\, dz\, ds = 0.$$

This choice of G is natural since we are considering a flux with a jumping probability that is a scalar multiple of the same jumping probability that moves things in the interior of the domain.

For this G we can still prove convergence but in a weaker sense.

THEOREM 3.20. *Let Ω be a bounded $C^{2+\alpha}$ domain for some $0 < \alpha < 1$, $g \in C^{1+\alpha,(1+\alpha)/2}(\overline{(\mathbb{R}^N \setminus \Omega)} \times [0,T])$, $v \in C^{2+\alpha,1+\alpha/2}(\overline{\Omega} \times [0,T])$ the solution of (3.7). Let J be as before and $G(x,\xi) = C_2 J(\xi)$, where C_2 is defined by (3.21). Let u_ε be the solution of (3.12). Then, for each $t \in [0,T]$,*

$$u_\varepsilon(x,t) \rightharpoonup v(x,t) \quad weakly^* \text{ in } L^\infty(\Omega) \text{ as } \varepsilon \to 0.$$

First, we prove that in this case F_ε goes to zero in the sense of measures.

LEMMA 3.21. *Let G be as in (3.20); then there exists a constant C independent of ε such that*

$$\int_0^T \int_\Omega |F_\varepsilon(x,s)| \, dx \, ds \le C.$$

Moreover,

$$F_\varepsilon \rightharpoonup 0 \quad \text{in the sense of measures as } \varepsilon \to 0.$$

That is, for any continuous function θ,

$$\int_0^T \int_\Omega F_\varepsilon(x,t)\theta(x,t) \, dx \, dt \to 0 \quad \text{as } \varepsilon \to 0.$$

PROOF. As $G(x,\xi) = C_2 J(\xi)$ and g and $\tilde{v}$ are smooth, taking again the coordinate system of Lemma 3.14, we obtain

$$F_\varepsilon(x,t) = \frac{1}{\varepsilon} \int_{\mathbb{R}^N \setminus \Omega} J_\varepsilon(x-y) \left(C_2 g(y,t) - \frac{y_N - x_N}{\varepsilon} g(\bar{x},t) \right)$$

$$-\frac{1}{\varepsilon} \int_{\mathbb{R}^N \setminus \Omega} J_\varepsilon(x-y) \sum_{i=1}^{N-1} \tilde{v}_{x_i}(\bar{x},t)\frac{y_i - x_i}{\varepsilon} \, dy$$

$$-\int_{\mathbb{R}^N \setminus \Omega} J_\varepsilon(x-y) \sum_{|\beta|=2} \frac{D^\beta \tilde{v}(\bar{x},t)}{2} \left(\left(\frac{y-\bar{x}}{\varepsilon}\right)^\beta - \left(\frac{x-\bar{x}}{\varepsilon}\right)^\beta \right) dy + O(\varepsilon^\alpha)$$

$$= \frac{1}{\varepsilon} \int_{\mathbb{R}^N \setminus \Omega} J_\varepsilon(x-y) \left(C_2 g(\bar{x},t) - \frac{y_N - x_N}{\varepsilon} g(\bar{x},t) \right)$$

$$-\frac{1}{\varepsilon} \int_{\mathbb{R}^N \setminus \Omega} J_\varepsilon(x-y) \sum_{i=1}^{N-1} \tilde{v}_{x_i}(\bar{x},t)\frac{y_i - x_i}{\varepsilon} \, dy + O(1)\chi_{\Omega_\varepsilon} + O(\varepsilon^\alpha).$$

Let

$$B_\varepsilon(x,t) := \int_{\mathbb{R}^N \setminus \Omega} J_\varepsilon(x-y) \left(C_2 g(\bar{x},t) - \frac{y_N - x_N}{\varepsilon} g(\bar{x},t) \right)$$

$$-\int_{\mathbb{R}^N \setminus \Omega} J_\varepsilon(x-y) \sum_{i=1}^{N-1} \tilde{v}_{x_i}(\bar{x},t)\frac{y_i - x_i}{\varepsilon} \, dy.$$

Proceeding in a similar way as in the proof of Lemma 3.14 we get for ε small,

$$\int_{\mathbb{R}^N \setminus \Omega} J_\varepsilon(x-y) \left(C_2 g(\bar{x},t) - \frac{y_N - x_N}{\varepsilon} g(\bar{x},t) \right)$$

$$= g(\bar{x},t) \int_{(\mathbb{R}^N \setminus \Omega) \cap \{|y_N - \bar{x}_N| \leq \kappa\varepsilon^2\}} J_\varepsilon(x-y) \left(C_2 - \frac{y_N - x_N}{\varepsilon} \right) dy$$

$$+ g(\bar{x},t) \int_{(\mathbb{R}^N \setminus \Omega) \cap \{y_N - \bar{x}_N > 0\}} J_\varepsilon(x-y) \left(C_2 - \frac{y_N - x_N}{\varepsilon} \right) dy$$

$$- g(\bar{x},t) \int_{(\mathbb{R}^N \setminus \Omega) \cap \{0 < y_N - \bar{x}_N < \kappa\varepsilon^2\}} J_\varepsilon(x-y) \left(C_2 - \frac{y_N - x_N}{\varepsilon} \right) dy$$

$$= C_1 g(\bar{x},t) \int_{\{z_N > s\}} J(z)(C_2 - z_N)\, dz + O(\varepsilon)\chi_{\Omega_\varepsilon}.$$

Moreover,

$$\int_{\mathbb{R}^N \setminus \Omega} J_\varepsilon(x-y) \sum_{i=1}^{N-1} \tilde{v}_{x_i}(\bar{x},t) \frac{y_i - x_i}{\varepsilon}\, dy$$

$$= \sum_{i=1}^{N-1} \tilde{v}_{x_i}(\bar{x},t) \int_{\{|y_N - \bar{x}_N| \leq \kappa\varepsilon^2\}} J_\varepsilon(x-y) \frac{y_i - x_i}{\varepsilon} dy$$

$$+ \sum_{i=1}^{N-1} \tilde{v}_{x_i}(\bar{x},t) \int_{\{y_N - \bar{x}_N > \kappa\varepsilon^2\}} J_\varepsilon(x-y) \frac{y_i - x_i}{\varepsilon} dy$$

$$= C_1 \sum_{i=1}^{N-1} \tilde{v}_{x_i}(\bar{x},t) \int_{\{z_N - s > \kappa\varepsilon\}} J(z) z_i dz + O(\varepsilon)\chi_{\Omega_\varepsilon}$$

$$= I_2 + O(\varepsilon)\chi_{\Omega_\varepsilon}.$$

As in Lemma 3.14 we have $I_2 = 0$. Therefore,

$$B_\varepsilon(x,t) = C_1 g(\bar{x},t) \int_{\{z_N > s\}} J(z)\, (C_2 - z_N)\, dz + O(\varepsilon)\chi_{\Omega_\varepsilon}.$$

Observe that B_ε is bounded and supported in Ω_ε. Hence

$$\int_0^t \int_\Omega |F_\varepsilon(x,\tau)|\, dx\, d\tau \leq \frac{1}{\varepsilon} \int_0^t \int_{\Omega_\varepsilon} |B_\varepsilon(x,\tau)|\, dx\, d\tau + Ct|\Omega_\varepsilon| + Ct|\Omega|\varepsilon^\alpha \leq C.$$

This proves the first assertion of the lemma.

Now, for a point $x \in \Omega_\varepsilon$, let us write

$$x = \bar{x} - \mu\eta(\bar{x}) \quad \text{with } 0 < \mu < \varepsilon.$$

For ε small and $0 < \mu < \varepsilon$, let dS_μ be the area element of the set $\{x \in \Omega :$ dist $(x, \partial\Omega) = \mu\}$. Then $dS_\mu = dS + O(\varepsilon)$, where dS is the area element of $\partial\Omega$.

Thus, taking $\mu = s\varepsilon$ we get, for any continuous test function θ,

$$\frac{1}{\varepsilon}\int_0^T\int_{\Omega_\varepsilon} B_\varepsilon(x,t)\theta(\bar{x},t)\,dx\,dt$$

$$= O(\varepsilon) + C_1\int_0^T\int_{\partial\Omega} g(\bar{x},t)\theta(\bar{x},t)\int_0^1\int_{\{z_N>s\}} J(z)\big(C_2-z_N\big)\,dz\,ds\,dS\,dt$$

$$= O(\varepsilon) \to 0 \quad \text{as } \varepsilon \to 0,$$

since we have chosen C_2 such that

$$\int_0^1\int_{\{z_N>s\}} J(z)\big(C_2-z_N\big)\,dz\,ds = 0.$$

With all these estimates, going back to F_ε, we have

$$F_\varepsilon(x,t) = \frac{1}{\varepsilon}B_\varepsilon(x,t) + O(1)\chi_{\Omega_\varepsilon} + O(\varepsilon^\alpha).$$

Thus, we obtain

$$\int_0^T\int_{\Omega_\varepsilon} F_\varepsilon(x,t)\theta(\bar{x},t)\,dx\,dt \to 0 \qquad \text{as } \varepsilon \to 0.$$

On the other hand, if $\sigma(r)$ is the modulus of continuity of θ,

$$\int_0^T\int_{\Omega_\varepsilon} F_\varepsilon(x,t)\theta(x,t)\,dx\,dt = \int_0^T\int_{\Omega_\varepsilon} F_\varepsilon(x,t)\theta(\bar{x},t)\,dx\,dt$$

$$+ \int_0^T\int_{\Omega_\varepsilon} F_\varepsilon(x,t)\big(\theta(x,t)-\theta(\bar{x},t)\big)\,dx\,dt$$

$$\leq \int_0^T\int_{\Omega_\varepsilon} F_\varepsilon(x,t)\theta(\bar{x},t)\,dx\,dt + C\sigma(\varepsilon)\int_0^T\int_{\Omega_\varepsilon} |F_\varepsilon(x,t)|\,dx\,dt \to 0 \quad \text{as } \varepsilon \to 0.$$

Finally, since $F_\varepsilon = O(\varepsilon^\alpha)$ in $\Omega\setminus\Omega_\varepsilon$,

$$\int_0^T\int_{\Omega\setminus\Omega_\varepsilon} F_\varepsilon(x,t)\theta(x,t)\,dx\,dt \to 0 \qquad \text{as } \varepsilon \to 0$$

and this ends the proof.   □

Next we prove that u_ε is uniformly bounded when G is given as in (3.20).

LEMMA 3.22. *Let G be as in (3.20). There exists a constant C independent of ε such that*

$$\|u_\varepsilon\|_{L^\infty(\overline{\Omega}\times[0,T])} \leq C.$$

PROOF. Again we use a comparison argument. Let us look for a supersolution. Take an auxiliary function v_1 as a solution of

$$(3.22) \quad \begin{cases} (v_1)_t - \Delta v_1 = h & \text{in } \Omega \times (0,T), \\[2mm] \dfrac{\partial v_1}{\partial \eta} = g_1 & \text{on } \partial\Omega \times (0,T), \\[2mm] v_1(\cdot, 0) = v_{10} & \text{in } \Omega, \end{cases}$$

for some smooth functions $h(x,t) \geq 1$, $v_{10}(x) \geq u_0(x)$ and

$$g_1(x,t) \geq \frac{2}{K}(C_2 + 1) \max_{\partial\Omega \times [0,T]} |g(x,t)| + 1 \qquad (K \text{ being as in } (3.17))$$

and such that v_1 has an extension $\tilde{v}_1$ that belongs to $C^{2+\alpha, 1+\alpha/2}(\mathbb{R}^N \times [0,T])$. Let M be an upper bound for v_1 in $\overline{\Omega} \times [0,T]$. As before, v_1 is a solution of

$$\begin{cases} (v_1)_t - L_\varepsilon v_1 = H(x,t,\varepsilon) & \text{in } \Omega \times (0,T), \\[2mm] v_1(x,0) = v_{10}(x) & \text{in } \Omega, \end{cases}$$

where H satisfies

$$H(x,t,\varepsilon) \geq \frac{g_1(\bar{x},t)}{\varepsilon} \int_{\mathbb{R}^N \setminus \Omega} J_\varepsilon(x-y)\eta(\bar{x}) \cdot \frac{y-x}{\varepsilon}\, dy - D_1 \int_{\mathbb{R}^N \setminus \Omega} J_\varepsilon(x-y)\, dy + \frac{1}{2}.$$

By Lemma 3.15, for $\varepsilon < \bar{\varepsilon}$,

$$H(x,t,\varepsilon) \geq \left(\frac{g_1(\bar{x},t)\, K}{\varepsilon} - D_1 \right) \int_{\mathbb{R}^N \setminus \Omega} J_\varepsilon(x-y)\, dy + \frac{1}{2}.$$

Let us recall that

$$F_\varepsilon(x,t) = \tilde{L}_\varepsilon(\tilde{v}) - \Delta\tilde{u} + \frac{C_2}{\varepsilon} \int_{\mathbb{R}^N \setminus \Omega} J_\varepsilon(x-y)g(y,t)\, dy$$

$$- \frac{1}{\varepsilon^2} \int_{\mathbb{R}^N \setminus \Omega} J_\varepsilon(x-y)\big(\tilde{v}(y,t) - \tilde{v}(x,t)\big)\, dy.$$

Then, proceeding once again as in Lemma 3.14, we have

$$\begin{aligned} |F_\varepsilon(x,t)| &\leq \frac{|g(\bar{x},t)|\, C_2}{\varepsilon} \int_{\mathbb{R}^N \setminus \Omega} J_\varepsilon(x-y)\, dy \\[2mm] &\quad + \frac{|g(\bar{x},t)|}{\varepsilon} \int_{\mathbb{R}^N \setminus \Omega} J_\varepsilon(x-y) \left| \eta(\bar{x}) \cdot \frac{y-x}{\varepsilon} \right| dy \\[2mm] &\quad + C\varepsilon^\alpha + C \int_{\mathbb{R}^N \setminus \Omega} J_\varepsilon(x-y)\, dy \\[2mm] &\leq \left(\frac{C_2 + 1}{\varepsilon} \max_{\partial\Omega \times [0,T]} |g(x,t)| + C \right) \int_{\mathbb{R}^N \setminus \Omega} J_\varepsilon(x-y)\, dy + C\varepsilon^\alpha \\[2mm] &\leq \left(\frac{g_1(\bar{x},t)\, K}{2\varepsilon} + C \right) \int_{\mathbb{R}^N \setminus \Omega} J_\varepsilon(x-y)\, dy + C\varepsilon^\alpha \end{aligned}$$

if $\varepsilon < \bar{\varepsilon}$, by our choice of g_1.

Therefore, for every ε small enough, we obtain

$$|F_\varepsilon(x,t)| \leq H(x,t,\varepsilon),$$

and, by a comparison argument, we conclude that

$$-M \leq -v_1(x,t) \leq u_\varepsilon(x,t) \leq v_1(x,t) \leq M,$$

for every $(x,t) \in \overline{\Omega} \times [0,T]$. This ends the proof. $\square$

Finally, we prove Theorem 3.20.

PROOF OF THEOREM 3.20. By Lemma 3.21 we have that

$$F_\varepsilon(x,t) \rightharpoonup 0 \quad \text{as } \varepsilon \to 0 \quad \text{in the sense of measures in } \Omega \times [0,T].$$

Assume first that $\psi \in C_0^{2+\alpha}(\Omega)$ and let $\tilde{\varphi}_\varepsilon$ be such that

$$\begin{cases} (\tilde{\varphi}_\varepsilon)_t - L_\varepsilon \tilde{\varphi}_\varepsilon = 0, \\ \tilde{\varphi}_\varepsilon(x,0) = \psi(x), \qquad x \in \Omega, \ t > 0. \end{cases}$$

Let $\tilde{\varphi}$ be a solution of

$$\begin{cases} \tilde{\varphi}_t - \Delta\tilde{\varphi} = 0 & \text{in } \Omega \times (0,T), \\[1mm] \dfrac{\partial\tilde{\varphi}}{\partial\eta} = 0 & \text{on } \partial\Omega \times (0,T), \\[1mm] \tilde{\varphi}(.,0) = \psi & \text{in } \Omega. \end{cases}$$

Then, by Theorem 3.16, we know that $\tilde{\varphi}_\varepsilon \to \tilde{\varphi}$ uniformly in $\Omega \times [0,T]$.

For a fixed $t > 0$ set $\varphi_\varepsilon(x,s) = \tilde{\varphi}_\varepsilon(x,t-s)$. Then φ_ε satisfies

$$\begin{cases} (\varphi_\varepsilon)_s + L_\varepsilon \varphi_\varepsilon = 0 & \text{for } x \in \Omega, \ s < t, \\ \varphi_\varepsilon(x,t) = \psi(x) & \text{for } x \in \Omega. \end{cases}$$

Analogously, if we set $\varphi(x,s) = \tilde{\varphi}(x,t-s)$, then φ satisfies

$$\begin{cases} \varphi_s + \Delta\varphi = 0 & \text{in } \Omega \times (0,t), \\[1mm] \dfrac{\partial\varphi}{\partial\eta} = 0 & \text{on } \partial\Omega \times (0,t), \\[1mm] \varphi(x,t) = \psi(x) & \text{in } \Omega. \end{cases}$$

Consequently, for $w_\varepsilon = u_\varepsilon - v$,

$$\int_\Omega w_\varepsilon(x,t)\,\psi(x)\,dx = \int_0^t \int_\Omega \frac{\partial w_\varepsilon}{\partial s}(x,s)\,\varphi_\varepsilon(x,s)\,dx\,ds$$

$$+ \int_0^t \int_\Omega \frac{\partial \varphi_\varepsilon}{\partial s}(x,s)\,w_\varepsilon(x,s)\,dx\,ds$$

$$= \int_0^t \int_\Omega L_\varepsilon(w_\varepsilon)(x,s)\varphi_\varepsilon(x,s)\,dx\,ds + \int_0^t \int_\Omega F_\varepsilon(x,s)\,\varphi_\varepsilon(x,s)\,dx\,ds$$

$$+ \int_0^t \int_\Omega \frac{\partial \varphi_\varepsilon}{\partial s}(x,s)\,w_\varepsilon(x,s)\,dx\,ds$$

$$= \int_0^t \int_\Omega L_\varepsilon(\varphi_\varepsilon)(x,s)w_\varepsilon(x,s)\,dx\,ds + \int_0^t \int_\Omega F_\varepsilon(x,s)\,\varphi_\varepsilon(x,s)\,dx\,ds$$

$$+ \int_0^t \int_\Omega \frac{\partial \varphi_\varepsilon}{\partial s}(x,s)\,w_\varepsilon(x,s)\,dx\,ds$$

$$= \int_0^t \int_\Omega F_\varepsilon(x,s)\varphi_\varepsilon(x,s)\,dx\,ds.$$

By Lemma 3.21,

$$\left| \int_0^t \int_\Omega F_\varepsilon(x,s)\varphi_\varepsilon(x,s)\,dx\,ds \right| \leq \left| \int_0^t \int_\Omega F_\varepsilon(x,s)\varphi(x,s)\,dx\,ds \right|$$

$$+ \sup_{0<s<t} \|\varphi_\varepsilon(x,s) - \varphi(x,s)\|_{L^\infty(\Omega)} \int_0^t \int_\Omega |F_\varepsilon(x,s)|\,dx\,ds \to 0$$

as $\varepsilon \to 0$. This proves the result when $\psi \in C_0^{2+\alpha}(\Omega)$.

We now deal with the general case. Let $\psi \in L^1(\Omega)$. Choose $\psi_n \in C_0^{2+\alpha}(\Omega)$ such that $\psi_n \to \psi$ in $L^1(\Omega)$. Then

$$\left| \int_\Omega w_\varepsilon(x,t)\,\psi(x)\,dx \right| \leq \left| \int_\Omega w_\varepsilon(x,t)\,\psi_n(x)\,dx \right| + \|\psi_n - \psi\|_{L^1(\Omega)} \|w_\varepsilon(.,t)\|_{L^\infty(\Omega)}.$$

By Lemma 3.22, $\{w_\varepsilon\}$ is uniformly bounded and hence the result follows. $\qquad\square$

Bibliographical notes

The results of this chapter are based on [68], [79] and [80].

CHAPTER 4

A nonlocal convection diffusion problem

In this chapter we analyze a nonlocal equation that takes into account convective and diffusive effects. We deal with

$$u_t(x,t) = (J * u - u)(x,t) + (G * (f(u)) - f(u))(x,t),$$

in the whole $\mathbb{R}^N$. Here we have added an extra term $G * (f(u)) - f(u)$ with a nonsymmetric kernel G. This will give a term of the form $\mathbf{b} \cdot \nabla f(u)$ when we rescale the problem obtaining in the limit a local convection-diffusion equation, that is, an equation of the form $v_t(x,t) = \Delta v(x,t) + \mathbf{b} \cdot \nabla f(v)(x,t)$.

For this model we also analyze its asymptotic behaviour as $t \to \infty$ under appropriate hypothesis on the nonlinearity f. We demonstrate a weakly nonlinear behaviour; that is, we show that under some conditions on f, the linear part dominates for large times.

4.1. A nonlocal model with a nonsymmetric kernel

We deal with the nonlocal evolution equation

(4.1)
$$\begin{cases} u_t(x,t) = (J * u - u)(x,t) + (G * (f(u)) - f(u))(x,t), \\ \\ u(x,0) = u_0(x), \qquad\qquad\qquad\qquad x \in \mathbb{R}^N, t > 0. \end{cases}$$

Let us first state our basic assumptions. The function J satisfies hypothesis (H) as in the previous chapters and G is nonnegative, not necessarily symmetric and satisfies

$$\int_{\mathbb{R}^N} G(x)\,dx = 1.$$

Moreover, we consider J and G smooth, $J, G \in \mathcal{S}(\mathbb{R}^N)$, the space of rapidly decreasing functions. To obtain a diffusion operator similar to the Laplacian we impose in addition that J satisfies

$$\frac{1}{2}\partial^2_{\xi_i \xi_i} \widehat{J}(0) = \frac{1}{2}\int_{\mathbb{R}^N} J(z)z_i^2\,dz = 1.$$

This implies that the Fourier transform of J satisfies

$$\widehat{J}(\xi) - 1 + \xi^2 \sim |\xi|^3, \quad \text{for } \xi \text{ close to } 0.$$

Here the notation $h \sim g$ means that there exist constants C_1 and C_2 such that $C_1 h \le g \le C_2 h$. We can consider more general kernels J with expansions in Fourier variables of the form $\widehat{J}(\xi) - 1 + A\,\xi^2 \sim |\xi|^3$. Since the results (and the proofs) are almost the same, we do not include the details for this more general case, but we comment on how the results are modified by the appearance of A.

The nonlinearity f is assumed to be nondecreasing with $f(0) = 0$ and locally Lipschitz continuous (a typical example considered below is $f(u) = |u|^{q-1}u$ with $q > 1$).

The equation in (4.1) has a diffusion operator $J * u - u$ and a nonlinear convective part given by $G * (f(u)) - f(u)$. Concerning this last term, if G is not symmetric, then individuals have greater probability of jumping in one direction than in others, provoking a convective effect. We call equation (4.1), a *nonlocal convection-diffusion equation*.

First, we prove existence and uniqueness of a solution $u \in C([0, \infty); L^1(\mathbb{R}^N)) \cap L^\infty([0, \infty); \mathbb{R}^N)$ with initial condition $u(\cdot, 0) = u_0 \in L^1(\mathbb{R}^N) \cap L^\infty(\mathbb{R}^N)$. Moreover, a contraction principle in $L^1(\mathbb{R}^N)$ is obtained; that is, if u and v are solutions of (4.1) corresponding to initial data $u_0, v_0 \in L^1(\mathbb{R}^N) \cap L^\infty(\mathbb{R}^N)$ respectively, then

$$\|u(t) - v(t)\|_{L^1(\mathbb{R}^N)} \le \|u_0 - v_0\|_{L^1(\mathbb{R}^N)}.$$

In addition,

$$\|u(t)\|_{L^\infty(\mathbb{R}^N)} \le \|u_0\|_{L^\infty(\mathbb{R}^N)}.$$

We have to emphasize again the lack of regularizing effect. This has already been observed for the linear problem $w_t = J * w - w$ in Chapter 1.

In [98], the authors prove that the solutions of the local convection-diffusion problem, $u_t = \Delta u + \mathbf{b} \cdot \nabla f(u)$, satisfy an estimate of the form

$$\|u(t)\|_{L^\infty(\mathbb{R}^N)} \le C(\|u_0\|_{L^1(\mathbb{R}^N)}) \, t^{-N/2}$$

for any initial datum $u_0 \in L^1(\mathbb{R}^N) \cap L^\infty(\mathbb{R}^N)$. Due to the absence of regularizing effect, in our nonlocal model, we cannot prove such type of inequality independently of the $L^\infty(\mathbb{R}^N)$-norm of the initial datum. Moreover, in the one-dimensional case with a suitable bound on the nonlinearity f that appears in the convective part we can prove that such an inequality does not hold in general. We remark again that the $L^1(\mathbb{R}^N) - L^\infty(\mathbb{R}^N)$ regularizing effect is not available for the linear equation, $w_t = J * w - w$.

Concerning rescaling, we obtain a solution of a standard convection-diffusion equation

$$(4.2) \qquad v_t(x, t) = \Delta v(x, t) + \mathbf{b} \cdot \nabla f(v)(x, t), \quad t > 0, \ x \in \mathbb{R}^N,$$

as the limit of solutions of rescaled problems such as (4.1) when the scaling parameter goes to zero. In fact, let us consider

$$J_\varepsilon(x) = \frac{1}{\varepsilon^N} J\left(\frac{x}{\varepsilon}\right), \quad G_\varepsilon(x) = \frac{1}{\varepsilon^N} G\left(\frac{x}{\varepsilon}\right),$$

and the solution $u_\varepsilon(x, t)$ of our convection-diffusion problem rescaled adequately,

$$(4.3) \quad \begin{cases} (u_\varepsilon)_t(x, t) = \dfrac{1}{\varepsilon^2} \displaystyle\int_{\mathbb{R}^N} J_\varepsilon(x - y)(u_\varepsilon(y, t) - u_\varepsilon(x, t)) \, dy \\[2mm] \qquad\qquad + \dfrac{1}{\varepsilon} \displaystyle\int_{\mathbb{R}^N} G_\varepsilon(x - y)(f(u_\varepsilon(y, t)) - f(u_\varepsilon(x, t))) \, dy, \\[2mm] u_\varepsilon(x, 0) = u_0(x), \qquad\qquad\qquad\qquad\qquad\qquad x \in \mathbb{R}^N, \ t > 0. \end{cases}$$

Remark that the scaling is different for the diffusive part of the equation, $J * u - u$, and for the convective part, $G * f(u) - f(u)$. The same different scaling properties can be observed for the local terms Δu and $\mathbf{b} \cdot \nabla f(u)$. With the above notation, for any $T > 0$, we have

$$\lim_{\varepsilon \to 0} \sup_{t \in [0,T]} \|u_\varepsilon - v\|_{L^2(\mathbb{R}^N)} = 0,$$

where $v(x,t)$ is the unique solution of the local convection-diffusion problem (4.2) with initial condition $v(x,0) = u_0(x) \in L^1(\mathbb{R}^N) \cap L^\infty(\mathbb{R}^N)$ and $\mathbf{b} = (b_1, \ldots, b_N)$ given by

$$b_j = \int_{\mathbb{R}^N} x_j \, G(x) \, dx, \qquad j = 1, \ldots, N.$$

This result also justifies the use of the name "nonlocal convection-diffusion problem" when we refer to (4.1).

From the hypotheses on J and G it follows that

$$|\widehat{G}(\xi) - 1 - i\mathbf{b} \cdot \xi| \le C|\xi|^2$$

and

$$|\widehat{J}(\xi) - 1 + \xi^2| \le C|\xi|^3$$

for every $\xi \in \mathbb{R}^N$. These bounds are exactly what we will use in the proof of the above convergence result.

Remark that when G is symmetric, then $b = 0$ and we obtain the heat equation in the limit. Of course the most interesting case is when $b \neq 0$ (this happens when G is not symmetric). Also we note that the conclusion of the theorem holds for other $L^p(\mathbb{R}^N)$-norms besides the $L^2(\mathbb{R}^N)$-norm, however the proof is more involved.

We can consider kernels J such that

$$A = \frac{1}{2} \int_{\mathbb{R}^N} J(z) z_i^2 \, dz \neq 1.$$

This gives the expansion $\widehat{J}(\xi) - 1 + A\xi^2 \sim |\xi|^3$, for ξ close to 0. In this case we arrive to a convection-diffusion equation with a multiple of the Laplacian as the diffusion operator, $v_t = A\Delta v + \mathbf{b} \cdot \nabla f(v)$.

Next, we study the asymptotic behaviour as $t \to \infty$ of solutions of (4.1). To this end we first analyze the decay of solutions taking into account only the diffusive part (the linear part) of the equation. These solutions have a similar decay rate as the one that holds for the heat equation; see Chapter 1 where the Fourier transform plays a key role. Using similar techniques, we prove the following result. Let $p \in [1, \infty]$; for any $u_0 \in L^1(\mathbb{R}^N) \cap L^\infty(\mathbb{R}^N)$ the solution $w(x,t)$ of the linear problem

$$(4.4) \quad \begin{cases} w_t(x,t) = (J * w - w)(x,t), & x \in \mathbb{R}^N,\ t > 0, \\ u(x,0) = u_0(x), & x \in \mathbb{R}^N, \end{cases}$$

satisfies the decay estimate

$$\|w(t)\|_{L^p(\mathbb{R}^N)} \le C(\|u_0\|_{L^1(\mathbb{R}^N)}, \|u_0\|_{L^\infty(\mathbb{R}^N)}) \, t^{-\frac{N}{2}(1 - \frac{1}{p})},$$

for t large.

We begin with the study of the asymptotic behaviour of the complete problem (4.1). To this end we have to impose some growth condition on f. Therefore, in the sequel we restrict ourselves to the nonlinearities f that are pure powers, that is,

$$(4.5) \qquad\qquad f(u) = |u|^{q-1}u, \quad \text{with} \quad q > 1.$$

The analysis is more involved than the one performed for the linear part and here we cannot use the Fourier transform directly (of course, by the presence of the nonlinear term). The strategy is to write a variation of constants formula for the solution and then to get estimates which imply that the nonlinear part decays faster than the linear one. For the local convection diffusion equation this analysis was performed by Escobedo and Zuazua in [98]. However, in the previously mentioned reference, energy estimates were used together with Sobolev inequalities to obtain decay bounds. These Sobolev inequalities are not available for the nonlocal model, since the linear part does not have any regularizing effect (see Remark 4.20). Therefore, we have to avoid the use of energy estimates and tackle the problem using a variant of the Fourier splitting method proposed by Schonbek to deal with local problems; see [142], [143] and [144].

The precise decay rate is given by

$$(4.6) \qquad\qquad \|u(t)\|_{L^p(\mathbb{R}^N)} \le C(\|u_0\|_{L^1(\mathbb{R}^N)}, \|u_0\|_{L^\infty(\mathbb{R}^N)}) \, t^{-\frac{N}{2}(1-\frac{1}{p})},$$

for every $p \in [1, \infty)$.

Finally, looking at the first order term in the asymptotic expansion of the solution, for $q > (N+1)/N$, we find that this leading order term is the same as the one that appears in the linear local heat equation. This is due to the fact that the nonlinear convection is of higher order and the radially symmetric diffusion leads to Gaussian kernels in the asymptotic regime.

Assume that f satisfies (4.5) with $q > (N+1)/N$ and let the initial condition u_0 belong to $L^1(\mathbb{R}^N, 1+|x|) \cap L^\infty(\mathbb{R}^N)$. For any $p \in [2, \infty)$ the following holds:

$$t^{-\frac{N}{2}(1-\frac{1}{p})}\|u(\cdot, t) - MG_t^2(\cdot)\|_{L^p(\mathbb{R}^N)} \le C(J, G, p, N)\,\alpha_q(t),$$

where

$$M = \int_{\mathbb{R}^N} u_0(x)\, dx$$

and

$$\alpha_q(t) = \begin{cases} t^{-\frac{1}{2}} & \text{if} \quad q \ge (N+2)/N, \\[2mm] t^{\frac{1-N(q-1)}{2}} & \text{if} \quad (N+1)/N < q < (N+2)/N. \end{cases}$$

Remark that we prove a weak nonlinear behaviour; in fact the decay rate and the first order term in the expansion are the same that appear in the linear model $w_t = J * w - w$.

As before, recall that the hypotheses on J imply that

$$\widehat{J}(\xi) - (1 - |\xi|^2) \sim |\xi|^3,$$

for ξ close to 0. This is the key property of J used in the proof of the results. Observe that when we have an expansion of the form

$$\widehat{J}(\xi) - (1 - A|\xi|^2) \sim |\xi|^3,$$

for ξ close to zero, we get as first order term a Gaussian profile of the form G_{At}^2.

Note also that $q = \frac{N+1}{N}$ is a critical exponent for the local convection-diffusion problem $v_t = \Delta v + \mathbf{b} \cdot \nabla(v^q)$; see [98]. When q is supercritical, $q > \frac{(N+1)}{N}$, for the local equation, it also gives an asymptotic simplification to the heat semigroup as $t \to \infty$.

The first order term in the asymptotic behaviour for critical or subcritical exponents $1 < q \leq \frac{(N+1)}{N}$ is open. One of the main difficulties that one has to face here is the absence of a self-similar profile due to the nonhomogeneous behaviour of the convolution kernels.

4.2. The linear semigroup revisited

In this section we analyze the asymptotic behaviour of the solutions of the equation

$$(4.7) \quad \begin{cases} w_t(x,t) = (J * w - w)(x,t), & x \in \mathbb{R}^N,\, t > 0, \\ w(x,0) = u_0(x), & x \in \mathbb{R}^N, \end{cases}$$

where J is nonnegative and compactly supported. As we have already mentioned, this equation shares many properties with the classical heat equation, $w_t = \Delta w$, such as: bounded stationary solutions are constants, a maximum principle holds for both of them and perturbations propagate with infinite speed; see [106]. However, there is no regularizing effect in general. In fact, the singularity of the source solution, that is, a solution of (4.7) with initial condition a delta measure, $u_0 = \delta_0$, remains with an exponential decay. This fundamental solution can be decomposed as

$$(4.8) \quad S(x,t) = e^{-t}\delta_0 + K_t(x),$$

where $K_t(x)$ is smooth; see Lemma 1.6. In this way we see that there is no regularizing effect since the solution w of (4.7) can be written as

$$(4.9) \quad w(t) = S(t) * u_0 = e^{-t}u_0 + K_t * u_0$$

with K_t smooth, which means that $w(\cdot, t)$ is as regular as u_0 is. This fact makes the analysis of (4.7) more involved.

In the following we give estimates on the regular part of the fundamental solution K_t whose Fourier transform is

$$(4.10) \quad \widehat{K}_t(\xi) = e^{t(\hat{J}(\xi)-1)} - e^{-t}.$$

The behaviour in the $L^p(\mathbb{R}^N)$-norms of K_t is obtained by analyzing the cases $p = \infty$ and $p = 1$. The case $p = 1$ follows by using the fact that the $L^1(\mathbb{R}^N)$-norm of the solutions of (4.7) does not increase. This analysis overlaps with the one performed in Chapter 1 but we include some details here for completeness. However, here we consider $J \in \mathcal{S}(\mathbb{R}^N)$ and hence all the derivatives of $\hat{J}$ (and of $\widehat{K}_t$) make sense. Therefore, we need less restrictive conditions than those of Chapter 1 (in particular, we do not impose $N \leq 3$ here).

The analysis of the behaviour of the gradient ∇K_t is more involved and differs from the calculations performed in Chapter 1. The behaviour in the $L^p(\mathbb{R}^N)$-norms with $2 \leq p \leq \infty$ follows from the estimates of the L^∞-norm and the L^2-norm, for which we use Plancherel's identity. However the case $1 \leq p < 2$ is more tricky. In order to evaluate the $L^1(\mathbb{R}^N)$-norm of ∇K_t we use the Carlson type inequality

$$(4.11) \qquad \|\varphi\|_{L^1(\mathbb{R}^N)} \leq C \|\varphi\|_{L^2(\mathbb{R}^N)}^{1-\frac{N}{2m}} \||x|^m \varphi\|_{L^2(\mathbb{R}^N)}^{\frac{N}{2m}},$$

which holds for $m > N/2$. The use of the above inequality with $\varphi = \nabla K_t$ implies that $|x|^m \nabla K_t$ belongs to $(L^2(\mathbb{R}^N))^N$. To guarantee this property and to obtain the decay rate in the $L^2(\mathbb{R}^N)$-norm of $|x|^m \nabla K_t$ we use that $J \in \mathcal{S}(\mathbb{R}^N)$ in Lemma 4.3.

The following lemma shows the decay rate in the $L^p(\mathbb{R}^N)$-norms of the kernel K_t for $1 \leq p \leq \infty$.

LEMMA 4.1. *Let J be such that $\widehat{J}(\xi) \in L^1(\mathbb{R}^N)$, $\nabla \widehat{J}(\xi) \in (L^2(\mathbb{R}^N))^N$ and*

$$\widehat{J}(\xi) - 1 + \xi^2 \sim |\xi|^3, \quad \partial_\xi \widehat{J}(\xi) \sim -\xi \quad as \ \xi \sim 0.$$

For any $p \geq 1$ there exists a positive constant $c(p, J)$ such that K_t, defined in (4.10), satisfies

$$(4.12) \qquad \|K_t\|_{L^p(\mathbb{R}^N)} \leq c(p, J) \, t^{-\frac{N}{2}(1-\frac{1}{p})}$$

for any $t > 0$ large enough.

REMARK 4.2. *A stronger inequality can be proved if $p = \infty$,*

$$\|K_t\|_{L^\infty(\mathbb{R}^N)} \leq Cte^{-\delta t}\|\widehat{J}\|_{L^1(\mathbb{R}^N)} + C\, t^{-N/2}$$

for some positive $\delta = \delta(J)$.

Moreover, for $p = 1$ we have

$$\|K_t\|_{L^1(\mathbb{R}^N)} \leq 2,$$

and for any $p \in [1, \infty]$,

$$\|S(t)\|_{L^p(\mathbb{R}^N) - L^p(\mathbb{R}^N)} \leq 3,$$

where S is given by (4.8).

PROOF OF LEMMA 4.1. We analyze the cases $p = \infty$ and $p = 1$; the others can be easily obtained applying Hölder's inequality.

Case $p = \infty$. Since $\widehat{K_t}(\xi) = e^{t(\widehat{J}(\xi)-1)} - e^{-t}$, we have

$$\|K_t\|_{L^\infty(\mathbb{R}^N)} \leq \frac{1}{(2\pi)^N} \int_{\mathbb{R}^N} |e^{t(\widehat{J}(\xi)-1)} - e^{-t}| d\xi.$$

Observe that the symmetry of J guarantees that $\widehat{J}$ is a real number. Let us choose $R > 0$ such that

$$(4.13) \qquad |\widehat{J}(\xi)| \leq 1 - \frac{|\xi|^2}{2} \quad \text{for all } |\xi| \leq R.$$

Once R is fixed, there exists $\delta = \delta(J)$, $0 < \delta < 1$, with

$$(4.14) \qquad |\widehat{J}(\xi)| \leq 1 - \delta \quad \text{for all } |\xi| \geq R.$$

Using that for any real numbers a and b the inequality

$$|e^a - e^b| \leq |a - b| \max\{e^a, e^b\}$$

holds, we obtain, for any $|\xi| \geq R$,

$$(4.15) \qquad |e^{t(\widehat{J}(\xi)-1)} - e^{-t}| \leq t|\widehat{J}(\xi)| \max\{e^{-t}, e^{t(\widehat{J}(\xi)-1)}\} \leq te^{-\delta t}|\widehat{J}(\xi)|.$$

Then the following integral decays exponentially:

$$\int_{|\xi|\geq R} |e^{t(\widehat{J}(\xi)-1)} - e^{-t}|d\xi \leq e^{-\delta t}t \int_{|\xi|\geq R} |\widehat{J}(\xi)|d\xi.$$

Using that this term is exponentially small, it remains to prove that

$$(4.16) \qquad I(t) = \int_{|\xi|\leq R} |e^{t(\widehat{J}(\xi)-1)} - e^{-t}|d\xi \leq Ct^{-N/2}.$$

To handle this case we use the following estimates:

$$|I(t)| \leq \int_{|\xi|\leq R} e^{t(\widehat{J}(\xi)-1)}d\xi + e^{-t}C \leq \int_{|\xi|\leq R} d\xi + e^{-t}C \leq C$$

and

$$|I(t)| \leq \int_{|\xi|\leq R} e^{t(\widehat{J}(\xi)-1)}d\xi + Ce^{-t} \leq \int_{|\xi|\leq R} e^{-\frac{t|\xi|^2}{2}} + Ce^{-t}$$

$$= t^{-N/2}\int_{|\eta|\leq Rt^{1/2}} e^{-\frac{|\eta|^2}{2}} + Ce^{-t} \leq C(1+t)^{-N/2}.$$

The last two estimates prove (4.16) and this finishes the analysis of this case.

Case $p = 1$. We denote by sgn_0 the function given by

$$\text{sgn}_0(r) = \begin{cases} -1 & \text{if } r < 0, \\ 0 & \text{if } r = 0, \\ 1 & \text{if } r > 0. \end{cases}$$

First we prove that the $L^1(\mathbb{R}^N)$-norm of the solutions of equation (4.4) does not increase. Multiplying equation (4.4) by $\text{sgn}_0(w(x,t))$ and integrating with respect to the space variable we obtain

$$\frac{d}{dt}\int_{\mathbb{R}^N} |w(x,t)|\,dx = \int_{\mathbb{R}^N}\int_{\mathbb{R}^N} J(x-y)w(y,t)\text{sgn}_0(w(x,t))\,dy\,ds$$

$$- \int_{\mathbb{R}^N} |w(x,t)|\,dx \leq \int_{\mathbb{R}^N}\int_{\mathbb{R}^N} J(x-y)|w(y,t)|\,dx\,dy - \int_{\mathbb{R}^N} |w(x,t)|\,dx \leq 0,$$

which shows that the $L^1(\mathbb{R}^N)$-norm does not increase. Hence, for any $u_0 \in L^1(\mathbb{R}^N)$,

$$\int_{\mathbb{R}^N} |e^{-t}u_0(x) + (K_t * u_0)(x)|\,dx \leq \int_{\mathbb{R}^N} |u_0(x)|\,dx.$$

Consequently,

$$\int_{\mathbb{R}^N} |(K_t * u_0)(x)|\,dx \leq 2\int_{\mathbb{R}^N} |u_0(x)|\,dx.$$

Choosing $(u_0)_n \in L^1(\mathbb{R}^N)$ such that $(u_0)_n \to \delta_0$ in $\mathcal{S}'(\mathbb{R}^N)$ we obtain in the limit that

$$\int_{\mathbb{R}^N} |K_t(x)|dx \leq 2.$$

This ends the proof of the L^1-case and the proof of the lemma is concluded. $\square$

The following lemma plays a key role when analyzing the decay of the complete problem (4.1). In the sequel $L^1(\mathbb{R}^N, a(x))$ denotes the following weighted space:

$$L^1(\mathbb{R}^N, a(x)) = \left\{\varphi : \mathbb{R}^N \to \mathbb{R} \text{ measurable } : \int_{\mathbb{R}^N} a(x)|\varphi(x)|dx < \infty\right\}.$$

LEMMA 4.3. *Let $p \geq 1$ and $J \in \mathcal{S}(\mathbb{R}^N)$. There exists a positive constant $C = C(p, J)$ depending on p and J such that*

$$\|K_t * \varphi - K_t\|_{L^p(\mathbb{R}^N)} \leq C t^{-\frac{N}{2}(1-\frac{1}{p})-\frac{1}{2}} \|\varphi\|_{L^1(\mathbb{R}^N, |x|)}$$

for all $\varphi \in L^1(\mathbb{R}^N, 1 + |x|)$ such that $\int_{\mathbb{R}^N} \varphi = 1$.

PROOF. Explicit computations show that

$$(K_t * \varphi - K_t)(x) = \int_{\mathbb{R}^N} \varphi(y)(K_t(x - y) - K_t(x)) \, dy$$

(4.17)

$$= \int_{\mathbb{R}^N} \varphi(y) \int_0^1 \nabla K_t(x - sy) \cdot (-y) \, ds \, dy.$$

We analyze the cases $p = 1$ and $p = \infty$; the other cases follow by interpolation.

For $p = \infty$ we have

$$(4.18) \qquad \|K_t * \varphi - K_t\|_{L^\infty(\mathbb{R}^N)} \leq \|\nabla K_t\|_{L^\infty(\mathbb{R}^N)} \int_{\mathbb{R}^N} |y| |\varphi(y)| \, dy.$$

In the case $p = 1$, by using (4.17), we get

$$\int_{\mathbb{R}^N} |(K_t * \varphi - K_t)(x)| \, dx \leq \int_{\mathbb{R}^N} \int_{\mathbb{R}^N} |y| |\varphi(y)| \int_0^1 |\nabla K_t(x - sy)| \, ds \, dy \, dx$$

(4.19)

$$= \int_{\mathbb{R}^N} |y| |\varphi(y)| \int_0^1 \int_{\mathbb{R}^N} |\nabla K_t(x - sy)| \, dx \, ds \, dy$$

$$= \int_{\mathbb{R}^N} |y| |\varphi(y)| \, dy \int_{\mathbb{R}^N} |\nabla K_t(x)| \, dx.$$

In view of (4.18) and (4.19) it is sufficient to prove that

$$\|\nabla K_t\|_{L^\infty(\mathbb{R}^N)} \leq C t^{-\frac{N}{2}-\frac{1}{2}}$$

and

$$\|\nabla K_t\|_{L^1(\mathbb{R}^N)} \leq C t^{-\frac{1}{2}}.$$

In the first case, with R and δ as in (4.13) and (4.14), by (4.15) we obtain

$$\|\nabla K_t\|_{L^\infty(\mathbb{R}^N)} \leq \frac{1}{(2\pi)^N} \int_{\mathbb{R}^N} |\xi| |e^{t(\widehat{J}(\xi)-1)} - e^{-t}| d\xi$$

$$= \frac{1}{(2\pi)^N} \int_{|\xi| \leq R} |\xi| |e^{t(\widehat{J}(\xi)-1)} - e^{-t}| d\xi$$

$$+ \frac{1}{(2\pi)^N} \int_{|\xi| \geq R} |\xi| |e^{t(\widehat{J}(\xi)-1)} - e^{-t}| d\xi$$

$$\leq \frac{1}{(2\pi)^N} \int_{|\xi| \leq R} |\xi| e^{-t|\xi|^2/2} d\xi$$

$$+ e^{-t} \frac{1}{(2\pi)^N} \int_{|\xi| \leq R} |\xi| d\xi + t \int_{|\xi| \geq R} |\xi| |\widehat{J}(\xi)| e^{-\delta t} d\xi$$

$$\leq Ct^{-\frac{N}{2}-\frac{1}{2}} + Ce^{-t} + Cte^{-\delta t}$$

$$\leq Ct^{-\frac{N}{2}-\frac{1}{2}},$$

provided that $|\xi|\widehat{J}(\xi)$ belongs to $L^1(\mathbb{R}^N)$.

In the second case it is enough to prove that the $L^1(\mathbb{R}^N)$-norm of $\partial_{x_1} K_t$ is controlled by $t^{-1/2}$. By Carlson's inequality (4.11), we get

$$\|\partial_{x_1} K_t\|_{L^1(\mathbb{R}^N)} \leq C \|\partial_{x_1} K_t\|_{L^2(\mathbb{R}^N)}^{1-\frac{N}{2m}} \||x|^m \partial_{x_1} K_t\|_{L^2(\mathbb{R}^N)}^{\frac{N}{2m}},$$

for any $m > N/2$.

Now the aim is to prove that, for any $t > 0$, we have

$$(4.20) \qquad \|\partial_{x_1} K_t\|_{L^2(\mathbb{R}^N)} \leq Ct^{-\frac{N}{4}-\frac{1}{2}}$$

and

$$(4.21) \qquad \||x|^m \partial_{x_1} K_t\|_{L^2(\mathbb{R}^N)} \leq Ct^{\frac{m-1}{2}-\frac{N}{4}}.$$

By Plancherel's identity, estimate (4.15) and using that $|\xi|\widehat{J}(\xi)$ belongs to $L^2(\mathbb{R}^N)$, we obtain

$$\|\partial_{x_1} K_t\|_{L^2(\mathbb{R}^N)}^2 = (2\pi)^{-\frac{N}{2}} \int_{\mathbb{R}^N} |\xi_1|^2 |e^{t(\widehat{J}(\xi)-1)} - e^{-t}|^2 d\xi$$

$$\leq 2(2\pi)^{-\frac{N}{2}} \int_{|\xi| \leq R} |\xi_1|^2 e^{-t|\xi|^2} d\xi + (2\pi)^{-\frac{N}{2}} e^{-2t} \int_{|\xi| \leq R} |\xi_1|^2 d\xi$$

$$+ (2\pi)^{-\frac{N}{2}} \int_{|\xi| \geq R} |\xi_1|^2 e^{-2\delta t} t^2 |\widehat{J}(\xi)|^2 d\xi$$

$$\leq Ct^{-\frac{N}{2}-\frac{1}{2}} + Ce^{-2t} + Ce^{-2\delta t} t^2$$

$$\leq Ct^{-\frac{N}{2}-\frac{1}{2}}.$$

This shows (4.20).

To prove (4.21), observe that

$$\||x|^m \partial_{x_1} K_t\|^2_{L^2(\mathbb{R}^N)} \leq C \int_{\mathbb{R}^N} (x_1^{2m} + \cdots + x_N^{2m})|\partial_{x_1} K_t(x)|^2 \, dx.$$

Thus, by symmetry it is sufficient to prove that

$$\int_{\mathbb{R}^N} |\partial_{\xi_1}^m(\xi_1 \widehat{K}_t(\xi))|^2 \, d\xi \leq C \, t^{m-1-\frac{N}{2}}$$

and

$$\int_{\mathbb{R}^N} |\partial_{\xi_2}^m(\xi_1 \widehat{K}_t(\xi))|^2 \, d\xi \leq C \, t^{m-1-\frac{N}{2}}.$$

Note that

$$|\partial_{\xi_1}^m(\xi_1 \widehat{K}_t(\xi))| = |\xi_1 \partial_{\xi_1}^m \widehat{K}_t(\xi) + m \partial_{\xi_1}^{m-1} \widehat{K}_t(\xi)| \leq |\xi||\partial_{\xi_1}^m \widehat{K}_t(\xi)| + m|\partial_{\xi_1}^{m-1} \widehat{K}_t(\xi)|$$

and

$$|\partial_{\xi_2}^m(\xi_1 \widehat{K}_t)| \leq |\xi||\partial_{\xi_2}^m \widehat{K}_t(\xi)|.$$

Hence we just have to prove that

$$\int_{\mathbb{R}^N} |\xi|^{2r}|\partial_{\xi_1}^n \widehat{K}_t(\xi)|^2 d\xi \leq C \, t^{n-r-\frac{N}{2}}, \qquad (r, n) \in \{(0, m-1), (1, m)\}.$$

Choosing $m = [N/2] + 1$ (the notation $[\cdot]$ stands for the floor function) the above inequality has to hold for $n = [N/2], [N/2] + 1$.

First we recall the following elementary identity:

$$\partial_{\xi_1}^n(e^g) = e^g \sum_{i_1 + 2i_2 + \cdots + ni_n = n} a_{i_1, \ldots, i_n} (\partial_{\xi_1}^1 g)^{i_1} (\partial_{\xi_1}^2 g)^{i_2} \cdots (\partial_{\xi_1}^n g)^{i_n},$$

where $a_{i_1, \ldots, i_n}$ are universal constants independent of g. Taking into account that

$$\widehat{K}_t(\xi) = e^{t(\widehat{J}(\xi)-1)} - e^{-t}$$

we have

$$\partial_{\xi_1}^n \widehat{K}_t(\xi) = e^{t(\widehat{J}(\xi)-1)} \sum_{i_1 + 2i_2 + \cdots + ni_n = n} a_{i_1, \ldots, i_n} t^{i_1 + \cdots + i_n} \prod_{j=1}^n [\partial_{\xi_1}^j \widehat{J}(\xi)]^{i_j}$$

and hence

$$|\partial_{\xi_1}^n \widehat{K}_t(\xi)|^2 \leq C \, e^{2t(\widehat{J}(\xi)-1)} \sum_{i_1 + 2i_2 + \cdots + ni_n = n} t^{2(i_1 + \cdots + i_n)} \prod_{j=1}^n [\partial_{\xi_1}^j \widehat{J}(\xi)]^{2i_j}.$$

Using that all the partial derivatives of $\widehat{J}$ decay faster than any polynomial in $|\xi|$ as $|\xi| \to \infty$, we obtain that

$$\int_{|\xi| > R} |\xi|^{2r}|\partial_{\xi_1}^n \widehat{K}_t(\xi)|^2 d\xi \leq C \, e^{-2\delta t} t^{2n},$$

where R and δ are chosen as in (4.13) and (4.14). Since $J \in \mathcal{S}(\mathbb{R}^N)$, $\widehat{J}(\xi)$ is smooth; consequently for all $|\xi| \leq R$,

$$|\partial_{\xi_1} \widehat{J}(\xi)| \leq C \, |\xi|$$

and

$$|\partial_{\xi_1}^j \widehat{J}(\xi)| \leq C, \qquad j = 2, \ldots, n.$$

Then for all $|\xi| \leq R$ we have

$$|\partial_{\xi_1}^n \widehat{K_t}(\xi)|^2 \leq C\, e^{-t|\xi|^2} \sum_{i_1+2i_2+\cdots+ni_n=n} t^{2(i_1+\cdots+i_n)}|\xi|^{2i_1}.$$

Finally, using that for any $l \geq 0$

$$\int_{|\xi| \leq R} e^{-t|\xi|^2} |\xi|^l d\xi \leq Ct^{-\frac{N}{2}-\frac{l}{2}},$$

we obtain

$$\int_{|\xi| \leq R} |\xi|^{2r} |\partial_{\xi_1}^n K_t(\xi)|^2 d\xi \leq Ct^{-\frac{N}{2}} \sum_{i_1+2i_2+\cdots+ni_n=n} t^{2p(i_1,\ldots,i_N)-r},$$

where

$$p(i_1,\ldots,i_n) = (i_1+\cdots+i_n) - \frac{i_1}{2} = \frac{i_1}{2} + i_2 + \cdots + i_n$$

$$\leq \frac{i_1 + 2i_2 + \cdots + ni_n}{2} = \frac{n}{2},$$

and the proof is complete. □

Next we prove a decay estimate that takes into account the linear semigroup applied to the convolution with a kernel G.

LEMMA 4.4. *Let* $1 \leq p \leq r \leq \infty$, $J \in \mathcal{S}(\mathbb{R}^N)$, $G \in L^1(\mathbb{R}^N, |x|)$ *and* $S(t) = e^{-t}\delta_0 + K_t$. *Then there exists a positive constant* $C = C(p, J, G)$ *such that the following estimate holds:*

$$(4.22) \quad \|S(t) * G * \varphi - S(t) * \varphi\|_{L^r(\mathbb{R}^N)} \leq Ct^{-\frac{N}{2}(\frac{1}{p}-\frac{1}{r})-\frac{1}{2}}(\|\varphi\|_{L^p(\mathbb{R}^N)} + \|\varphi\|_{L^r(\mathbb{R}^N)})$$

for all $\varphi \in L^p(\mathbb{R}^N) \cap L^r(\mathbb{R}^N)$.

REMARK 4.5. In fact, the following stronger inequality holds:

$$\|S(t) * G * \varphi - S(t) * \varphi\|_{L^r(\mathbb{R}^N)} \leq C\, t^{-\frac{N}{2}(\frac{1}{p}-\frac{1}{r})-\frac{1}{2}} \|\varphi\|_{L^p(\mathbb{R}^N)} + C\, e^{-t}\|\varphi\|_{L^r(\mathbb{R}^N)}.$$

PROOF. We have

$$S(t) * G * \varphi - S(t) * \varphi = e^{-t}(G * \varphi - \varphi) + K_t * G * \varphi - K_t * \varphi.$$

The first term in the above right hand side satisfies

$$e^{-t}\|G * \varphi - \varphi\|_{L^r(\mathbb{R}^N)} \leq e^{-t}(\|G\|_{L^1(\mathbb{R}^N)}\|\varphi\|_{L^r(\mathbb{R}^N)} + \|\varphi\|_{L^r(\mathbb{R}^N)}) \leq 2e^{-t}\|\varphi\|_{L^r(\mathbb{R}^N)}.$$

For the second term, by Lemma 4.3 we get that K_t satisfies

$$\|K_t * G - K_t\|_{L^a(\mathbb{R}^N)} \leq C(r, J)\|G\|_{L^1(\mathbb{R}^N, |x|)} t^{-\frac{N}{2}(1-\frac{1}{a})-\frac{1}{2}}$$

for all $t \geq 0$, where a is such that $1/r = 1/a + 1/p - 1$. From this, using Young's inequality we complete the proof. □

4.3. Existence and uniqueness of the convection problem

In this section the previous results and estimates on the linear semigroup are used to prove the existence and uniqueness of the solution of the nonlinear problem (4.1). The proof is based on the variation of constants formula.

First, we state what we understand by a solution.

DEFINITION 4.6. A function $u \in C([0, \infty); L^1(\mathbb{R}^N)) \cap L^\infty([0, \infty); \mathbb{R}^N)$ is a *solution* of (4.1) if it satisfies

$$(4.23) \qquad u(t) = S(t) * u_0 + \int_0^t S(t-s) * (G * (f(u)) - f(u))(s) \, ds,$$

where $(S(t))_{t \geq 0}$ is the semigroup solution of problem (4.4).

Now we are ready to prove existence and uniqueness.

THEOREM 4.7. *For any $u_0 \in L^1(\mathbb{R}^N) \cap L^\infty(\mathbb{R}^N)$ there exists a unique global solution of problem (4.1). Moreover, if u and v are solutions of (4.1) corresponding to initial data $u_0, v_0 \in L^1(\mathbb{R}^N) \cap L^\infty(\mathbb{R}^N)$ respectively, then the following contraction property holds:*

$$\|u(t) - v(t)\|_{L^1(\mathbb{R}^N)} \leq \|u_0 - v_0\|_{L^1(\mathbb{R}^N)}$$

for any $t \geq 0$. In addition,

$$\|u(t)\|_{L^\infty(\mathbb{R}^N)} \leq \|u_0\|_{L^\infty(\mathbb{R}^N)}$$

for any $t \geq 0$.

PROOF. For the initial condition $u_0 \in L^1(\mathbb{R}^N) \cap L^\infty(\mathbb{R}^N)$, consider the integral equation associated with (4.1), as in Definition 4.6,

$$(4.24) \qquad u(t) = S(t) * u_0 + \int_0^t S(t-s) * (G * (f(u)) - f(u))(s) \, ds,$$

the functional

$$\mathcal{T}_{u_0}(u)(t) = S(t) * u_0 + \int_0^t S(t-s) * (G * (f(u)) - f(u))(s) \, ds$$

and the space

$$X_{t_0} = C([0, t_0]; L^1(\mathbb{R}^N)) \cap L^\infty([0, t_0]; \mathbb{R}^N)$$

endowed with the norm

$$|||u||| = \sup_{t \in [0, t_0]} \left(\|u(t)\|_{L^1(\mathbb{R}^N)} + \|u(t)\|_{L^\infty(\mathbb{R}^N)} \right).$$

We prove that $\mathcal{T}_{u_0}$ is a contraction in the ball of radius R of X_{t_0}, B_R, if t_0 is small enough.

Step I. Local Existence. Let $M = \max\{\|u_0\|_{L^1(\mathbb{R}^N)}, \|u_0\|_{L^\infty(\mathbb{R}^N)}\}$ and $p = 1$ or ∞. Then, using the results of Lemma 4.1 we obtain

$$\|\mathcal{T}_{u_0}(u)(t)\|_{L^p(\mathbb{R}^N)} \leq \|S(t) * u_0\|_{L^p(\mathbb{R}^N)}$$

$$+ \int_0^t \|S(t-s) * G * (f(u)) - S(t-s) * f(u)\|_{L^p(\mathbb{R}^N)} \, ds$$

$$\leq \left(e^{-t} + \|K_t\|_{L^1(\mathbb{R}^N)}\right) \|u_0\|_{L^p(\mathbb{R}^N)}$$

$$+ \int_0^t 2(e^{-(t-s)} + \|K_{t-s}\|_{L^1(\mathbb{R}^N)})\|f(u)(s)\|_{L^p(\mathbb{R}^N)} \, ds$$

$$\leq 3 \|u_0\|_{L^p(\mathbb{R}^N)} + 6 \, t_0 f(R) \leq 3M + 6 \, t_0 f(R).$$

This implies that

$$\||\mathcal{T}_{u_0}(u)\|| \leq 6M + 12 \, t_0 f(R).$$

Choosing $R = 12M$ and t_0 such that $12 \, t_0 f(R) < 6M$ we get that $\mathcal{T}_{u_0}(B_R) \subset B_R$.

Let us take u and v in B_R. Then, for $p = 1$ or ∞, we have

$$\|\mathcal{T}_{u_0}(u)(t) - \mathcal{T}_{u_0}(v)(t)\|_{L^p(\mathbb{R}^N)}$$

$$\leq \int_0^t \|(S(t-s) * G - S(t-s)) * (f(u) - f(v))\|_{L^p(\mathbb{R}^N)} ds$$

$$\leq 6 \int_0^t \|f(u)(s) - f(v)(s)\|_{L^p(\mathbb{R}^N)} \, ds$$

$$\leq C \int_0^t \|u(s) - v(s)\|_{L^p(\mathbb{R}^N)} \, ds$$

$$\leq C \, t_0 \, \||u - v\||.$$

Choosing t_0 small we obtain that $\mathcal{T}_{u_0}$ is a contraction in B_R and then there exists a unique local solution u of (4.24).

Step II. Global existence. To prove global existence of the solutions we have to guarantee that both the $L^1(\mathbb{R}^N)$ and $L^\infty(\mathbb{R}^N)$-norms of the solutions do not blow up in finite time. We apply the following lemma to control the $L^\infty(\mathbb{R}^N)$-norm of the solutions.

LEMMA 4.8. *Let $\theta \in L^1(\mathbb{R}^N)$ and K be a nonnegative function with mass one. Then, for any $\mu \geq 0$, we have*

(4.25) $$\int_{\theta(x) > \mu} \int_{\mathbb{R}^N} K(x - y)\theta(y) \, dy \, dx \leq \int_{\theta(x) > \mu} \theta(x) \, dx$$

and

(4.26) $$\int_{\theta(x) < -\mu} \int_{\mathbb{R}^N} K(x - y)\theta(y) \, dy \, dx \geq \int_{\theta(x) < -\mu} \theta(x) \, dx.$$

PROOF. First of all we point out that we only have to show (4.25). Indeed, once it is proved, then (4.26) follows immediately applying (4.25) to the function $-\theta$.

First, we prove estimate (4.25) for $\mu = 0$ and then we apply this case to obtain the case $\mu \neq 0$.

For $\mu = 0$ we have the following inequalities:

$$\int_{\theta(x)>0} \int_{\mathbb{R}^N} K(x-y)\theta(y) \, dy \, dx$$

$$\leq \int_{\theta(x)>0} \int_{\theta(y)>0} K(x-y)\theta(y) \, dy \, dx$$

$$= \int_{\theta(y)>0} \theta(y) \int_{\theta(x)>0} K(x-y) \, dx \, dy$$

$$\leq \int_{\theta(y)>0} \theta(y) \int_{\mathbb{R}^N} K(x-y) \, dx \, dy$$

$$= \int_{\theta(y)>0} \theta(y) \, dy.$$

Let us analyze the case $\mu > 0$. In this case the inequality

$$\int_{\theta(x)>\mu} \theta(x) \, dx \leq \int_{\mathbb{R}^N} |\theta(x)| \, dx$$

shows that the set $\{x \in \mathbb{R}^N : \theta(x) > \mu\}$ has finite measure. Then, by the first step, we get

$$\int_{\theta(x)>\mu} \int_{\mathbb{R}^N} K(x-y)\theta(y) \, dy \, dx$$

$$= \int_{\theta(x)>\mu} \int_{\mathbb{R}^N} K(x-y)(\theta(y) - \mu) \, dy \, dx + \int_{\theta(x)>\mu} \mu \, dx$$

$$\leq \int_{\theta(x)>\mu} (\theta(x) - \mu) \, dx + \int_{\theta(x)>\mu} \mu \, dx$$

$$= \int_{\theta(x)>\mu} \theta(x) \, dx.$$

This completes the proof of (4.25).   □

Control of the L^1-norm. We multiply equation (4.1) by $\text{sgn}_0(u(x,t))$ and integrate in $\mathbb{R}^N$ to obtain the estimate

$$\frac{d}{dt} \int_{\mathbb{R}^N} |u(x,t)|\, dx = \int_{\mathbb{R}^N} \int_{\mathbb{R}^N} J(x-y)u(y,t)\text{sgn}_0(u(x,t))\, dy\, dx - \int_{\mathbb{R}^N} |u(x,t)|dx$$

$$+ \int_{\mathbb{R}^N} \int_{\mathbb{R}^N} G(x-y)f(u(y,t))\text{sgn}_0(u(x,t))dydx - \int_{\mathbb{R}^N} \text{sgn}_0(f(u(x,t)))u(x,t)\, dx$$

$$\leq \int_{\mathbb{R}^N} \int_{\mathbb{R}^N} J(x-y)|u(y,t)|\, dy\, dx - \int_{\mathbb{R}^N} |u(x,t)|\, dx$$

$$+ \int_{\mathbb{R}^N} \int_{\mathbb{R}^N} G(x-y)|f(u(y,t))|\, dy\, dx - \int_{\mathbb{R}^N} |f(u(x,t))|\, dx$$

$$= \int_{\mathbb{R}^N} |u(y,t)| \int_{\mathbb{R}^N} J(x-y)\, dx\, dy - \int_{\mathbb{R}^N} |u(x,t)|\, dx$$

$$+ \int_{\mathbb{R}^N} |f(u(y,t))| \int_{\mathbb{R}^N} G(x-y)\, dx\, dy - \int_{\mathbb{R}^N} |f(u(x,t))|\, dx \leq 0,$$

which shows that the $L^1(\mathbb{R}^N)$-norm of $u(\cdot,t)$ does not increase in time.

Control of the L^∞-norm. Let us denote $m = \|u_0\|_{L^\infty(\mathbb{R}^N)}$. Multiplying the equation in (4.1) by $\text{sgn}_0(u-m)^+$ and integrating in the x variable we get

$$\frac{d}{dt} \int_{\mathbb{R}^N} (u(x,t)-m)^+ dx = I_1(t) + I_2(t),$$

where

$$I_1(t) = \int_{\mathbb{R}^N} \int_{\mathbb{R}^N} J(x-y)u(y,t)\text{sgn}_0(u(x,t)-m)^+\, dy\, dx$$

$$- \int_{\mathbb{R}^N} u(x,t)\text{sgn}_0(u(x,t)-m)^+\, dx$$

and

$$I_2(t) = \int_{\mathbb{R}^N} \int_{\mathbb{R}^N} G(x-y)f(u(y,t))\text{sgn}_0(u(x,t)-m)^+\, dy\, dx$$

$$- \int_{\mathbb{R}^N} f(u(x,t))\text{sgn}_0(u(x,t)-m)^+\, dx.$$

We claim that both I_1 and I_2 are negative. Thus $(u(x,t)-m)^+ = 0$ a.e. $x \in \mathbb{R}^N$ and then $u(x,t) \leq m$ for all $t > 0$ and a.e. $x \in \mathbb{R}^N$.

In the case of I_1, applying Lemma 4.8 with $K = J$, $\theta = u(t)$ and $\mu = m$ we obtain

$$\int_{\mathbb{R}^N} \int_{\mathbb{R}^N} J(x-y)u(y,t)\text{sgn}_0(u(x,t)-m)^+\, dy\, dx$$

$$= \int_{u(x)>m} \int_{\mathbb{R}^N} J(x-y)u(y,t)\, dy\, dx$$

$$\leq \int_{u(x)>m} u(x,t)\, dx.$$

To handle I_2 we proceed in a similar manner. Applying Lemma 4.8 with

$$\theta(x) = f(u(x,t)) \quad \text{and} \quad \mu = f(m)$$

we get

$$\int_{f(u(x,t))>f(m)} \int_{\mathbb{R}^N} G(x-y)f(u(y,t)) \, dy \, dx$$
$$\leq \int_{f(u(x,t))>f(m)} f(u(x,t)) \, dx.$$

Using that f is a nondecreasing function, we rewrite this inequality in an equivalent form to obtain the desired inequality

$$\int_{\mathbb{R}^N} \int_{\mathbb{R}^N} G(x-y)f(u(y,t)) \text{sgn}_0(u(x,t)-m)^+ \, dy \, dx$$

$$= \int_{u(x,t) \geq m} \int_{\mathbb{R}^N} G(x-y)f(u(y,t)) \, dy \, dx$$

$$= \int_{f(u(x,t)) \geq f(m)} \int_{\mathbb{R}^N} G(x-y)f(u(y,t)) \, dy \, dx$$

$$\leq \int_{u(x,t) \geq m} f(u(x,t)) \, dx.$$

In a similar way, by using inequality (4.26), we have

$$\frac{d}{dt} \int_{\mathbb{R}^N} (u(x,t)+m)^- \, dx \leq 0,$$

which implies that $u(x,t) \geq -m$ for all $t > 0$ and a.e. $x \in \mathbb{R}^N$.

We conclude that $\|u(t)\|_{L^\infty(\mathbb{R}^N)} \leq \|u_0\|_{L^\infty(\mathbb{R}^N)}$.

Step III. Uniqueness and contraction property. Let us consider two solutions u and v corresponding to initial data u_0 and v_0 respectively. We prove that for any $t > 0$,

$$\frac{d}{dt} \int_{\mathbb{R}^N} |u(x,t) - v(x,t)| \, dx \leq 0.$$

To this end, we multiply by $\text{sgn}_0(u(x,t) - v(x,t))$ the equation satisfied by $u - v$, and using the symmetry of J, the positivity of J and G and that their masses are equal to one, we obtain

$$\frac{d}{dt} \int_{\mathbb{R}^N} |u(x,t) - v(x,t)| \, dx$$

$$= \int_{\mathbb{R}^N} \int_{\mathbb{R}^N} J(x-y)(u(y,t) - v(y,t))\mathrm{sgn}_0(u(x,t) - v(x,t)) \, dx \, dy$$

$$- \int_{\mathbb{R}^N} \int_{\mathbb{R}^N} |u(x,t) - v(x,t)| \, dx$$

$$+ \int_{\mathbb{R}^N} \int_{\mathbb{R}^N} G(x-y)(f(u(y,t)) - f(v(y,t)))\mathrm{sgn}_0(u(x,t) - v(x,t)) \, dx \, dy$$

$$- \int_{\mathbb{R}^N} |f(u(x,t)) - f(v(x,t))| \, dx$$

$$\leq \int_{\mathbb{R}^N} \int_{\mathbb{R}^N} J(x-y)|u(y,t) - v(y,t)| \, dx \, dy - \int_{\mathbb{R}^N} |u(x,t) - v(x,t)| \, dx$$

$$+ \int_{\mathbb{R}^N} \int_{\mathbb{R}^N} G(x-y)|f(u(y,t)) - f(v(y,t))| \, dx \, dy$$

$$- \int_{\mathbb{R}^N} |f(u(x,t)) - f(v(x,t))| \, dx = 0.$$

Thus we get the uniqueness of the solutions and the contraction property

$$\|u(t) - v(t)\|_{L^1(\mathbb{R}^N)} \leq \|u_0 - v_0\|_{L^1(\mathbb{R}^N)}.$$

This ends the proof of Theorem 4.7. $\square$

In contrast to what happens for the local convection-diffusion problem (see [98]), due to the lack of regularizing effect, the $L^\infty(\mathbb{R})$-norm is not bounded for positive times when we consider initial conditions in $L^1(\mathbb{R})$.

PROPOSITION 4.9. *Let $N = 1$ and $|f(u)| \leq C|u|^q$ with $1 \leq q < 2$. If $u(t)$ is the solution of (4.1), then*

$$\sup_{u_0 \in L^1(\mathbb{R})} \sup_{t \in [0,1]} \frac{t^{\frac{1}{2}} \|u(t)\|_{L^\infty(\mathbb{R})}}{\|u_0\|_{L^1(\mathbb{R})}} = \infty.$$

PROOF. Assume by contradiction that

(4.27) $$\sup_{u_0 \in L^1(\mathbb{R})} \sup_{t \in [0,1]} \frac{t^{\frac{1}{2}} \|u(t)\|_{L^\infty(\mathbb{R})}}{\|u_0\|_{L^1(\mathbb{R})}} = M < \infty.$$

Using the representation formula (4.24) we get

$$\|u(1)\|_{L^\infty(\mathbb{R})} \geq \|S(1) * u_0\|_{L^\infty(\mathbb{R})} - \left\| \int_0^1 S(1-s) * (G * (f(u)) - f(u))(s) \, ds \right\|_{L^\infty(\mathbb{R})}.$$

By Lemma 4.4 the last term can be bounded as follows:

$$\left\| \int_0^1 S(1-s) * (G * (f(u)) - f(u))(s) \, ds \right\|_{L^\infty(\mathbb{R})}$$

$$\leq \int_0^1 (1-s)^{-\frac{1}{2}} \|f(u(s))\|_{L^\infty(\mathbb{R})} \, ds$$

$$\leq C \int_0^1 \|u(s)\|_{L^\infty(\mathbb{R})}^q ds \leq C M^q \|u_0\|_{L^1(\mathbb{R})}^q \int_0^1 s^{-\frac{q}{2}} ds$$

$$\leq C M^q \|u_0\|_{L^1(\mathbb{R})}^q,$$

provided that $q < 2$. This implies that the $L^\infty(\mathbb{R})$-norm of the solution at time $t = 1$ satisfies

$$\|u(1)\|_{L^\infty(\mathbb{R})} \geq \|S(1) * u_0\|_{L^\infty(\mathbb{R})} - C M^q \|u_0\|_{L^1(\mathbb{R})}^q$$

$$\geq e^{-1} \|u_0\|_{L^\infty(\mathbb{R})} - \|K_1\|_{L^\infty(\mathbb{R})} \|u_0\|_{L^1(\mathbb{R})} - C M^q \|u_0\|_{L^1(\mathbb{R})}^q$$

$$\geq e^{-1} \|u_0\|_{L^\infty(\mathbb{R})} - C \|u_0\|_{L^1(\mathbb{R})} - C M^q \|u_0\|_{L^1(\mathbb{R})}^q.$$

Choosing now a sequence $u_{0,\varepsilon}$ with $\|u_{0,\varepsilon}\|_{L^1(\mathbb{R})} = 1$ and $\|u_{0,\varepsilon}\|_{L^\infty(\mathbb{R})} \to \infty$ we obtain that

$$\|u_{0,\varepsilon}(1)\|_{L^\infty(\mathbb{R})} \to \infty,$$

a contradiction with the assumption (4.27). The proof of the result is now completed. □

4.4. Rescaling the kernels. Convergence to the local convection-diffusion problem

In this section we prove the convergence of solutions of the nonlocal problem to solutions of the local convection-diffusion equation when we rescale the kernels and let the scaling parameter go to zero.

As in the previous sections, we begin with the analysis of the linear part. Remark that in Chapter 1 we have shown the convergence of the approximations to the solutions of the local diffusion problem in the L^∞-norm. Now we show the convergence in the L^2-norm.

LEMMA 4.10. *Assume that $u_0 \in L^2(\mathbb{R}^N)$. Let w_ε be the solution of*

$$(4.28) \quad \begin{cases} (w_\varepsilon)_t(x,t) = \dfrac{1}{\varepsilon^2} \displaystyle\int_{\mathbb{R}^N} J_\varepsilon(x-y)(w_\varepsilon(y,t) - w_\varepsilon(x,t)) \, dy, \\ w_\varepsilon(x,0) = u_0(x), \qquad\qquad\qquad x \in \mathbb{R}^N, \, t > 0, \end{cases}$$

and w the solution of

$$(4.29) \quad \begin{cases} w_t(x,t) = \Delta w(x,t) & \text{in } \mathbb{R}^N \times (0,\infty), \\ w(x,0) = u_0(x) & \text{in } \mathbb{R}^N. \end{cases}$$

Then, for any positive T,

$$\lim_{\varepsilon \to 0} \sup_{t \in [0,T]} \|w_\varepsilon - w\|_{L^2(\mathbb{R}^N)} = 0.$$

PROOF. Taking the Fourier transform in (4.28) we get

$$\frac{d}{dt}\widehat{w_\varepsilon}(\xi, t) = \frac{1}{\varepsilon^2}\left(\widehat{J_\varepsilon}(\xi)\widehat{w_\varepsilon}(\xi, t) - \widehat{w_\varepsilon}(\xi, t)\right).$$

Therefore,

$$\widehat{w_\varepsilon}(\xi, t) = \exp\left(t\frac{\widehat{J_\varepsilon}(\xi) - 1}{\varepsilon^2}\right)\widehat{u_0}(\xi).$$

Since $\widehat{J_\varepsilon}(\xi) = \widehat{J}(\varepsilon\xi)$, we get

$$\widehat{w_\varepsilon}(\xi, t) = \exp\left(t\frac{\widehat{J}(\varepsilon\xi) - 1}{\varepsilon^2}\right)\widehat{u_0}(\xi).$$

By Plancherel's identity, using the well-known formula for solutions of (4.29),

$$\widehat{w}(\xi, t) = e^{-t\xi^2}\widehat{u_0}(\xi),$$

we get

$$\|w_\varepsilon(t) - w(t)\|_{L^2(\mathbb{R}^N)}^2 = (2\pi)^{-\frac{N}{2}}\int_{\mathbb{R}^N}\left|e^{t\frac{\widehat{J}(\varepsilon\xi)-1}{\varepsilon^2}} - e^{-t\xi^2}\right|^2 |\widehat{u_0}(\xi)|^2 \, d\xi.$$

Taking R as in (4.13) we split the integral in two parts according to the sets $\{\xi \in \mathbb{R}^N : |\xi| \geq R/\varepsilon\}$ and $\{\xi \in \mathbb{R}^N : |\xi| \leq R/\varepsilon\}$. Then, for δ as in (4.14),

(4.30)
$$\int_{|\xi|\geq R/\varepsilon}\left|e^{t\frac{\widehat{J}(\varepsilon\xi)-1}{\varepsilon^2}} - e^{-t\xi^2}\right|^2 |\widehat{u_0}(\xi)|^2 d\xi$$

$$\leq \int_{|\xi|\geq R/\varepsilon}\left(e^{-\frac{t\delta}{\varepsilon^2}} + e^{\frac{-tR^2}{\varepsilon^2}}\right)^2 |\widehat{u_0}(\xi)|^2 d\xi$$

$$\leq \left(e^{-\frac{t\delta}{\varepsilon^2}} + e^{\frac{-tR^2}{\varepsilon^2}}\right)^2 \|u_0\|_{L^2(\mathbb{R}^N)}^2 \underset{\varepsilon \to 0}{\to} 0.$$

To treat the integral on the set $\{\xi \in \mathbb{R}^N : |\xi| \leq R/\varepsilon\}$ we use the fact that on this set we have

(4.31)
$$\left|e^{t\frac{\widehat{J}(\varepsilon\xi)-1}{\varepsilon^2}} - e^{-t\xi^2}\right| \leq t\left|\frac{\widehat{J}(\varepsilon\xi) - 1}{\varepsilon^2} + \xi^2\right|\max\left\{e^{t\frac{\widehat{J}(\varepsilon\xi)-1}{\varepsilon^2}}, e^{-t\xi^2}\right\}$$

$$\leq t\left|\frac{\widehat{J}(\varepsilon\xi) - 1}{\varepsilon^2} + \xi^2\right|\max\left\{e^{-\frac{t\xi^2}{2}}, e^{-t\xi^2}\right\}$$

$$\leq t\left|\frac{\widehat{J}(\varepsilon\xi) - 1}{\varepsilon^2} + \xi^2\right|e^{-\frac{t\xi^2}{2}}.$$

Thus,

$$\int_{|\xi| \le R/\varepsilon} \left| e^{t \frac{\widehat{J}(\varepsilon\xi)-1}{\varepsilon^2}} - e^{-t\xi^2} \right|^2 |\widehat{u_0}(\xi)|^2 d\xi$$

$$\le \int_{|\xi| \le R/\varepsilon} e^{-t|\xi|^2} t^2 \left| \frac{\widehat{J}(\varepsilon\xi)-1}{\varepsilon^2} + \xi^2 \right|^2 |\widehat{u_0}(\xi)|^2 d\xi$$

$$\le \int_{|\xi| \le R/\varepsilon} e^{-t\xi^2} t^2 |\xi|^4 \left| \frac{\widehat{J}(\varepsilon\xi)-1+\varepsilon^2\xi^2}{\varepsilon^2\xi^2} \right|^2 |\widehat{u_0}(\xi)|^2 d\xi.$$

Since $|\widehat{J}(\xi) - 1| \le C|\xi|^2$ for all $\xi \in \mathbb{R}^N$,

$$(4.32) \qquad \left| \frac{\widehat{J}(\varepsilon\xi)-1+\varepsilon^2\xi^2}{\varepsilon^2\xi^2} \right| \le \frac{C}{\varepsilon^2|\xi|^2} \varepsilon^2|\xi|^2 \le C.$$

Using this bound and that $e^{-|s|}s^2 \le C$, we get that

$$\sup_{t\in[0,T]} \int_{|\xi| \le R/\varepsilon} \left| e^{t \frac{\widehat{J}(\varepsilon\xi)-1}{\varepsilon^2}} - e^{-t\xi^2} \right|^2 |\widehat{u_0}(\xi)|^2 d\xi$$

$$\le C \int_{\mathbb{R}^N} \left| \frac{\widehat{J}(\varepsilon\xi)-1+\varepsilon^2|\xi|^2}{\varepsilon^2|\xi|^2} \right|^2 |\widehat{u_0}(\xi)|^2 \chi_{\{|\xi| \le R/\varepsilon\}} d\xi.$$

By inequality (4.32) together with the fact that

$$\lim_{\varepsilon \to 0} \frac{\widehat{J}(\varepsilon\xi)-1+\varepsilon^2|\xi|^2}{\varepsilon^2|\xi|^2} = 0$$

and that $\widehat{u_0} \in L^2(\mathbb{R}^N)$, by Dominated Convergence Theorem, we obtain

$$(4.33) \qquad \lim_{\varepsilon \to 0} \sup_{t\in[0,T]} \int_{|\xi| \le R/\varepsilon} \left| e^{t \frac{\widehat{J}(\varepsilon\xi)-1}{\varepsilon^2}} - e^{-t\xi^2} \right|^2 |\widehat{u_0}(\xi)|^2 d\xi = 0.$$

We conclude that

$$\lim_{\varepsilon \to 0} \sup_{t\in[0,T]} \|w_\varepsilon(t) - w(t)\|_{L^2(\mathbb{R}^N)}^2 = 0,$$

as we wanted to prove.   □

Next lemma provides us with a uniform (independent of ε) decay for the nonlocal convective part. Remember that $G_\varepsilon(x) = \frac{1}{\varepsilon^N} G\left(\frac{x}{\varepsilon}\right)$.

LEMMA 4.11. *Let $S_\varepsilon(t)$ be the linear semigroup associated to (4.28). There exists a positive constant $C = C(J, G)$ such that*

$$\left\| \left(\frac{S_\varepsilon(t) * G_\varepsilon - S_\varepsilon(t)}{\varepsilon} \right) * \varphi \right\|_{L^2(\mathbb{R}^N)} \le C t^{-\frac{1}{2}} \|\varphi\|_{L^2(\mathbb{R}^N)}$$

for all $t > 0$ large and $\varphi \in L^2(\mathbb{R}^N)$, uniformly on $\varepsilon > 0$.

PROOF. We denote by $\Phi_\varepsilon(x, t)$ the function

$$\Phi_\varepsilon(x, t) = \frac{(S_\varepsilon(t) * G_\varepsilon)(x) - S_\varepsilon(t)(x)}{\varepsilon}.$$

Then by the definition of S_ε and G_ε we obtain

$$\Phi_\varepsilon(x,t) = \frac{1}{(2\pi)^N} \int_{\mathbb{R}^N} e^{ix\cdot\xi} e^{\frac{t(\hat{J}(\varepsilon\xi)-1)}{\varepsilon^2}} \frac{\widehat{G}(\xi\varepsilon)-1}{\varepsilon} d\xi$$

$$= \varepsilon^{-N-1} \frac{1}{(2\pi)^N} \int_{\mathbb{R}^N} e^{i\varepsilon^{-1}x\cdot\xi} e^{\frac{t(\hat{J}(\xi)-1)}{\varepsilon^2}} (\widehat{G}(\xi)-1) d\xi$$

$$= \varepsilon^{-N-1} \Phi_1(x\varepsilon^{-1}, t\varepsilon^{-2}).$$

At this point, we observe that, for $\varepsilon = 1$, Lemma 4.4 implies

$$\|\Phi_1(t) * \varphi\|_{L^2(\mathbb{R}^N)} \le Ct^{-\frac{1}{2}} \|\varphi\|_{L^2(\mathbb{R}^N)}.$$

Hence

$$\|\Phi_\varepsilon(t) * \varphi\|_{L^2(\mathbb{R}^N)} = \varepsilon^{-N-1} \|\Phi_1(\varepsilon^{-1}\cdot, t\varepsilon^{-2}) * \varphi\|_{L^2(\mathbb{R}^N)}$$

$$= \varepsilon^{-1} \|[\Phi_1(t\varepsilon^{-2}) * \varphi(\varepsilon\cdot)](\varepsilon^{-1}\cdot)\|_{L^2(\mathbb{R}^N)}$$

$$= \varepsilon^{-1+\frac{N}{2}} \|\Phi_1(t\varepsilon^{-2}) * \varphi(\varepsilon\cdot)\|_{L^2(\mathbb{R}^N)}$$

$$\le C\varepsilon^{-1+\frac{N}{2}} (t\varepsilon^{-2})^{-\frac{1}{2}} \|\varphi(\varepsilon\cdot)\|_{L^2(\mathbb{R}^N)}$$

$$\le Ct^{-\frac{1}{2}} \|\varphi\|_{L^2(\mathbb{R}^N)}$$

and the proof is finished. $\qquad\square$

LEMMA 4.12. *Let $T > 0$, $M > 0$ and $\alpha \in (0,1)$. There exists a constant $C > 0$ such that*

$$(4.34) \qquad \sup_{t\in[0,T]} \int_0^t \left\| \left(\frac{S_\varepsilon(s) * G_\varepsilon - S_\varepsilon(s)}{\varepsilon} - \mathbf{b}\cdot\nabla G_s^2 \right) * \varphi(s) \right\|_{L^2(\mathbb{R}^N)} ds$$

$$\le CM(\varepsilon^{\frac{\alpha}{2}} + \varepsilon^{1-\alpha}),$$

uniformly for all $\varphi \in L^\infty(0,T;L^2(\mathbb{R}^N))$ such that $\|\varphi\|_{L^\infty(0,T;L^2(\mathbb{R}^N))} \le M$. Consequently,

$$\lim_{\varepsilon\to 0} \sup_{t\in[0,T]} \int_0^t \left\| \left(\frac{S_\varepsilon(s) * G_\varepsilon - S_\varepsilon(s)}{\varepsilon} - \mathbf{b}\cdot\nabla G_s^2 \right) * \varphi(s) \right\|_{L^2(\mathbb{R}^N)} ds = 0.$$

Here $\mathbf{b} = (b_1, \ldots, b_N)$ is given by

$$b_j = \int_{\mathbb{R}^N} x_j G(x)\, dx, \qquad j = 1, \ldots, N.$$

PROOF. Let us denote by $I_\varepsilon(t)$ the quantity

$$I_\varepsilon(t) = \int_0^t \left\| \left(\frac{S_\varepsilon(s) * G_\varepsilon - S_\varepsilon(s)}{\varepsilon} - \mathbf{b}\cdot\nabla G_s^2 \right) * \varphi(s) \right\|_{L^2(\mathbb{R}^N)} ds.$$

Choose $\alpha \in (0,1)$. Then

$$I_\varepsilon(t) \le \begin{cases} I_{1,\varepsilon} & \text{if } t \le \varepsilon^\alpha, \\ I_{1,\varepsilon} + I_{2,\varepsilon}(t) & \text{if } t \ge \varepsilon^\alpha, \end{cases}$$

where

$$I_{1,\varepsilon} = \int_0^{\varepsilon^\alpha} \left\| \left(\frac{S_\varepsilon(s) * G_\varepsilon - S_\varepsilon(s)}{\varepsilon} - \mathbf{b} \cdot \nabla G_s^2 \right) * \varphi(s) \right\|_{L^2(\mathbb{R}^N)} ds$$

and

$$I_{2,\varepsilon}(t) = \int_{\varepsilon^\alpha}^t \left\| \left(\frac{S_\varepsilon(s) * G_\varepsilon - S_\varepsilon(s)}{\varepsilon} - \mathbf{b} \cdot \nabla G_s^2 \right) * \varphi(s) \right\|_{L^2(\mathbb{R}^N)} ds.$$

By Lemma 4.11, the first term $I_{1,\varepsilon}$ satisfies

$$
\begin{aligned}
I_{1,\varepsilon} &\leq \int_0^{\varepsilon^\alpha} \left\| \left(\frac{S_\varepsilon(s) * G_\varepsilon - S_\varepsilon(s)}{\varepsilon} \right) * \varphi \right\|_{L^2(\mathbb{R}^N)} ds \\
&\quad + \int_0^{\varepsilon^\alpha} \| \mathbf{b} \cdot \nabla G_s^2 * \varphi \|_{L^2(\mathbb{R}^N)} \, ds \\
&\leq C \int_0^{\varepsilon^\alpha} s^{-\frac{1}{2}} \| \varphi(s) \|_{L^2(\mathbb{R}^N)} ds \\
&\quad + C \int_0^{\varepsilon^\alpha} \| \nabla G_s^2 \|_{L^1(\mathbb{R}^N)} \| \varphi(s) \|_{L^2(\mathbb{R}^N)} ds \\
&\leq CM \int_0^{\varepsilon^\alpha} s^{-\frac{1}{2}} \, ds = 2CM\varepsilon^{\frac{\alpha}{2}}.
\end{aligned}
$$

(4.35)

To bound $I_{2,\varepsilon}(t)$, by Plancherel's identity, we have

$I_{2,\varepsilon}(t)$

$$= (2\pi)^{-\frac{N}{2}} \int_{\varepsilon^\alpha}^t \left\| \left(e^{s(\widehat{J}(\varepsilon\xi)-1)/\varepsilon^2} \left(\frac{\widehat{G}(\varepsilon\xi)-1}{\varepsilon} \right) - i\,\mathbf{b} \cdot \xi e^{-s|\xi|^2} \right) \widehat{\varphi}(s) \right\|_{L_\xi^2(\mathbb{R}^N)} ds$$

$$\leq (2\pi)^{-\frac{N}{2}} \int_{\varepsilon^\alpha}^t \left\| \left(e^{s(\widehat{J}(\varepsilon\xi)-1)/\varepsilon^2} - e^{-s|\xi|^2} \right) \left(\frac{\widehat{G}(\varepsilon\xi)-1}{\varepsilon} \right) \widehat{\varphi}(s) \right\|_{L_\xi^2(\mathbb{R}^N)} ds$$

$$+ (2\pi)^{-\frac{N}{2}} \int_{\varepsilon^\alpha}^t \left\| e^{-s|\xi|^2} \left(\frac{\widehat{G}(\varepsilon\xi)-1}{\varepsilon} - i\,\mathbf{b} \cdot \xi \right) \widehat{\varphi}(s) \right\|_{L_\xi^2(\mathbb{R}^N)} ds$$

$$= (2\pi)^{-\frac{N}{2}} \left(\int_{\varepsilon^\alpha}^t R_{1,\varepsilon}(s) \, ds + \int_{\varepsilon^\alpha}^t R_{2,\varepsilon}(s) \, ds \right),$$

where

$$R_{1,\varepsilon}(s) = \left\| \left(e^{s(\widehat{J}(\varepsilon\xi)-1)/\varepsilon^2} - e^{-s|\xi|^2} \right) \left(\frac{\widehat{G}(\varepsilon\xi)-1}{\varepsilon} \right) \widehat{\varphi}(s) \right\|_{L_\xi^2(\mathbb{R}^N)}$$

and

$$R_{2,\varepsilon}(s) = \left\| e^{-s|\xi|^2} \left(\frac{\widehat{G}(\varepsilon\xi)-1}{\varepsilon} - i\,\mathbf{b} \cdot \xi \right) \widehat{\varphi}(s) \right\|_{L_\xi^2(\mathbb{R}^N)}.$$

Next, upper bounds for $R_{1,\varepsilon}$ and $R_{2,\varepsilon}$ are obtained. Observe that $R_{1,\varepsilon}$ satisfies

$$(R_{1,\varepsilon})^2(s) \leq 2((R_{3,\varepsilon})^2(s) + (R_{4,\varepsilon})^2(s)),$$

where

$$(R_{3,\varepsilon})^2(s) = \int_{|\xi| \leq R/\varepsilon} \left(e^{s(\widehat{J}(\varepsilon\xi)-1)/\varepsilon^2} - e^{-s|\xi|^2} \right)^2 \left| \frac{\widehat{G}(\varepsilon\xi) - 1}{\varepsilon} \right|^2 |\widehat{\varphi}(\xi,s)|^2 d\xi$$

and

$$(R_{4,\varepsilon})^2(s) = \int_{|\xi| \geq R/\varepsilon} \left(e^{s(\widehat{J}(\varepsilon\xi)-1)/\varepsilon^2} - e^{-s|\xi|^2} \right)^2 \left| \frac{\widehat{G}(\varepsilon\xi) - 1}{\varepsilon} \right|^2 |\widehat{\varphi}(\xi,s)|^2 d\xi.$$

For $R_{3,\varepsilon}$ we proceed as in the proof of Lemma 4.11 by choosing δ and R as in (4.13) and (4.14). Using estimate (4.31) and the facts that $|\widehat{G}(\xi) - 1| \leq C|\xi|$ and $|\widehat{J}(\xi) - 1 + \xi^2| \leq C|\xi|^3$ for every $\xi \in \mathbb{R}^N$, we get

$$(R_{3,\varepsilon})^2(s) \leq C \int_{|\xi| \leq R/\varepsilon} e^{-s|\xi|^2} s^2 \left| \frac{\widehat{J}(\varepsilon\xi) - 1 + \xi^2\varepsilon^2}{\varepsilon^2} \right|^2 |\xi|^2 |\widehat{\varphi}(\xi,s)|^2 d\xi$$

$$\leq C \int_{|\xi| \leq R/\varepsilon} e^{-s|\xi|^2} s^2 \left[\frac{(\varepsilon\xi)^3}{\varepsilon^2} \right]^2 |\xi|^2 |\widehat{\varphi}(\xi,s)|^2 d\xi$$

$$= C \int_{|\xi| \leq R/\varepsilon} e^{-s|\xi|^2} s^2 \varepsilon^2 |\xi|^8 |\widehat{\varphi}(\xi,s)|^2 d\xi$$

$$\leq \varepsilon^2 s^{-2} \int_{\mathbb{R}^N} e^{-s|\xi|^2} s^4 |\xi|^8 |\widehat{\varphi}(\xi,s)|^2 d\xi$$

$$\leq C\varepsilon^{2-2\alpha} \int_{\mathbb{R}^N} |\widehat{\varphi}(\xi,s)|^2 d\xi$$

$$\leq C\varepsilon^{2-2\alpha} M^2.$$

In the case of $R_{4,\varepsilon}$, we use that $|\widehat{G}(\xi)| \leq 1$ and we proceed as in the proof of (4.30), to obtain

$$(R_{4,\varepsilon})^2(s) \leq \int_{|\xi| \geq R/\varepsilon} \left(e^{-\frac{s\delta}{\varepsilon^2}} + e^{-\frac{sR^2}{\varepsilon^2}} \right)^2 \varepsilon^{-2} |\widehat{\varphi}(\xi,s)|^2 d\xi$$

$$\leq \left(e^{-\frac{\delta}{\varepsilon^{2-\alpha}}} + e^{-\frac{R^2}{\varepsilon^{2-\alpha}}} \right)^2 \varepsilon^{-2} \int_{|\xi| \geq R/\varepsilon} |\widehat{\varphi}(\xi,s)|^2 d\xi$$

$$\leq M^2 \left(e^{-\frac{\delta}{\varepsilon^{2-\alpha}}} + e^{-\frac{R^2}{\varepsilon^{2-\alpha}}} \right)^2 \varepsilon^{-2}$$

$$\leq CM^2 \varepsilon^{2-2\alpha}$$

for sufficiently small ε. Then

$$(4.36) \qquad \int_{\varepsilon^\alpha}^t R_{1,\varepsilon}(s)ds \leq CTM\varepsilon^{1-\alpha}.$$

The second term can be estimated in a similar way. Using that

$$|\widehat{G}(\xi) - 1 - i\mathbf{b} \cdot \xi| \le C|\xi|^2$$

for every $\xi \in \mathbb{R}^N$, we have

$$(R_{2,\varepsilon})^2(s) \le \int_{\mathbb{R}^N} e^{-2s|\xi|^2} \left| \frac{\widehat{G}(\varepsilon\xi) - 1 - i\,\mathbf{b} \cdot \xi\varepsilon}{\varepsilon} \right|^2 |\widehat{\varphi}(\xi, s)|^2 d\xi$$

$$\le C \int_{\mathbb{R}^N} e^{-2s|\xi|^2} \left[\frac{(\xi\varepsilon)^2}{\varepsilon} \right]^2 |\widehat{\varphi}(\xi, s)|^2 d\xi$$

$$= C \int_{\mathbb{R}^N} e^{-2s|\xi|^2} \varepsilon^2 |\xi|^4 |\widehat{\varphi}(\xi, s)|^2 d\xi$$

$$= C\varepsilon^2 s^{-2} \int_{\mathbb{R}^N} e^{-2s|\xi|^2} s^2 |\xi|^4 |\widehat{\varphi}(\xi, s)|^2 d\xi$$

$$\le C\varepsilon^{2(1-\alpha)} \int_{\mathbb{R}^N} |\widehat{\varphi}(\xi, s)|^2 d\xi$$

$$\le CM^2 \varepsilon^{2(1-\alpha)},$$

and we conclude that

(4.37) $$\int_{\varepsilon^\alpha}^t R_{2,\varepsilon}(s)\,ds \le CTM\varepsilon^{1-\alpha}.$$

Now, by (4.35), (4.36) and (4.37) we obtain that

$$\sup_{t \in [0,T]} I_\varepsilon(t) \le CM(\varepsilon^{\frac{\alpha}{2}} + \varepsilon^{1-\alpha}) \to 0, \qquad \text{as} \quad \varepsilon \to 0,$$

which finishes the proof.   □

Now we are ready to prove the convergence result.

THEOREM 4.13. *With the above notation, for any $T > 0$, we have*

$$\lim_{\varepsilon \to 0} \sup_{t \in [0,T]} \|u_\varepsilon(t) - v(t)\|_{L^2(\mathbb{R}^N)} = 0,$$

where u_ε is the solution of (4.1) with the kernels J and G rescaled with ε as above, and $v(x,t)$ is the unique solution of the local convection-diffusion problem (4.2) with initial condition $v(x,0) = u_0(x) \in L^1(\mathbb{R}^N) \cap L^\infty(\mathbb{R}^N)$ and $\mathbf{b} = (b_1, \ldots, b_N)$ given by

$$b_j = \int_{\mathbb{R}^N} x_j\, G(x)\, dx, \qquad j = 1, \ldots, N.$$

PROOF. First we write the two problems in the semigroup formulation,

$$u_\varepsilon(t) = S_\varepsilon(t) * u_0 + \int_0^t \frac{S_\varepsilon(t-s) * G_\varepsilon - S_\varepsilon(t-s)}{\varepsilon} * f(u_\varepsilon(s))\, ds$$

and

$$v(t) = G_t^2 * u_0 + \int_0^t \mathbf{b} \cdot \nabla G_{t-s}^2 * f(v(s))\, ds.$$

Then

(4.38) $$\sup_{t\in[0,T]}\|u_\varepsilon(t)-v(t)\|_{L^2(\mathbb{R}^N)} \leq \sup_{t\in[0,T]} I_{1,\varepsilon}(t) + \sup_{t\in[0,T]} I_{2,\varepsilon}(t),$$

where

$$I_{1,\varepsilon}(t) = \|S_\varepsilon(t)*u_0 - G_t^2*u_0\|_{L^2(\mathbb{R}^N)}$$

and

$$I_{2,\varepsilon}(t) = \left\| \int_0^t \frac{S_\varepsilon(t-s)*G_\varepsilon - S_\varepsilon(t-s)}{\varepsilon} * f(u_\varepsilon(s))\,ds \right.$$

$$\left. - \int_0^t \mathbf{b}\cdot\nabla G_{t-s}^2 * f(v(s))\,ds \right\|_{L^2(\mathbb{R}^N)}.$$

In view of Lemma 4.10 we have

$$\sup_{t\in[0,T]} I_{1,\varepsilon}(t) \to 0 \qquad \text{as } \varepsilon\to 0.$$

So it remains to analyze the second term $I_{2,\varepsilon}$. To this end, we split it again

$$I_{2,\varepsilon}(t) \leq I_{3,\varepsilon}(t) + I_{4,\varepsilon}(t),$$

where

$$I_{3,\varepsilon}(t) = \int_0^t \left\| \frac{S_\varepsilon(t-s)*G_\varepsilon - S_\varepsilon(t-s)}{\varepsilon} * \big(f(u_\varepsilon(s)) - f(v(s))\big) \right\|_{L^2(\mathbb{R}^N)} ds$$

and

$$I_{4,\varepsilon}(t) = \int_0^t \left\| \left(\frac{S_\varepsilon(t-s)*G_\varepsilon - S_\varepsilon(t-s)}{\varepsilon} - \mathbf{b}\cdot\nabla G_{t-s}^2 \right) * f(v(s)) \right\|_{L^2(\mathbb{R}^N)} ds.$$

Using Young's inequality, Lemma 4.11 and that from the hypotheses we have a uniform bound for u_ε and u in terms of $\|u_0\|_{L^1(\mathbb{R}^N)}$, $\|u_0\|_{L^\infty(\mathbb{R}^N)}$, we obtain

$$I_{3,\varepsilon}(t) \leq \int_0^t \frac{\|f(u_\varepsilon(s)) - f(v(s))\|_{L^2(\mathbb{R}^N)}}{|t-s|^{\frac{1}{2}}}\,ds$$

(4.39) $$\leq \|f(u_\varepsilon)-f(v)\|_{L^\infty(0,T;L^2(\mathbb{R}^N))} \int_0^t \frac{1}{|t-s|^{\frac{1}{2}}}\,ds$$

$$\leq 2T^{1/2}\|u_\varepsilon-v\|_{L^\infty(0,T;L^2(\mathbb{R}^N))} C(\|u_0\|_{L^1(\mathbb{R}^N)}, \|u_0\|_{L^\infty(\mathbb{R}^N)}).$$

By Lemma 4.12, choosing $\alpha = 2/3$ in (4.34), we get

(4.40) $$\sup_{t\in[0,T]} I_{4,\varepsilon} \leq C\varepsilon^{\frac{1}{3}}\|f(v)\|_{L^\infty(0,T;L^2(\mathbb{R}^N))} \leq \varepsilon^{\frac{1}{3}} C(\|u_0\|_{L^1(\mathbb{R}^N)}, \|u_0\|_{L^\infty(\mathbb{R}^N)}).$$

By (4.38), (4.39) and (4.40) we get

$$\|u_\varepsilon-v\|_{L^\infty(0,T;L^2(\mathbb{R}^N))} \leq \|I_{1,\varepsilon}\|_{L^\infty(0,T;L^2(\mathbb{R}^N))}$$

$$+ T^{\frac{1}{2}} C(\|u_0\|_{L^1(\mathbb{R}^N)}, \|u_0\|_{L^\infty(\mathbb{R}^N)})\|u_\varepsilon-v\|_{L^\infty(0,T;L^2(\mathbb{R}^N))}$$

$$+ C\varepsilon^{\frac{1}{3}} C(\|u_0\|_{L^1(\mathbb{R}^N)}, \|u_0\|_{L^\infty(\mathbb{R}^N)}).$$

Taking $T = T_0$ sufficiently small, depending on $\|u_0\|_{L^1(\mathbb{R}^N)}$ and $\|u_0\|_{L^\infty(\mathbb{R}^N)}$, we have

$$\|u_\varepsilon - v\|_{L^\infty(0,T;L^2(\mathbb{R}^N))}$$

$$\leq C(\|u_0\|_{L^1(\mathbb{R}^N)}, \|u_0\|_{L^\infty(\mathbb{R}^N)})(\varepsilon^{\frac{1}{3}} + \|I_{1,\varepsilon}\|_{L^\infty(0,T;L^2(\mathbb{R}^N))}) \to 0$$

as $\varepsilon \to 0$.

Using the same argument in any interval $[\tau, \tau+T_0]$, the stability of the solutions of the equation (4.3) in the $L^2(\mathbb{R}^N)$-norm and having in mind that

$$\|u_\varepsilon(\tau)\|_{L^1(\mathbb{R}^N)} + \|u_\varepsilon(\tau)\|_{L^\infty(\mathbb{R}^N)} \leq \|u_0\|_{L^1(\mathbb{R}^N)} + \|u_0\|_{L^\infty(\mathbb{R}^N)}$$

for any time $\tau > 0$, we obtain

$$\lim_{\varepsilon \to 0} \sup_{t \in [0,T]} \|u_\varepsilon - v\|_{L^2(\mathbb{R}^N)} = 0,$$

and the proof is complete.   □

4.5. Long time behaviour of the solutions

The aim of this chapter is to obtain the first term in the asymptotic expansion of the solution u of (4.1). The main ingredient of the proofs is the following lemma inspired by the Fourier splitting method introduced by Schonbek (see [142], [143] and [144]).

LEMMA 4.14. *Let R and δ be such that the function $\widehat{J}$ satisfies*

$$(4.41) \qquad\qquad \widehat{J}(\xi) \leq 1 - \frac{|\xi|^2}{2}, \quad |\xi| \leq R,$$

and

$$(4.42) \qquad\qquad \widehat{J}(\xi) \leq 1 - \delta, \quad |\xi| \geq R.$$

Let us assume that the function $u : [0,\infty) \times \mathbb{R}^N \to \mathbb{R}$ satisfies the differential inequality

$$(4.43) \qquad \frac{d}{dt} \int_{\mathbb{R}^N} |u(x,t)|^2 \, dx \leq c \int_{\mathbb{R}^N} (J * u - u)(x,t) u(x,t) \, dx,$$

for any $t > 0$. Then, for any $1 \leq r < \infty$ there exists a constant $a = rN/c\delta$ such that

$$\int_{\mathbb{R}^N} |u(x,at)|^2 dx \leq \frac{\|u(0)\|_{L^2(\mathbb{R}^N)}^2}{(t+1)^{rN}}$$

$$(4.44)$$

$$+ \frac{rN\omega_N(2\delta)^{\frac{N}{2}}}{(t+1)^{rN}} \int_0^t (s+1)^{rN-\frac{N}{2}-1} \|u(as)\|_{L^1(\mathbb{R}^N)}^2 ds$$

for all positive times t, where ω_N is the volume of the unit ball in $\mathbb{R}^N$. In particular,

$$(4.45) \quad \|u(at)\|_{L^2(\mathbb{R}^N)} \leq \frac{\|u(0)\|_{L^2(\mathbb{R}^N)}}{(t+1)^{\frac{rN}{2}}} + \frac{(2\omega_N)^{\frac{1}{2}}(2\delta)^{\frac{N}{4}}}{(t+1)^{\frac{N}{4}}} \|u\|_{L^\infty(0,\infty;L^1(\mathbb{R}^N))}.$$

REMARK 4.15. Condition (4.41) can be replaced by $\widehat{J}(\xi) \leq 1 - A|\xi|^2$ for $|\xi| \leq R$, but omitting the constant A in the proof we simplify some formulas.

REMARK 4.16. The differential inequality (4.43) can be written in the following form:

$$\frac{d}{dt}\int_{\mathbb{R}^N}|u(x,t)|^2\,dx \le -\frac{c}{2}\int_{\mathbb{R}^N}\int_{\mathbb{R}^N}J(x-y)(u(x,t)-u(y,t))^2\,dx\,dy.$$

This is the nonlocal version of the energy method used in [**98**]. However, in the nonlocal case, exactly the same inequalities as used in [**98**] cannot be applied.

PROOF. Let R and δ be as in (4.41) and (4.42). We set $a = \frac{rN}{c\delta}$ and consider the set

$$A(t) = \left\{\xi \in \mathbb{R}^N : |\xi| \le M(t) = \left(\frac{2rN}{c(t+a)}\right)^{1/2}\right\}.$$

Inequality (4.43) gives

(4.46)
$$\frac{d}{dt}\int_{\mathbb{R}^N}|u(x,t)|^2 dx \le c(2\pi)^{-N}\int_{\mathbb{R}^N}(\widehat{J}(\xi)-1)|\widehat{u}(\xi,t)|^2 d\xi$$

$$\le c(2\pi)^{-N}\int_{\mathbb{R}^N\setminus A(t)}(\widehat{J}(\xi)-1)|\widehat{u}(\xi,t)|^2 d\xi.$$

We now claim that

(4.47)
$$c(\widehat{J}(\xi)-1) \le -\frac{rN}{t+a}, \qquad \text{for every } \xi \in \mathbb{R}^N\setminus A(t).$$

In fact, using the hypotheses (4.41) and (4.42) on the function $\widehat{J}$, we have, for any $|\xi| \ge R$,

$$c(\widehat{J}(\xi)-1) \le -c\delta = -\frac{rN}{a} \le -\frac{rN}{t+a}$$

and

$$c(\widehat{J}(\xi)-1) \le -\frac{c|\xi|^2}{2} \le -\frac{c}{2}\frac{2rN}{c(t+a)} = -\frac{rN}{t+a}$$

for all $\xi \in \mathbb{R}^N\setminus A(t)$ with $|\xi| \le R$. Hence (4.47) holds.

Substituting (4.47) in (4.46), by Plancherel's identity, we obtain

$$\frac{d}{dt}\int_{\mathbb{R}^N}|u(x,t)|^2 dx \le -\frac{rN}{t+a}(2\pi)^{-N}\int_{\mathbb{R}^N\setminus A(t)}|\widehat{u}(\xi,t)|^2 d\xi$$

$$\le -\frac{rN}{t+a}(2\pi)^{-N}\int_{\mathbb{R}^N}|\widehat{u}(\xi,t)|^2 d\xi + \frac{rN}{t+a}(2\pi)^{-N}\int_{|\xi|\le M(t)}|\widehat{u}(\xi,t)|^2 d\xi$$

$$\le -\frac{rN}{t+a}\int_{\mathbb{R}^N}|u(x,t)|^2 dx + \frac{rN}{t+a}M(t)^N\omega_N\|\widehat{u}(t)\|^2_{L^\infty(\mathbb{R}^N)}$$

$$\le -\frac{rN}{t+a}\int_{\mathbb{R}^N}|u(x,t)|^2 dx + \frac{rN}{t+a}\left[\frac{2rN}{c(t+a)}\right]^{\frac{N}{2}}\omega_N\|u(t)\|^2_{L^1(\mathbb{R}^N)}.$$

This implies that

$$\frac{d}{dt}\left[(t+a)^{rN}\int_{\mathbb{R}^N}|u(x,t)|^2 dx\right]$$

$$= (t+a)^{rN}\left[\frac{d}{dt}\int_{\mathbb{R}^N}|u(x,t)|^2 dx\right] + rN(t+a)^{rN-1}\int_{\mathbb{R}^N}|u(x,t)|^2 dx$$

$$\leq (t+a)^{rN-\frac{N}{2}-1} rN\left(\frac{2rN}{c}\right)^{\frac{N}{2}}\omega_N\|u(t)\|^2_{L^1(\mathbb{R}^N)}.$$

Integrating the last inequality with respect to the time variable we get

$$(t+a)^{rN}\int_{\mathbb{R}^N}|u(x,t)|^2 dx - a^{rN}\int_{\mathbb{R}^N}|u(x,0)|^2 dx$$

$$\leq rN\omega_N\left(\frac{2rN}{c}\right)^{\frac{N}{2}}\int_0^t (s+a)^{rN-\frac{N}{2}-1}\|u(s)\|^2_{L^1(\mathbb{R}^N)} ds$$

and hence

$$\int_{\mathbb{R}^N}|u(x,t)|^2\, dx \leq \frac{a^{rN}}{(t+a)^{rN}}\int_{\mathbb{R}^N}|u(x,0)|^2\, dx$$

$$+\frac{rN\omega_N}{(t+a)^{rN}}\left(\frac{2rN}{c}\right)^{\frac{N}{2}}\int_0^t (s+a)^{rN-\frac{N}{2}-1}\|u(s)\|^2_{L^1(\mathbb{R}^N)}\, ds.$$

Replacing t by ta we have

$$\int_{\mathbb{R}^N}|u(x,at)|^2\, dx$$

$$\leq \frac{\|u(0)\|^2_{L^2(\mathbb{R}^N)}}{(t+1)^{rN}} + \frac{rN\omega_N}{(t+1)^{rN}a^{rN}}\left(\frac{2rN}{c}\right)^{\frac{N}{2}}\int_0^{at}(s+a)^{rN-\frac{N}{2}-1}\|u(s)\|^2_{L^1(\mathbb{R}^N)}\, ds$$

$$= \frac{\|u(0)\|^2_{L^2(\mathbb{R}^N)}}{(t+1)^{rN}} + \frac{rN\omega_N}{(t+1)^{rN}}\left(\frac{2rN}{ca}\right)^{\frac{N}{2}}\int_0^t (s+1)^{rN-\frac{N}{2}-1}\|u(as)\|^2_{L^1(\mathbb{R}^N)}\, ds$$

$$= \frac{\|u(0)\|^2_{L^2(\mathbb{R}^N)}}{(t+1)^{rN}} + \frac{rN\omega_N(2\delta)^{\frac{N}{2}}}{(t+1)^{rN}}\int_0^t (s+1)^{rN-\frac{N}{2}-1}\|u(as)\|^2_{L^1(\mathbb{R}^N)}\, ds,$$

which proves (4.44).

Estimate (4.45) is obtained as follows:

$$\int_{\mathbb{R}^N} |u(x, at)|^2 dx \leq \frac{\|u(0)\|_{L^2(\mathbb{R}^N)}^2}{(t+1)^{rN}}$$

$$+ \frac{rN\omega_N (2\delta)^{\frac{N}{2}}}{(t+1)^{rN}} \|u\|_{L^\infty([0,\infty); L^1(\mathbb{R}^N))}^2 \int_0^t (s+1)^{rN-\frac{N}{2}-1} ds$$

$$\leq \frac{\|u(0)\|_{L^2(\mathbb{R}^N)}^2}{(t+1)^{rN}} + \frac{2\omega_N (2\delta)^{\frac{N}{2}}}{(t+1)^{\frac{N}{2}}} \|u\|_{L^\infty([0,\infty); L^1(\mathbb{R}^N))}^2.$$

This ends the proof. □

LEMMA 4.17. *Let* $2 \leq p < \infty$. *For any function* $u : \mathbb{R}^N \to \mathbb{R}$, $I(u)$ *defined by*

$$I(u) = \int_{\mathbb{R}^N} (J * u - u)(x)|u(x)|^{p-1}\mathrm{sgn}_0(u(x)) \, dx$$

satisfies

$$I(u) \quad \leq \frac{4(p-1)}{p^2} \int_{\mathbb{R}^N} (J * |u|^{p/2} - |u|^{p/2})(x)|u(x)|^{p/2} \, dx$$

$$= -\frac{2(p-1)}{p^2} \int_{\mathbb{R}^N} \int_{\mathbb{R}^N} J(x-y)(|u(y)|^{p/2} - |u(x)|^{p/2})^2 \, dx \, dy.$$

REMARK 4.18. This result is a nonlocal counterpart of the well-known identity

$$\int_{\mathbb{R}^N} \Delta v \, |v|^{p-1}\mathrm{sgn}_0(v) \, dx = -\frac{4(p-1)}{p^2} \int_{\mathbb{R}^N} |\nabla(|v|^{p/2})|^2 \, dx.$$

PROOF. Using the symmetry of J, $I(u)$ can be written in the following manner:

$$I(u) \quad = \int_{\mathbb{R}^N} \int_{\mathbb{R}^N} J(x-y)(u(y) - u(x))|u(x)|^{p-1}\mathrm{sgn}_0(u(x)) \, dx \, dy$$

$$= \int_{\mathbb{R}^N} \int_{\mathbb{R}^N} J(x-y)(u(x) - u(y))|u(y)|^{p-1}\mathrm{sgn}_0(u(y)) \, dx \, dy.$$

Thus

$$I(u) = -\frac{1}{2} \int_{\mathbb{R}^N} \int_{\mathbb{R}^N} J(x-y)(u(x) - u(y))$$

$$\times (|u(x)|^{p-1}\mathrm{sgn}_0(u(x)) - |u(y)|^{p-1}\mathrm{sgn}_0(u(y))) \, dx \, dy.$$

Using the inequality,

$$||\alpha|^{p/2} - |\beta|^{p/2}|^2 \leq \frac{p^2}{4(p-1)}(\alpha - \beta)(|\alpha|^{p-1}\mathrm{sgn}_0(\alpha) - |\beta|^{p-1}\mathrm{sgn}_0(\beta)),$$

which holds for all real numbers α and β and for every $2 \leq p < \infty$, we obtain that $I(u)$ can be bounded from above as follows:

$$I(u) \leq -\frac{4(p-1)}{2p^2} \int_{\mathbb{R}^N} \int_{\mathbb{R}^N} J(x-y)(|u(y)|^{p/2} - |u(x)|^{p/2})^2 \, dx \, dy$$

$$= -\frac{4(p-1)}{2p^2} \int_{\mathbb{R}^N} \int_{\mathbb{R}^N} J(x-y)(|u(y)|^p - 2|u(y)|^{p/2}|u(x)|^{p/2} + |u(x)|^p) \, dx \, dy$$

$$= \frac{4(p-1)}{p^2} \int_{\mathbb{R}^N} (J * |u|^{p/2} - |u|^{p/2})(x)|u(x)|^{p/2} \, dx,$$

and the proof is finished. $\qquad\qquad\qquad\qquad\qquad\qquad\qquad\qquad\qquad\square$

Now we are ready to proceed with the proof of the decay.

THEOREM 4.19. *Assume that f satisfies (4.5) with $q > 1$ and*

$$u_0 \in L^1(\mathbb{R}^N) \cap L^\infty(\mathbb{R}^N).$$

Then, for every $p \in [1, \infty)$ the solution of the equation (4.1) satisfies

$$(4.48) \qquad \|u(t)\|_{L^p(\mathbb{R}^N)} \leq C(\|u_0\|_{L^1(\mathbb{R}^N)}, \|u_0\|_{L^\infty(\mathbb{R}^N)})t^{-\frac{N}{2}(1-\frac{1}{p})}.$$

PROOF. Let u be the solution of the nonlocal convection-diffusion problem. Then, by the same arguments used to control the $L^1(\mathbb{R}^N)$-norm in the proof of the existence result, we obtain

$$\frac{d}{dt} \int_{\mathbb{R}^N} |u(x,t)|^p dx = p \int_{\mathbb{R}^N} (J * u - u)(x,t)|u(x,t)|^{p-1}\mathrm{sgn}_0(u(x,t)) \, dx$$

$$+ \int_{\mathbb{R}^N} (G * f(u) - f(u))(x,t)|u(x,t)|^{p-1}\mathrm{sgn}_0(u(x,t)) \, dx$$

$$\leq p \int_{\mathbb{R}^N} (J * u - u)(x,t)|u(x,t)|^{p-1}\mathrm{sgn}_0(u(x,t)) \, dx.$$

By Lemma 4.17 we get that the $L^p(\mathbb{R}^N)$-norm of the solution u satisfies the following differential inequality:

$$(4.49) \quad \frac{d}{dt} \int_{\mathbb{R}^N} |u(x,t)|^p \, dx \leq \frac{4(p-1)}{p} \int_{\mathbb{R}^N} (J * |u|^{p/2} - |u|^{p/2})(x,t)|u(x,t)|^{p/2} \, dx.$$

First, let us consider $p = 2$. Consequently,

$$\frac{d}{dt} \int_{\mathbb{R}^N} |u(x,t)|^2 \, dx \leq 2 \int_{\mathbb{R}^N} (J * |u| - |u|)(x,t)|u(x,t)| \, dx.$$

Applying Lemma 4.14 with $|u|$, $c = 2$, $r = 1$ and using that $\|u\|_{L^\infty(0,\infty;L^1(\mathbb{R}^N))} \leq \|u_0\|_{L^1(\mathbb{R}^N)}$, we obtain

$$\left\|u\left(\frac{tN}{2\delta}\right)\right\|_{L^2(\mathbb{R}^N)} \leq \frac{\|u_0\|_{L^2(\mathbb{R}^N)}}{(t+1)^{\frac{1}{2}}} + \frac{(2\omega_N)^{\frac{1}{2}}(2\delta)^{\frac{N}{4}}}{(t+1)^{\frac{N}{4}}}\|u\|_{L^\infty(0,\infty;L^1(\mathbb{R}^N))}$$

$$\leq \frac{\|u_0\|_{L^2(\mathbb{R}^N)}}{(t+1)^{\frac{N}{2}}} + \frac{(2\omega_N)^{\frac{1}{2}}(2\delta)^{\frac{N}{4}}}{(t+1)^{\frac{N}{4}}}\|u_0\|_{L^1(\mathbb{R}^N)}$$

$$\leq \frac{C(J, \|u_0\|_{L^1(\mathbb{R}^N)}, \|u_0\|_{L^\infty(\mathbb{R}^N)})}{(t+1)^{\frac{N}{4}}},$$

which proves (4.48) in the case $p = 2$. Since the $L^1(\mathbb{R}^N)$-norm of the solutions of (4.1) does not increase, $\|u(t)\|_{L^1(\mathbb{R}^N)} \leq \|u_0\|_{L^1(\mathbb{R}^N)}$, by Hölder's inequality we get the desired decay rate (4.48) in any $L^p(\mathbb{R}^N)$-norm with $p \in [1,2]$.

In the following, using an inductive argument, we prove the result for any $r = 2^m$, with $m \geq 1$ an integer. By Hölder's inequality this implies the decay in the $L^p(\mathbb{R}^N)$-norm for any $2 < p < \infty$.

Let us choose $r = 2^m$ with $m \geq 1$ and assume that

$$\|u(t)\|_{L^r(\mathbb{R}^N)} \leq Ct^{-\frac{N}{2}(1-\frac{1}{r})}$$

holds for some positive constant $C = C(J, \|u_0\|_{L^1(\mathbb{R}^N)}, \|u_0\|_{L^\infty(\mathbb{R}^N)})$ and for every positive time t. We want to show an analogous estimate for $p = 2r = 2^{m+1}$.

By (4.49) with $p = 2r$ we obtain the following differential inequality:

$$\frac{d}{dt}\int_{\mathbb{R}^N}|u(x,t)|^{2r}dx \leq \frac{4(2r-1)}{2r}\int_{\mathbb{R}^N}(J*|u|^r - |u|^r)(x,t)|u(x,t)|^r dx.$$

Applying Lemma 4.14 with $|u|^r$, $c(r) = 2(2r-1)/r$ and $a = \frac{rN}{c(r)\delta}$, we get

$$\int_{\mathbb{R}^N}|u(at)|^{2r} \leq \frac{\|u_0^r\|_{L^2(\mathbb{R}^N)}^2}{(t+1)^{rN}} + \frac{N\omega_N(2\delta)^{\frac{N}{2}}}{(t+1)^{rN}}\int_0^t(s+1)^{rN-\frac{N}{2}-1}\|u^r(as)\|_{L^1(\mathbb{R}^N)}^2 ds$$

$$\leq \frac{\|u_0\|_{L^{2r}(\mathbb{R}^N)}^{2r}}{(t+1)^{rN}} + \frac{C}{(t+1)^{rN}}\int_0^t(s+1)^{rN-\frac{N}{2}-1}\|u(as)\|_{L^r(\mathbb{R}^N)}^{2r} ds$$

$$\leq \frac{C(J, \|u_0\|_{L^1(\mathbb{R}^N)}, \|u_0\|_{L^\infty(\mathbb{R}^N)})}{(t+1)^N}\left(1 + \int_0^t(s+1)^{rN-\frac{N}{2}-1}(s+1)^{-Nr(1-\frac{1}{r})}ds\right)$$

$$\leq \frac{C}{(t+1)^{Nr}}(1 + (t+1)^{\frac{N}{2}}) \leq C(t+1)^{\frac{N}{2}-Nr},$$

and then

$$\|u(at)\|_{L^{2r}(\mathbb{R}^N)} \leq C(J, \|u_0\|_{L^1(\mathbb{R}^N)}, \|u_0\|_{L^\infty(\mathbb{R}^N)})(t+1)^{-\frac{N}{2}(1-\frac{1}{2r})},$$

which finishes the proof. $\qquad\square$

Let us conclude this section with a remark concerning the applicability of energy methods to studying nonlocal problems.

REMARK 4.20. If we want to use energy estimates to get decay rates (for example in $L^2(\mathbb{R}^N)$), we easily arrive at

$$\frac{d}{dt}\int_{\mathbb{R}^N}|w(x,t)|^2\,dx = -\frac{1}{2}\int_{\mathbb{R}^N}\int_{\mathbb{R}^N}J(x-y)(w(x,t)-w(y,t))^2\,dx\,dy$$

when we deal with a solution of the linear equation, $w_t = J*w - w$, and to

$$\frac{d}{dt}\int_{\mathbb{R}^N}|u(x,t)|^2\,dx \leq -\frac{1}{2}\int_{\mathbb{R}^N}\int_{\mathbb{R}^N}J(x-y)(u(x,t)-u(y,t))^2\,dx\,dy$$

if we consider the complete convection-diffusion problem. However, we cannot go further since no inequality of the form

$$\left(\int_{\mathbb{R}^N}|u(x)|^p\,dx\right)^{\frac{2}{p}} \leq C\int_{\mathbb{R}^N}\int_{\mathbb{R}^N}J(x-y)(u(x)-u(y))^2\,dx\,dy$$

is true for $p > 2$.

4.6. Weakly nonlinear behaviour

In this section we find the leading order term in the asymptotic expansion of the solution to (4.1). We use ideas from [**98**] showing that the nonlinear term decays faster than the linear part.

THEOREM 4.21. *Suppose f satisfies (4.5) with $q > (N+1)/N$ and the initial condition u_0 belongs to $L^1(\mathbb{R}^N, 1+|x|)\cap L^\infty(\mathbb{R}^N)$. For any $p \in [2,\infty)$ the following inequality holds:*

$$t^{-\frac{N}{2}(1-\frac{1}{p})}\|u(t)-MG_t^2\|_{L^p(\mathbb{R}^N)} \leq C(J,G,p,N)\,\alpha_q(t),$$

where

$$M = \int_{\mathbb{R}^N}u_0(x)\,dx$$

and

$$\alpha_q(t) = \begin{cases} t^{-\frac{1}{2}} & \text{if } q \geq (N+2)/N, \\ t^{\frac{1-N(q-1)}{2}} & \text{if } (N+1)/N < q < (N+2)/N. \end{cases}$$

Now we give a result that extends to nonlocal diffusion problems the result of [**96**] in the case of the heat equation. The proof is close to those of Chapter 1.

LEMMA 4.22. *Let $J \in \mathcal{S}(\mathbb{R}^N)$ such that*

$$\widehat{J}(\xi) - 1 + |\xi|^2 \sim |\xi|^3, \qquad \xi \sim 0.$$

Then, for every $p \in [2,\infty)$, there exists some positive constant $C = C(p,J)$ such that

$$\|S(t)*\varphi - MG_t^2\|_{L^p(\mathbb{R}^N)} \leq Ce^{-t}\|\varphi\|_{L^p(\mathbb{R}^N)}$$

(4.50)

$$+ C\|\varphi\|_{L^1(\mathbb{R}^N,|x|)}t^{-\frac{N}{2}(1-\frac{1}{p})-\frac{1}{2}}, \qquad t > 0,$$

for every $\varphi \in L^1(\mathbb{R}^N, 1+|x|)$ with $M = \int_{\mathbb{R}^N}\varphi(x)\,dx$.

PROOF. We write $S(t) = e^{-t}\delta_0 + K_t$. Then it is sufficient to show that

$$\|K_t * \varphi - MK_t\|_{L^p(\mathbb{R}^N)} \le C\|\varphi\|_{L^1(\mathbb{R}^N, |x|)} t^{-\frac{N}{2}(1-\frac{1}{p})-\frac{1}{2}}$$

and

$$t^{\frac{N}{2}(1-\frac{1}{p})}\|K_t - G_t^2\|_{L^p(\mathbb{R}^N)} \le Ct^{-\frac{1}{2}}.$$

The first estimate follows by Lemma 4.3. The second uses the hypotheses on $\widehat{J}$. The details can be fixed working as in Chapter 1.   □

REMARK 4.23. If we consider a condition such as $\widehat{J}(\xi) - 1 + A|\xi|^2 \sim |\xi|^3$ for $\xi \sim 0$, we obtain as profile a modified Gaussian G_{At}^2.

REMARK 4.24. The case $p \in [1, 2)$ is more subtle. The analysis performed in the previous sections to handle the case $p = 1$ can also be extended to cover this case when the dimension N satisfies $1 \le N \le 3$. Indeed in this case, if J satisfies $\widehat{J}(\xi) \sim 1 - A|\xi|^s$, $\xi \sim 0$, then we need $s = 2$ to obtain the Gaussian profile and also s has to be grater than $[N/2] + 1$ to apply the methods we use.

Now we are ready to prove that the same expansion holds for solutions of the complete problem (4.1) when $q > \frac{(N+1)}{N}$.

PROOF OF THEOREM 4.21. In view of (4.50) it is sufficient to show that

$$t^{-\frac{N}{2}(1-\frac{1}{p})}\|u(t) - S(t) * u_0\|_{L^p(\mathbb{R}^N)} \le Ct^{-\frac{N}{2}(q-1)+\frac{1}{2}}.$$

Using the representation (4.24) we get that

$$\|u(t) - S(t) * u_0\|_{L^p(\mathbb{R}^N)} \le \int_0^t \|[S(t-s) * G - S(t-s)] * |u(s)|^{q-1}u(s)\|_{L^p(\mathbb{R}^N)} \, ds.$$

We now estimate the right hand side term as follows. We split it in two parts, one in which we integrate on $(0, t/2)$ and the other where we integrate on $(t/2, t)$. Concerning the second term, by Lemma 4.4 and Theorem 4.19 we have

$$\int_{t/2}^t \|(S(t-s) * G - S(t-s)) * |u(s)|^{q-1}u(s)\|_{L^p(\mathbb{R}^N)} \, ds$$

$$\le C(J, G) \int_{t/2}^t (t-s)^{-\frac{1}{2}}\|u(s)\|_{L^{pq}(\mathbb{R}^N)}^q \, ds$$

$$\le C(J, G, \|u_0\|_{L^1(\mathbb{R}^N)}, \|u_0\|_{L^\infty(\mathbb{R}^N)}) \int_{t/2}^t (t-s)^{-\frac{1}{2}}s^{-\frac{N}{2}(q-\frac{1}{p})} \, ds$$

$$\le Ct^{-\frac{N}{2}(q-\frac{1}{p})+\frac{1}{2}} \le Ct^{-\frac{N}{2}(1-\frac{1}{p})}t^{-\frac{N}{2}(q-1)+\frac{1}{2}}.$$

To bound the first term we proceed as follows:

$$\int_0^{t/2} \left\| (S(t-s) * G - S(t-s)) * |u(s)|^{q-1} u(s) \right\|_{L^p(\mathbb{R}^N)} ds$$

$$\leq C(p, J, G) \int_0^{t/2} (t-s)^{-\frac{N}{2}(1-\frac{1}{p})-\frac{1}{2}} \left(\left\| |u(s)|^q \right\|_{L^1(\mathbb{R}^N)} + \left\| |u(s)|^q \right\|_{L^p(\mathbb{R}^N)} \right) ds$$

$$\leq C t^{-\frac{N}{2}(1-\frac{1}{p})-\frac{1}{2}} \left(\int_0^{t/2} \|u(s)\|_{L^q(\mathbb{R}^N)}^q ds + \int_0^{t/2} \|u(s)\|_{L^{pq}(\mathbb{R}^N)}^q ds \right)$$

$$= C t^{-\frac{N}{2}(1-\frac{1}{p})-\frac{1}{2}} (I_1(t) + I_2(t)).$$

By Theorem 4.19, we have

$$I_1(t) \leq \int_0^{t/2} \|u(s)\|_{L^q(\mathbb{R}^N)}^q ds \leq C(\|u_0\|_{L^1(\mathbb{R}^N)}, \|u_0\|_{L^\infty(\mathbb{R}^N)}) \int_0^{t/2} (1+s)^{-\frac{N}{2}(q-1)} ds,$$

and an explicit computation of the last integral shows that

$$t^{-\frac{1}{2}} \int_0^{t/2} (1+s)^{-\frac{N}{2}(q-1)} ds \leq C t^{-\frac{N}{2}(q-1)+\frac{1}{2}}.$$

Arguing in the same way for I_2 we get

$$t^{-\frac{1}{2}} I_2(t) \leq C(\|u_0\|_{L^1(\mathbb{R}^N)}, \|u_0\|_{L^\infty(\mathbb{R}^N)}) t^{-\frac{1}{2}} \int_0^{t/2} (1+s)^{-\frac{Nq}{2}(1-\frac{1}{pq})} ds$$

$$\leq C(\|u_0\|_{L^1(\mathbb{R}^N)}, \|u_0\|_{L^\infty(\mathbb{R}^N)}) t^{-\frac{N}{2}(q-\frac{1}{p})+\frac{1}{2}}$$

$$\leq C(\|u_0\|_{L^1(\mathbb{R}^N)}, \|u_0\|_{L^\infty(\mathbb{R}^N)}) t^{-\frac{N}{2}(q-1)+\frac{1}{2}}.$$

This ends the proof. □

Bibliographical notes

The results of this chapter rely on [120].

CHAPTER 5

The Neumann problem for a nonlocal nonlinear diffusion equation

In this chapter we turn our the attention to nonlinear equations with Neumann boundary conditions. We study the nonlocal problem

(5.1)
$$\begin{cases} z_t(x,t) = \int_\Omega J(x-y)(u(y,t) - u(x,t))\, dy, & x \in \Omega,\ t > 0, \\[2mm] z(x,t) \in \gamma(u(x,t)), & x \in \Omega,\ t > 0, \\[2mm] z(x,0) = z_0(x), & x \in \Omega, \end{cases}$$

where Ω is a bounded domain, $z_0 \in L^1(\Omega)$, γ is a maximal monotone graph in $\mathbb{R}^2$ such that $0 \in \gamma(0)$ and $J : \mathbb{R}^N \to \mathbb{R}$ is a nonnegative continuous radial function with compact support, $J(0) > 0$ and $\int_{\mathbb{R}^N} J(x)dx = 1$; that is, J satisfies hypothesis (H) introduced in Chapter 1 and it has compact support.

Problem (5.1) is the nonlocal version of the local problem

(5.2)
$$\begin{cases} w_t - \Delta v = 0 & \text{in } \Omega \times (0, +\infty), \\[2mm] w \in \gamma(v) & \text{in } \Omega \times (0, +\infty), \\[2mm] \dfrac{\partial v}{\partial \eta} = 0 & \text{on } \partial\Omega \times (0, +\infty), \\[2mm] w(\cdot, 0) = w_0(\cdot) & \text{in } \Omega. \end{cases}$$

Problems of this type appear in many applications, as for instance, in modelling diffusion in porous media or in the so-called fast diffusion equation ([150]), which corresponds to γ of the form $\gamma(r) = |r|^{m-1}r$, $m > 0$. Moreover, since γ may be multivalued, problems of type (5.2) appear in various phenomena with changes of phase such as the multiphase Stefan problem ([90]), for which

$$\gamma(r) = \begin{cases} r - 1 & \text{if } r < 0, \\ [-1, 0] & \text{if } r = 0, \\ r & \text{if } r > 0, \end{cases}$$

and in the weak formulation of the mathematical model of the so-called Hele-Shaw problem (see [93] and [97]), for which

$$\gamma(r) = \begin{cases} 0 & \text{if } r < 0, \\ [0, 1] & \text{if } r = 0, \\ 1 & \text{if } r > 0. \end{cases}$$

The case where the domain of γ is different from $\mathbb{R}$ is also treated and corresponds to obstacle problems (see [**9**] for the local problem).

Solutions of (5.1) will be understood in the following sense.

DEFINITION 5.1. A *solution* of (5.1) in $[0, T]$ is a function $z \in W^{1,1}(0, T; L^1(\Omega))$ which satisfies $z(x, 0) = z_0(x)$, a.e. $x \in \Omega$, and for which there exists a function $u \in L^2(0, T; L^2(\Omega))$, $z \in \gamma(u)$ a.e. in $\Omega \times (0, T)$, such that

$$z_t(x, t) = \int_{\Omega} J(x - y)(u(y, t) - u(x, t)) \, dy \quad \text{a.e. in } \Omega \times (0, T).$$

REMARK 5.2. Observe that if $z \in W^{1,1}(0, T; L^1(\Omega))$, then $z \in C([0, T] : L^1(\Omega))$ and then $z(x, 0)$ makes sense.

For this type of solution we prove existence and uniqueness for initial conditions in $L^1(\Omega)$. Moreover, when γ is a continuous function, we find the asymptotic behaviour of the solutions: they converge as $t \to \infty$ to the mean value of the initial condition.

5.1. Existence and uniqueness of solutions

This section deals with the existence and uniqueness of solutions for the nonlocal Neumann problem (5.1). We start with some notation and preliminary results.

5.1.1. Notation and preliminaries. We collect some preliminaries and notation that will be used in the sequel. For a maximal monotone graph η in $\mathbb{R} \times \mathbb{R}$ and $r \in \mathbb{N}$ we denote by η_r the *Yosida approximation* of η, given by $\eta_r = r(I - (I + \frac{1}{r}\eta)^{-1})$. The function η_r is maximal monotone and Lipschitz continuous. We recall the definition of the *main section* η^0 of η,

$$\eta^0(s) := \begin{cases} \text{the element of minimal absolute value of } \eta(s) \text{ if } \eta(s) \neq \emptyset, \\[2mm] +\infty \quad \text{if } [s, +\infty) \cap D(\eta) = \emptyset, \\[2mm] -\infty \quad \text{if } (-\infty, s] \cap D(\eta) = \emptyset, \end{cases}$$

where $D(\eta)$ denotes the effective domain of η. The following properties hold: if $s \in D(\eta)$, $|\eta_r(s)| \leq |\eta^0(s)|$ and $\eta_r(s) \to \eta^0(s)$ as $r \to +\infty$, and if $s \notin D(\eta)$, $|\eta_r(s)| \to +\infty$ as $r \to +\infty$.

We will use the following notation: $\eta_- := \inf \mathrm{R}(\eta)$ and $\eta_+ := \sup \mathrm{R}(\eta)$, where $\mathrm{R}(\eta)$ denotes the range of η. If $0 \in D(\eta)$ then $j_\eta(r) = \int_0^r \eta^0(s) ds$ defines a convex lower semicontinuous function such that $\eta = \partial j_\eta$. If j_η^* is the Legendre transform of j_η, then $\eta^{-1} = \partial j_\eta^*$.

As a consequence of Proposition 3.4 we have the following Poincaré type inequality.

PROPOSITION 5.3. *Given J and Ω, there exists $\beta_1 = \beta_1(J, \Omega) > 0$ such that*

$$(5.3) \qquad \beta_1 \int_{\Omega} \left| u - \frac{1}{|\Omega|} \int_{\Omega} u \right|^2 \leq \frac{1}{2} \int_{\Omega} \int_{\Omega} J(x - y)(u(y) - u(x))^2 \, dy \, dx,$$

for every $u \in L^2(\Omega)$.

In order to obtain a generalized Poincaré type inequality we need the following result.

PROPOSITION 5.4. *Let $\Omega \subset \mathbb{R}^N$ be a bounded open set and $k > 0$. There exists a constant $C > 0$ such that, for any $K \subset \Omega$ with $|K| > k$,*

$$(5.4) \qquad \|u\|_{L^2(\Omega)} \leq C \left(\left\| u - \frac{1}{|\Omega|} \int_\Omega u \right\|_{L^2(\Omega)} + \left| \int_K u \right| \right), \qquad \forall\, u \in L^2(\Omega).$$

PROOF. Assume that the conclusion does not hold. Then, for every $n \in \mathbb{N}$ there exists $K_n \subset \Omega$ with $|K_n| > k$, and $u_n \in L^2(\Omega)$ satisfying

$$(5.5) \qquad \|u_n\|_{L^2(\Omega)} \geq n \left(\left\| u_n - \frac{1}{|\Omega|} \int_\Omega u_n \right\|_{L^2(\Omega)} + \left| \int_{K_n} u_n \right| \right), \qquad \forall\, n \in \mathbb{N}.$$

We normalize u_n by $\|u_n\|_{L^2(\Omega)} = 1$ for all $n \in \mathbb{N}$, and consequently we can assume that

$$(5.6) \qquad u_n \rightharpoonup u \qquad \text{weakly in } L^2(\Omega).$$

Moreover, by (5.5), we have

$$(5.7) \qquad \left\| u_n - \frac{1}{|\Omega|} \int_\Omega u_n \right\|_{L^2(\Omega)} \leq \frac{1}{n}, \quad \text{and} \quad \left| \int_{K_n} u_n \right| \leq \frac{1}{n}, \quad \forall\, n \in \mathbb{N}.$$

Hence

$$u_n - \frac{1}{|\Omega|} \int_\Omega u_n \to 0 \qquad \text{in } L^2(\Omega),$$

and by (5.6) we get $u(x) = \frac{1}{|\Omega|} \int_\Omega u = \alpha$ for almost all $x \in \Omega$, and $u_n \to \alpha$ strongly in $L^2(\Omega)$. Since $\|u_n\|_{L^2(\Omega)} = 1$ for each $n \in \mathbb{N}$, $\alpha \neq 0$. On the other hand, (5.7) implies

$$\lim_{n \to \infty} \int_{K_n} u_n = 0.$$

Since χ_{K_n} is bounded in $L^2(\Omega)$, we can extract a subsequence, still denoted by χ_{K_n}, such that

$$\chi_{K_n} \rightharpoonup \phi \qquad \text{weakly in } L^2(\Omega).$$

Moreover, ϕ is nonnegative and satisfies

$$k \leq \lim_{n \to \infty} |K_n| = \lim_{n \to \infty} \int_\Omega \chi_{K_n} = \int_\Omega \phi.$$

Now, since $u_n \to \alpha$ strongly in $L^2(\Omega)$ and $\chi_{K_n} \rightharpoonup \phi$ weakly in $L^2(\Omega)$, we have

$$0 = \lim_{n \to \infty} \int_{K_n} u_n = \lim_{n \to \infty} \int_\Omega \chi_{K_n} u_n = \alpha \int_\Omega \phi \neq 0,$$

a contradiction. $\square$

To simplify the notation we define the linear self-adjoint operator

$$A : L^2(\Omega) \to L^2(\Omega)$$

by

$$Au(x) = \int_\Omega J(x - y)(u(y) - u(x))\, dy, \qquad x \in \Omega.$$

As a consequence of the above results we have the next proposition, which plays the role of the classical generalized Poincaré inequality for Sobolev spaces.

PROPOSITION 5.5. *Let $\Omega \subset \mathbb{R}^N$ be a bounded open set and $k > 0$. There exists a constant $C = C(J, \Omega, k)$ such that, for any $K \subset \Omega$ with $|K| > k$,*

$$(5.8) \qquad \|u\|_{L^2(\Omega)} \leq C \left(\left(-\int_\Omega Au\, u \right)^{1/2} + \|u\|_{L^2(K)} \right), \qquad \forall\, u \in L^2(\Omega).$$

Using the above result we obtain the following lemma. Its proof follows the steps of the proof of [**8**, Lemma 4.2] that uses the classical generalized Poincaré inequality for Sobolev spaces.

LEMMA 5.6. *Let γ be a maximal monotone graph in $\mathbb{R}^2$ such that $0 \in \gamma(0)$. Let $\{u_n\}_{n\in\mathbb{N}} \subset L^2(\Omega)$ and $\{z_n\}_{n\in\mathbb{N}} \subset L^1(\Omega)$ such that, for every $n \in \mathbb{N}$, $z_n \in \gamma(u_n)$ a.e. in Ω.*

Suppose that

(i) *if $\gamma_+ = +\infty$, then there exists $M > 0$ such that*

$$\int_\Omega z_n^+ < M, \qquad \forall\, n \in \mathbb{N};$$

(ii) *if $\gamma_+ < +\infty$, then there exist $M \in \mathbb{R}$ and $h > 0$ such that*

$$\int_\Omega z_n < M < \gamma_+ |\Omega|, \qquad \forall\, n \in \mathbb{N}$$

and

$$\int_{\{x\in\Omega:z_n(x)<-h\}} |z_n,| < \frac{\gamma_+|\Omega| - M}{4}, \qquad \forall\, n \in \mathbb{N}.$$

Then there exists a constant C $(C = C(M, \Omega)$ in case (i), and $C = C(M, \Omega, \gamma, h)$ in case (ii)) such that

$$(5.9) \qquad \|u_n^+\|_{L^2(\Omega)} \leq C \left(\left(-\int_\Omega Au_n^+\, u_n^+ \right)^{1/2} + 1 \right), \qquad \forall\, n \in \mathbb{N}.$$

Suppose that

(iii) *if $\gamma_- = -\infty$, then there exists $M > 0$ such that*

$$\int_\Omega z_n^- < M, \qquad \forall\, n \in \mathbb{N};$$

(iv) *if $\gamma_- > -\infty$, then there exist $M \in \mathbb{R}$ and $h > 0$ such that*

$$\int_\Omega z_n > M > \gamma_- |\Omega|, \qquad \forall\, n \in \mathbb{N}$$

and

$$\int_{\{x\in\Omega:z_n(x)>h\}} z_n < \frac{M - \gamma_-|\Omega|}{4}, \qquad \forall\, n \in \mathbb{N}.$$

Then there exists a constant $\tilde{C}$ ($\tilde{C} = \tilde{C}(M,\Omega)$ in case (iii), and $\tilde{C} = \tilde{C}(M,\Omega,\gamma,h)$ in case (iv)) such that

$$(5.10) \qquad \|u_n^-\|_{L^2(\Omega)} \leq \tilde{C}\left(\left(-\int_\Omega Au_n^- \, u_n^-\right)^{1/2} + 1\right), \qquad \forall \, n \in \mathbb{N}.$$

PROOF. Let us only prove (5.9), the proof of (5.10) being similar. First, consider the case $\gamma_+ = +\infty$. Then, by assumption, there exists $M > 0$ such that

$$\int_\Omega z_n^+ < M, \qquad \forall \, n \in \mathbb{N}.$$

For $n \in \mathbb{N}$, let $K_n = \left\{ x \in \Omega : z_n^+(x) < \frac{2M}{|\Omega|} \right\}$. Then

$$0 \leq \int_{K_n} z_n^+ = \int_\Omega z_n^+ - \int_{\Omega \setminus K_n} z_n^+ \leq M - (|\Omega| - |K_n|)\frac{2M}{|\Omega|} = |K_n|\frac{2M}{|\Omega|} - M.$$

Therefore,

$$|K_n| \geq \frac{|\Omega|}{2},$$

and

$$\|u_n^+\|_{L^2(K_n)} \leq |K_n|^{1/2} \sup \gamma^{-1}\left(\frac{2M}{|\Omega|}\right).$$

By Proposition 5.5,

$$\|u_n^+\|_{L^2(\Omega)} \leq \tilde{C}(J,\Omega)\left(\left(-\int_\Omega Au_n^+ \, u_n^+\right)^{1/2} + |\Omega|^{1/2} \sup \gamma^{-1}\left(\frac{2M}{|\Omega|}\right)\right), \qquad \forall \, n \in \mathbb{N},$$

and consequently (5.9) holds in this case.

Now, consider the case $\gamma_+ < +\infty$. Let

$$\delta = \gamma_+ |\Omega| - M.$$

By assumption, for every $n \in \mathbb{N}$, we have

$$(5.11) \qquad \int_\Omega z_n < \gamma_+|\Omega| - \delta.$$

For $n \in \mathbb{N}$, let $K_n = \left\{ x \in \Omega : z_n(x) < \gamma_+ - \frac{\delta}{2|\Omega|} \right\}$. By (5.11),

$$\int_{K_n} z_n = \int_\Omega z_n - \int_{\Omega \setminus K_n} z_n < -\frac{\delta}{2} + |K_n|\left(\gamma_+ - \frac{\delta}{2|\Omega|}\right).$$

Moreover,

$$\int_{K_n} z_n = -\int_{K_n \cap \{x \in \Omega : z_n < -h\}} |z_n| + \int_{K_n \cap \{x \in \Omega : z_n \geq -h\}} z_n \geq -\frac{\delta}{4} - h|K_n|.$$

Therefore,

$$|K_n|\left(h - \frac{\delta}{2|\Omega|} + \gamma_+\right) \geq \frac{\delta}{4}.$$

Hence $|K_n| > 0$, $h - \frac{\delta}{2|\Omega|} + \gamma_+ > 0$ and

$$|K_n| \geq \frac{\delta}{4\left(h - \frac{\delta}{2|\Omega|} + \gamma_+\right)}.$$

Consequently,

$$\|u_n^+\|_{L^2(K_n)} \leq |K_n|^{1/2} \sup \gamma^{-1} \left(\gamma_+ - \frac{\delta}{2|\Omega|} \right).$$

Then, by Proposition 5.5,

$$\|u_n^+\|_{L^2(\Omega)} \leq \tilde{C}(J, \Omega, \gamma, h) \left(\left(- \int_\Omega A u_n^+ \, u_n^+ \right)^{1/2} + |\Omega|^{1/2} \sup \gamma^{-1} \left(\gamma_+ - \frac{\delta}{2|\Omega|} \right) \right).$$

This ends the proof of (5.9).   □

Finally, we state the following monotonicity result. Its proof is straightforward.

LEMMA 5.7. *Let* $T: \mathbb{R} \to \mathbb{R}$ *be a nondecreasing function. For every* $u \in L^2(\Omega)$ *such that* $T(u) \in L^2(\Omega)$, *we have*

$$- \int_\Omega A u(x) \, T(u(x)) \, dx = - \int_\Omega \int_\Omega J(x - y)(u(y) - u(x)) \, dy \, T(u(x)) \, dx$$

$$= \frac{1}{2} \int_\Omega \int_\Omega J(x - y) \left(T(u(y)) - T(u(x)) \right) (u(y) - u(x)) \, dy \, dx.$$

In particular, we have

$$- \int_\Omega A u(x) \, u(x) \, dx = \frac{1}{2} \int_\Omega \int_\Omega J(x - y)(u(y) - u(x))^2 \, dy \, dx.$$

5.1.2. Mild solutions and contraction principle. In this subsection we obtain a mild solution of the problem (5.1) by studying the associated integral operator.

Given a maximal monotone graph γ in $\mathbb{R}^2$ such that $0 \in \gamma(0)$, $\gamma_- < \gamma_+$, we consider the problem

$$(S_\phi^\gamma) \qquad \gamma(u) - Au \ni \phi \quad \text{in } \Omega.$$

DEFINITION 5.8. *Let* $\phi \in L^1(\Omega)$. *A pair of functions* $(u, z) \in L^2(\Omega) \times L^1(\Omega)$ *is a* solution *of problem* (S_ϕ^γ) *if* $z(x) \in \gamma(u(x))$ *a.e.* $x \in \Omega$ *and* $z(x) - Au(x) = \phi(x)$ *a.e.* $x \in \Omega$; *that is,*

$$z(x) - \int_\Omega J(x - y)(u(y) - u(x)) \, dy = \phi(x) \qquad \text{a.e. } x \in \Omega.$$

With respect to the uniqueness of solutions of problem (S_ϕ^γ), we have the following maximum principle.

THEOREM 5.9.

(i) *Let* $\phi_1 \in L^1(\Omega)$ *and let* (u_1, z_1) *be a subsolution of* $(S_{\phi_1}^\gamma)$; *that is,* $z_1(x) \in \gamma(u_1(x))$ *a.e.* $x \in \Omega$ *and* $z_1(x) - Au_1(x) \leq \phi_1(x)$ *a.e.* $x \in \Omega$. *Let* $\phi_2 \in L^1(\Omega)$ *and let* (u_2, z_2) *be a supersolution of* $(S_{\phi_2}^\gamma)$; *that is,* $z_2(x) \in \gamma(u_2(x))$ *a.e.* $x \in \Omega$ *and* $z_2(x) - Au_2(x) \geq \phi_2(x)$ *a.e.* $x \in \Omega$. *Then*

$$\int_\Omega (z_1 - z_2)^+ \leq \int_\Omega (\phi_1 - \phi_2)^+.$$

Moreover, if $\phi_1 \leq \phi_2$, $\phi_1 \neq \phi_2$, *then* $u_1(x) \leq u_2(x)$ *a.e.* $x \in \Omega$.

(ii) *Let $\phi \in L^1(\Omega)$, and let (u_1, z_1), (u_2, z_2) be two solutions of (S_ϕ^γ). Then $z_1 = z_2$ a.e. and there exists a constant c such that $u_1 = u_2 + c$ a.e.*

PROOF. To prove (i), let (u_1, z_1) be a subsolution of $(S_{\phi_1}^\gamma)$ and (u_2, z_2) a supersolution of $(S_{\phi_2}^\gamma)$. Then

$$-(Au_1(x) - Au_2(x)) + z_1(x) - z_2(x) \leq \phi_1(x) - \phi_2(x).$$

Let $T_k(r) = k \wedge (r \vee (-k))$, $k \geq 0$, $r \in \mathbb{R}$. Multiplying the above inequality by $\frac{1}{k} T_k^+(u_1 - u_2 + k \operatorname{sgn}_0^+(z_1 - z_2))$ and integrating we get,

$$\int_\Omega (z_1 - z_2) \frac{1}{k} T_k^+(u_1 - u_2 + k \operatorname{sgn}_0^+(z_1 - z_2))$$

$$- \int_\Omega (Au_1(x) - Au_2(x)) \frac{1}{k} T_k^+(u_1(x) - u_2(x) + k \operatorname{sgn}_0^+(z_1(x) - z_2(x)))\, dx$$

(5.12)

$$\leq \int_\Omega (\phi_1 - \phi_2) \frac{1}{k} T_k^+(u_1 - u_2 + k \operatorname{sgn}_0^+(z_1 - z_2))$$

$$\leq \int_\Omega (\phi_1(x) - \phi_2(x))^+\, dx.$$

Let us write $u = u_1 - u_2$ and $z = \operatorname{sgn}_0^+(z_1 - z_2)$; then, by the monotonicity result stated in Lemma 5.7,

$$\lim_{k \to 0} \int_\Omega (Au_1(x) - Au_2(x)) \frac{1}{k} T_k^+(u_1(x) - u_2(x) + k \operatorname{sgn}_0^+(z_1(x) - z_2(x)))\, dx$$

$$= \lim_{k \to 0} \int_\Omega Au(x) \frac{1}{k} T_k^+(u(x) + kz(x))\, dx$$

$$= \lim_{k \to 0} \int_\Omega A(u + kz)(x) \frac{1}{k} T_k^+(u(x) + kz(x))\, dx \leq 0.$$

Therefore, taking limit as k goes to 0 in (5.12), we obtain

$$\int_\Omega (z_1 - z_2)^+ \leq \int_\Omega (\phi_1 - \phi_2)^+.$$

Let us now suppose that $\phi_1 \leq \phi_2$, $\phi_1 \neq \phi_2$. By the previous calculations we know that $z_1 \leq z_2$. Since

$$\int_\Omega z_1 \leq \int_\Omega \phi_1 < \int_\Omega \phi_2 \leq \int_\Omega z_2,$$

$z_1 \neq z_2$. Going back to (5.12), if $u = u_1 - u_2$, we get

$$- \int_\Omega Au(x) T_k^+(u(x))\, dx = 0,$$

and therefore

$$- \int_\Omega Au(x) u^+(x)\, dx = 0.$$

Consequently, by Lemma 5.7, there exists a null set $C \subset \Omega \times \Omega$ such that

(5.13) $\quad J(x - y)(u^+(y) - u^+(x))(u(y) - u(x)) = 0 \quad$ for all $(x, y) \in \Omega \times \Omega \setminus C$.

Let B be a null subset of Ω such that if $x \notin B$, the section $C_x = \{y \in \Omega : (x, y) \in C\}$ is null. Let $x \notin B$; if $u(x) > 0$ then, since there exists $r_0 > 0$ such that $J(z) > 0$ for every z such that $|z| \leq r_0$, by a compactness argument and having in mind (5.13), it is easy to see that $u(y) = u(x) > 0$ for all $y \notin C_x$. Therefore $u_1(y) > u_2(y)$ for all $y \notin C_x$ in Ω and consequently $z_1(y) \geq z_2(y)$ a.e. in Ω, which contradicts the inequalities $z_1 \leq z_2$, $z_1 \neq z_2$.

Let us now prove (ii). As (u_i, z_i) are solutions of (S_ϕ^γ), we have that

$$-(Au_1(x) - Au_2(x)) + z_1(x) - z_2(x) = 0.$$

By (i), $z_1 = z_2$ a.e. Consequently,

$$0 = -(Au_1(x) - Au_2(x)) = -A(u_1 - u_2)(x).$$

Therefore, multiplying the above equation by $u = u_1 - u_2$ and integrating we obtain

$$\frac{1}{2} \int_\Omega \int_\Omega J(x - y)(u(y) - u(x))^2 \, dy \, dx = 0.$$

Then, by (5.3), it follows that u is constant a.e. in Ω. $\qquad\square$

In particular we have the following result.

COROLLARY 5.10. *Let $k > 0$ and $u \in L^2(\Omega)$ such that*

$$ku - Au \geq 0 \quad a.e. \ in \ \Omega;$$

then $u \geq 0$ a.e. in Ω.

PROOF. Since (u, ku) is a supersolution of (S_0^γ), where $\gamma(r) = kr$, and $(0, 0)$ is a subsolution of (S_0^γ), by Theorem 5.9, the result follows. $\qquad\square$

To study the existence of solutions of problem (S_ϕ^γ) we start with the following two lemmas, for which we need to recall the following definition (see Appendix A). For $u, v \in L^1(\Omega)$,

$$u \ll v \quad \text{if and only if} \quad \int_\Omega j(u) \, dx \leq \int_\Omega j(v) \, dx, \quad \forall j \in J_0,$$

where

$$J_0 = \{j : \mathbb{R} \to [0, +\infty], \text{ convex and lower semicontinuos with } j(0) = 0\}.$$

LEMMA 5.11. *Suppose $\gamma : \mathbb{R} \to \mathbb{R}$ is a nondecreasing Lipschitz continuous function with $\gamma(0) = 0$ and $\gamma_- < \gamma_+$. Let $\phi \in C(\overline{\Omega})$ such that $\gamma_- < \phi < \gamma_+$. Then there exists a solution $(u, \gamma(u))$ of the problem (S_ϕ^γ). Moreover, $\gamma(u) \ll \phi$.*

PROOF. Since $\gamma_- < \phi < \gamma_+$ and $\phi \in C(\overline{\Omega})$, we can find $c_1 \leq c_2$ such that

(5.14) $$\gamma_- < \gamma(c_1) \leq \phi(x) \leq \gamma(c_2) < \gamma_+ \quad \forall x \in \Omega.$$

Since γ is a nondecreasing Lipschitz continuous function, there exists $k > 0$ for which the function $s \mapsto ks - \gamma(s)$ is nondecreasing. By induction we can find a sequence $\{u_i\} \subset L^2(\Omega)$ such that

(5.15)
$$u_0 = c_1, \qquad u_i \leq u_{i+1} \leq c_2,$$

$$ku_{i+1} - Au_{i+1} = \phi - \gamma(u_i) + ku_i, \qquad \forall i \in \mathbb{N}.$$

Since $k > 0$, and A is self-adjoint, it is easy to see that k does not belong to the spectrum of A; then there exists $u_1 \in L^2(\Omega)$ such that

$$ku_1 - Au_1 = \phi - \gamma(c_1) + kc_1.$$

Then, by (5.14), we have

$$ku_1 - Au_1 = \phi - \gamma(c_1) + kc_1 \geq kc_1 = kc_1 - Ac_1.$$

Hence, from Corollary 5.10 we get that $u_0 = c_1 \leq u_1$. Analogously, there exists u_2 such that

$$ku_2 - Au_2 = \phi - \gamma(u_1) + ku_1.$$

As $c_1 \leq u_1$, we get

$$ku_2 - Au_2 \geq \phi - \gamma(c_1) + kc_1 = ku_1 - Au_1.$$

Again by Corollary 5.10, $u_1 \leq u_2$, and by induction we obtain a sequence u_i such that $u_i \leq u_{i+1}$. On the other hand, since the function $s \mapsto ks - \gamma(s)$ is nondecreasing and $c_1 \leq c_2$, by (5.14),

$$kc_2 - Ac_2 \geq \phi - \gamma(c_2) + kc_2 \geq \phi - \gamma(c_1) + kc_1 = ku_1 - Au_1.$$

Applying again Corollary 5.10, we get $c_2 \geq u_1$, and by an inductive argument we deduce that $u_i \leq c_2$ for all $i \in \mathbb{N}$. Hence (5.15) holds. Consequently, there exists $u \in L^\infty(\Omega)$ such that $u(x) = \lim_{i \to +\infty} u_i(x)$ a.e. in Ω. Taking limits in (5.15), we obtain that

$$ku - Au = \phi - \gamma(u) + ku,$$

and $(u, \gamma(u))$ is a solution of problem (S_ϕ^γ), that is,

(5.16) $$\gamma(u) - Au = \phi.$$

Let

$$P_0 = \{q \in C^\infty(\mathbb{R}) : 0 \leq q' \leq 1,\ \mathrm{supp}(q')\ \text{is compact and}\ 0 \notin \mathrm{supp}(q)\}.$$

Finally, given $p \in P_0$, multiplying (5.16) by $p(\gamma(u))$, and integrating in Ω, we get

$$\int_\Omega \gamma(u(x))p(\gamma(u(x)))\,dx - \int_\Omega Au(x)p(\gamma(u(x)))\,dx = \int_\Omega \phi(x)p(\gamma(u(x)))\,dx.$$

By Lemma 5.7, the second term in the above equality is nonnegative; therefore

$$\int_\Omega \gamma(u(x))p(\gamma(u(x)))\,dx \leq \int_\Omega \phi(x)p(\gamma(u(x)))\,dx.$$

By Proposition A.42, we conclude that $\gamma(u) \ll \phi$. $\qquad\square$

LEMMA 5.12. *Suppose γ is a maximal monotone graph in $\mathbb{R}^2$, $]-\infty, 0] \subset D(\gamma)$, $0 \in \gamma(0)$, $\gamma_- < \gamma_+$. Let $\tilde{\gamma}(s) = \gamma(s)$ if $s < 0$, $\tilde{\gamma}(s) = 0$ if $s \geq 0$. Assume that $\tilde{\gamma}$ is Lipschitz continuous in $]-\infty, 0]$. Let $\phi \in C(\overline{\Omega})$ such that $\gamma_- < \phi < \gamma_+$. Then there exists a solution (u, z) of (S_ϕ^γ). Moreover, $z \ll \phi$.*

PROOF. If $\gamma_- < 0$, we consider c_1 such that $\gamma(c_1) = \{m_1\}$, $\gamma_- < m_1 < 0$ with $m_1 \leq \phi$. And if $\gamma_- = 0$, we take $c_1 = m_1 = 0$. Let γ_r, $r \in \mathbb{N}$, be the Yosida approximation of γ and define the maximal monotone graph

$$\gamma^r(s) = \begin{cases} \gamma(s) & \text{if } s < 0, \\[2mm] \gamma_r(s) & \text{if } s \geq 0. \end{cases}$$

Observe that γ^r is a nondecreasing Lipschitz continuous function with $\gamma^r(0) = 0$, $\gamma^r \leq \gamma^{r+1}$, $\gamma_- = \gamma_-^r < \phi < \gamma_+^r$, for r large enough, and that it converges in the sense of maximal monotone graphs to γ. From the previous lemma, for each γ^r we obtain a solution (u_r, z_r) of $(S_\phi^{\gamma^r})$; that is, $z_r = \gamma^r(u_r)$ a.e. and

$$(5.17) \qquad\qquad z_r - Au_r = \phi.$$

Moreover, $z_r \ll \phi$, $z_r \geq m_1$, and $u_r \geq c_1$. Let

$$\hat{z}_r(x) = \begin{cases} z_r(x) & \text{if } u_r(x) \leq 0, \\ \gamma_{r+1}(u_r(x)) & \text{if } u_r(x) > 0. \end{cases}$$

Since γ_r is nondecreasing,

$$\hat{z}_r \geq z_r,$$

and also,

$$\hat{z}_r \in \gamma^{r+1}(u_r).$$

Therefore, $(u_r, \hat{z}_r)$ is a supersolution of $(S_\phi^{\gamma^{r+1}})$. Using Theorem 5.9, we obtain that

$$\hat{z}_r \geq z_{r+1}.$$

Now, if $\hat{z}_r = z_r$ then

$$z_r \geq z_{r+1},$$

and if $\hat{z}_r \neq z_r$, by Theorem 5.9,

$$u_r \geq u_{r+1}.$$

So, there exists a monotone nonincreasing subsequence of $\{u_r\}$, also denoted by $\{u_r\}$, with $u_r \geq \hat{c}_1$, or there exists a monotone nonincreasing subsequence of $\{z_r\}$, also denoted by $\{z_r\}$, with $z_r \geq m_1$. In the first case, we have that

$$u_r \to u \quad \text{in } L^2(\Omega),$$

and, since $z_r \ll \phi$, by Proposition A.44, we have

$$z_r \to z \quad \text{weakly in } L^1(\Omega).$$

In the second case, we obtain

$$(5.18) \qquad\qquad z_r \to z \quad \text{in } L^1(\Omega).$$

In fact, since $z_r \ll \phi$, we get that

$$(5.19) \qquad\qquad z_r \to z \quad \text{in } L^2(\Omega).$$

In this second case, multiplying (5.17) by $u_r - u_s$ and integrating we get

$$-\int_\Omega Au_r(u_r - u_s) = \int_\Omega \phi\,(u_r - u_s) - \int_\Omega z_r(u_r - u_s).$$

Moreover,

$$-\int_\Omega Au_s(u_r - u_s) = \int_\Omega \phi\,(u_r - u_s) - \int_\Omega z_s(u_r - u_s).$$

Hence, since $\int_\Omega z_r = \int_\Omega z_s$,

$$-\int_\Omega A(u_r - u_s)(u_r - u_s) = -\int_\Omega (z_r - z_s)(u_r - u_s)$$

$$= -\int_\Omega (z_r - z_s)\left(u_r - \frac{1}{|\Omega|}\int_\Omega u_r - \left(u_s - \frac{1}{|\Omega|}\int_\Omega u_s\right)\right).$$

By Proposition 5.3,

$$\beta_1 \left\| \left(u_r - \frac{1}{|\Omega|} \int_\Omega u_r \right) - \left(u_s - \frac{1}{|\Omega|} \int_\Omega u_s \right) \right\|_{L^2(\Omega)} \leq \|z_r - z_s\|_{L^2(\Omega)}.$$

From (5.19) we get

$$u_r - \frac{1}{|\Omega|} \int_\Omega u_r \to w \quad \text{in } L^2(\Omega).$$

Let us see that $\left\{ \frac{1}{|\Omega|} \int_\Omega u_r \right\}$ is bounded. If not, we can assume, passing to a subsequence if necessary, that it converges to $-\infty$. Then $u_r \to -\infty$ a.e. in Ω. On the other hand, since $z_r \in \gamma^r(u_r)$ and $\gamma^r \to \gamma$, by (5.18) and the above convergence on u_r, $z = \gamma_-$ a.e. in Ω. Consequently, $\int_\Omega \phi = \int_\Omega z = |\Omega|\gamma_-$ which contradicts the inequality $\phi > \gamma_-$. Thus, $\left\{ \frac{1}{|\Omega|} \int_\Omega u_r \right\}$ is bounded and there exists a subsequence of $\{u_r\}$, also denoted by $\{u_r\}$, such that

$$u_r \to u \quad \text{in } L^2(\Omega).$$

Therefore, in both cases, $z \in \gamma(u)$ a.e. in Ω, $z \ll \phi$, and, taking limit in

$$z_r - A u_r \ni \phi,$$

we obtain

$$z - A u \ni \phi,$$

which concludes the proof. $\square$

With this lemma in mind we proceed to extend the result to general monotone graphs.

THEOREM 5.13. *Suppose γ is a maximal monotone graph in $\mathbb{R}^2$, $0 \in \gamma(0)$ and $\gamma_- < \gamma_+$. Let $\phi \in C(\overline{\Omega})$ such that $\gamma_- < \phi < \gamma_+$. Then there exists a solution (u, z) of (S_ϕ^γ). Moreover, $z \ll \phi$.*

PROOF. Let γ_r be the Yosida approximation of γ and consider the maximal monotone graph

$$\gamma^r(s) = \begin{cases} \gamma(s) & \text{if } s > 0, \\ \gamma_r(s) & \text{if } s \leq 0. \end{cases}$$

Observe that γ^r satisfies the hypothesis of Lemma 5.12, $\gamma_-^r < \phi < \gamma_+^r$ for r large enough, $\gamma^r \geq \gamma^{r+1}$ and converges in the sense of maximal monotone graphs to γ. From the previous lemma, for each γ^r we obtain a solution (u_r, z_r) of $(S_\phi^{\gamma^r})$, $z_r \ll \phi$. Now, we can proceed similarly to the previous lemma to conclude the proof. $\square$

The natural space to study the problem (5.1) from the point of view of Nonlinear Semigroup Theory is $L^1(\Omega)$. In this space we define the operator

$$B^\gamma := \left\{ (z, \hat{z}) \in L^1(\Omega) \times L^1(\Omega) : \exists u \in L^2(\Omega) \, \text{s.t.} \, (u, z) \text{ is a solution of } (S_{z+\hat{z}}^\gamma) \right\}.$$

In other words, $\hat{z} \in B^\gamma(z)$ if and only if there exists $u \in L^2(\Omega)$ such that $z(x) \in \gamma(u(x))$ a.e. in Ω, and

$$(5.20) \qquad - \int_\Omega J(x - y)(u(y) - u(x)) \, dy = \hat{z}(x), \quad \text{a.e. } x \in \Omega.$$

The operator B^γ allows us to rewrite (5.1) as the following abstract Cauchy problem in $L^1(\Omega)$:

(5.21)
$$\begin{cases} z'(t) + B^\gamma(z(t)) \ni 0, & t \in (0, T), \\ z(0) = z_0. \end{cases}$$

A direct consequence of Theorems 5.9 and 5.13 is the following result.

COROLLARY 5.14. *Suppose γ is a maximal monotone graph in $\mathbb{R}^2$, $0 \in \gamma(0)$. Then the operator B^γ is T-accretive in $L^1(\Omega)$ and satisfies*

$$\{\phi \in C(\overline{\Omega}) : \gamma_- < \phi < \gamma_+\} \subset \mathrm{R}(I + B^\gamma).$$

The following theorem is a consequence of the above result.

THEOREM 5.15. *Let $T > 0$ and $z_{i0} \in L^1(\Omega)$, $i = 1, 2$. Let z_i be a solution in $[0, T]$ of (5.1) with initial datum z_{i0}, $i = 1, 2$. Then*

(5.22)
$$\int_\Omega (z_1(t) - z_2(t))^+ \leq \int_\Omega (z_{10} - z_{20})^+$$

for almost every $t \in (0, T)$.

PROOF. Let $(u_i(t), z_i(t))$ be solutions of (5.1) with initial data z_{0i}, $i = 1, 2$. Then, since they are strong solutions of (5.21) and A is T-accretive, (5.22) follows from Nonlinear Semigroup Theory (see Theorem A.56). □

In the next result we characterize $\overline{D(B^\gamma)}^{L^1(\Omega)}$.

THEOREM 5.16. *Suppose γ is a maximal monotone graph in $\mathbb{R}^2$. Then*

$$\overline{D(B^\gamma)}^{L^1(\Omega)} = \{z \in L^1(\Omega) : \gamma_- \leq z \leq \gamma_+\}.$$

PROOF. It is obvious that

$$\overline{D(B^\gamma)}^{L^1(\Omega)} \subset \{z \in L^1(\Omega) : \gamma_- \leq z \leq \gamma_+\}.$$

To obtain the other inclusion, it is enough to take $\phi \in C(\overline{\Omega})$ satisfying $\gamma_- < \phi < \gamma_+$ and to prove that $\phi \in \overline{D(B^\gamma)}^{L^1(\Omega)}$. Let $a, b \in \mathbb{R}$ such that $\gamma_- < a < \phi < b < \gamma_+$. Now, by Theorem 5.13, for any $n \in \mathbb{N}$, there exists $v_n := \left(I + \frac{1}{n}B^\gamma\right)^{-1}\phi \in D(B^\gamma)$, that is, $(v_n, n(\phi - v_n)) \in B^\gamma$, and thus there exists $u_n \in L^2(\Omega)$ such that $v_n \in \gamma(u_n)$ a.e. in Ω and

(5.23)
$$v_n(x) - \frac{1}{n} \int_\Omega J(x - y)(u_n(y) - u_n(x))\, dy = \phi(x) \qquad \forall\, x \in \Omega.$$

Moreover, $v_n \ll \phi$. Then

(5.24)
$$-\infty < \inf \gamma^{-1}(a) \leq u_n \leq \sup \gamma^{-1}(b) < +\infty.$$

Hence, from (5.23) and (5.24) it follows that $v_n \to \phi$ in $L^1(\Omega)$. □

As a consequence of the above results we have the following theorem concerning mild solutions (see Appendix A).

THEOREM 5.17. *Suposse γ is a maximal monotone graph in $\mathbb{R}^2$. Let $T > 0$ and $z_0 \in L^1(\Omega)$ satisfying $\gamma_- \leq z_0 \leq \gamma_+$. Then there exists a unique mild solution of (5.21). Moreover $z \ll z_0$.*

PROOF. For $n \in \mathbb{N}$, let $\varepsilon = T/n$, and consider a subdivision

$$t_0 = 0 < t_1 < \cdots < t_{n-1} < T = t_n$$

with $t_i - t_{i-1} = \varepsilon$. Let $z_0^\varepsilon \in C(\overline{\Omega})$ with

$$\gamma_- < z_0^\varepsilon < \gamma_+$$

and

$$\|z_0^\varepsilon - z_0\|_{L^1(\Omega)} \leq \varepsilon.$$

By Theorem 5.13, for n large enough, there exists a solution $(u_i^\varepsilon, z_i^\varepsilon)$ of

(5.25) $$\gamma(u_i^\varepsilon) - \varepsilon A u_i^\varepsilon \ni z_{i-1}^\varepsilon$$

for $i = 1, \ldots, n$, with

(5.26) $$z_i^\varepsilon \ll z_{i-1}^\varepsilon.$$

That is, there exists a unique solution $z_i^\varepsilon \in L^1(\Omega)$ of the time discretized scheme associated with (5.21),

$$z_i^\varepsilon + \varepsilon B^\gamma z_i^\varepsilon \ni \varepsilon z_{i-1}^\varepsilon, \qquad i = 1, \ldots, n.$$

Therefore, if we define $z_\varepsilon(t)$ by

$$\begin{cases} z_\varepsilon(0) = z_0^\varepsilon, \\ z_\varepsilon(t) = z_i^\varepsilon, & \text{for } t \in]t_{i-1}, t_i], \quad i = 1, \ldots, n, \end{cases}$$

it is an ε-approximate solution of problem (5.21).

By Theorem A.29, on account of Corollary 5.14 and Theorem 5.16, the problem (5.21) has a unique mild solution $z(t) \in C([0, T] : L^1(\Omega))$, obtained as $z(t) = L^1(\Omega)\text{-}\lim_{\varepsilon \to 0} z_\varepsilon(t)$ uniformly for $t \in [0, T]$. Finally, from (5.26) we get $z \ll z_0$. □

By Crandall-Liggett's Theorem (Theorem A.31), the mild solution obtained above is given by the exponential formula,

(5.27) $$e^{-tB^\gamma} z_0 = \lim_{n \to \infty} \left(I + \frac{t}{n} B^\gamma \right)^{-n} z_0.$$

The nonlinear contraction semigroup e^{-tB^γ} generated by the operator $-B^\gamma$ will be denoted in the sequel by $(S(t))_{t \geq 0}$.

In principle, it is not clear how these mild solutions have to be interpreted with respect to (5.1). In the next subsection we will see that they coincide with the solutions defined in the introduction.

5.1.3. Existence of solutions. In this subsection we prove that the mild solution of (5.21) is in fact a solution of problem (5.1).

THEOREM 5.18. *Let $z_0 \in L^1(\Omega)$ such that $\gamma_- \leq z_0 \leq \gamma_+$, $\gamma_- < \frac{1}{|\Omega|} \int_\Omega z_0 < \gamma_+$ and $\int_\Omega j_\gamma^*(z_0) < +\infty$. Then there exists a unique solution of (5.1) in $[0, T]$ for every $T > 0$. Moreover, $z \ll z_0$.*

PROOF. We divide the proof into three steps.

Step 1. First, suppose that

$$(5.28) \qquad \text{there exist } c_1, c_2 \text{ such that } c_1 \leq c_2, \, m_1 \in \gamma(c_1), m_2 \in \gamma(c_2)$$
$$\text{and } \gamma_- < m_1 \leq z_0 \leq m_2 < \gamma_+.$$

Let $z(t)$ be the mild solution of (5.21) given by Theorem 5.17. We show that z is a solution of problem (5.1).

For $n \in \mathbb{N}$, let $\varepsilon = T/n$, and consider a subdivision $t_0 = 0 < t_1 < \cdots < t_{n-1} < T = t_n$ with $t_i - t_{i-1} = \varepsilon$. Then it follows that

$$(5.29) \qquad z(t) = L^1(\Omega)\text{-}\lim_\varepsilon z_\varepsilon(t) \qquad \text{uniformly for } t \in [0, T],$$

where $z_\varepsilon(t)$ is given, for ε small enough, by

$$(5.30) \qquad \begin{cases} z_\varepsilon(t) = z_0 & \text{for } t \in \,]-\infty, 0], \\ z_\varepsilon(t) = z_i^n, & \text{for } t \in \,]t_{i-1}, t_i], \quad i = 1, \ldots, n, \end{cases}$$

where $(u_i^n, z_i^n) \in L^2(\Omega) \times L^1(\Omega)$ is the solution of

$$(5.31) \qquad -Au_i^n + \frac{z_i^n - z_{i-1}^n}{\varepsilon} = 0, \quad i = 1, 2, \ldots, n.$$

Moreover, $z_i^n \ll z_0$. Hence $\gamma_- < m_1 \leq z_i^n \leq m_2 < \gamma_+$ and consequently,

$$\inf \gamma^{-1}(m_1) \leq u_i^n \leq \sup \gamma^{-1}(m_2).$$

Therefore, if we write $u_\varepsilon(t) = u_i^n$, $t \in \,]t_{i-1}, t_i]$, $i = 1, \ldots, n$, we can assume that

$$(5.32) \qquad u_\varepsilon \rightharpoonup u \quad \text{weakly in } L^2(0, T; L^2(\Omega)) \text{ as } \varepsilon \to 0^+.$$

Since $z_\varepsilon \in \gamma(u_\varepsilon)$ a.e. in $\Omega \times (0, T))$ and $z_\varepsilon \to z$ in $L^1(0, T; L^1(\Omega))$, having in mind (5.32), we obtain that $z \in \gamma(u)$ a.e. in $\Omega \times (0, T)$. On the other hand, from (5.31),

$$\frac{z_\varepsilon(t) - z_\varepsilon(t - \varepsilon)}{\varepsilon} \rightharpoonup z_t \quad \text{weakly in } L^2(0, T; L^2(\Omega)) \text{ as } \varepsilon \to 0^+.$$

Step 2. Now let $z_0 \in L^1(\Omega)$ such that

$$\gamma_- \leq z_0 \leq \gamma_+, \quad \gamma_-|\Omega| < \int_\Omega z_0 < \gamma_+|\Omega|, \quad \int_\Omega j_\gamma^*(z_0) < +\infty,$$

and assume that

$$(5.33) \qquad \text{there exist } c_1 \text{ and } m_1 \in \gamma(c_1) \text{ with } \gamma_- < m_1 \leq z_0$$
$$\text{and (5.28) is not satisfied.}$$

We take $m_1 < 0$ if $\gamma_- < 0$. Let $z_{0n} \in L^\infty(\Omega)$, $z_{0n} \geq m_1$,

$$z_{0n} \nearrow z_0 \quad \text{as } n \text{ goes to } +\infty,$$

such that $\int_\Omega z_{0n} < \int_\Omega z_{0n+1}$ and $z_{0n} \leq m_2(n) < \gamma_+$, $m_2(n) \in \gamma(c_2(n))$ for some $c_2(n)$. By Step 1, there exists a solution z_n of problem (5.1) with initial datum z_{0n}, which is the mild solution of (5.21) with initial datum z_{0n} and satisfies $z_n \ll z_{0n}$. It is obvious that

$$\lim_{n\to\infty} z_n = z \quad \text{in } C([0,T] : L^1(\Omega)),$$

being z the mild solution of (5.21) with initial datum z_0, moreover $z \ll z_0$. Next we prove that z is the solution of (5.1).

Since z_n is a solution of problem (5.1) with initial datum z_{0n}, there exists $u_n \in L^2(0,T;L^2(\Omega))$, $z_n \in \gamma(u_n)$ a.e. in $\Omega \times (0,T)$, such that

$$(5.34) \qquad\qquad (z_n)_t - Au_n = 0.$$

Moreover, by Theorem 5.9, we can assume that

$$(5.35) \qquad\qquad u_n \text{ is nondecreasing in } n.$$

Multiplying (5.34) by u_n, we obtain

$$(5.36) \qquad \frac{d}{dt} \int_\Omega \left(\int_0^{z_n(t)} (\gamma^{-1})^0(s)ds \right) = \int_\Omega Au_n(t)u_n(t)dt$$

in $\mathcal{D}'(0,T)$. Indeed, since $u_n(t) \in \gamma^{-1}(z_n(t)) = \partial j_\gamma^*(z_n(t))$,

$$(z_n(t+\tau) - z_n(t))u_n(t) \leq \int_{z_n(t)}^{z_n(t+\tau)} (\gamma^{-1})^0(s)ds \quad \text{for all } \tau.$$

Consequently,

$$\int_\Omega (z_n)_t(t)u_n(t) = \frac{d}{dt} \int_\Omega \left(\int_0^{z_n(t)} (\gamma^{-1})^0(s)ds \right)$$

and (5.36) holds.

Integrating now (5.36) between 0 and T we get

$$(5.37) \qquad\qquad -\int_0^T \int_\Omega Au_n(t)\, u_n(t)\, dt \leq \int_\Omega j_\gamma^*(z_0).$$

Let us show that $\{u_n\}$ is bounded in $L^2(0,T;L^2(\Omega))$. In the case $\gamma_+ = +\infty$, let

$$M = \sup_{t\in[0,T]} \int_\Omega z^+(t) + 1$$

and $n_0 \in \mathbb{N}$ such that

$$\sup_{t\in[0,T]} \int_\Omega (z_n)^+(t) < M, \qquad \forall n \geq n_0.$$

In the case $\gamma_+ < +\infty$, since we have conservation of mass, there exist $M \in \mathbb{R}$ and $n_0 \in \mathbb{N}$ such that, for all $n \geq n_0$,

$$\sup_{t\in[0,T]} \int_\Omega z_n(t) < M < \gamma_+|\Omega|.$$

Moreover, since $z_n \ll z_{0n}$, we have that $z_n \geq m_1$. On the other hand,

$$(5.38) \quad \sup_{t\in[0,T]} \int_{\{x\in\Omega : z_n(t)(x) < -4(m_1^2+1)|\Omega|/(\gamma_+|\Omega|-M)\}} |z_n(t)| < \frac{\gamma_+|\Omega| - M}{4}, \; \forall n \in \mathbb{N}.$$

Indeed, (5.38) is obvious if $m_1 \geq 0$, and if $m_1 < 0$ it follows from the fact that $z_n \geq m_1$. Therefore, in both cases, by Lemma 5.6, there exists $C > 0$ such that

$$(5.39) \quad \|(u_n(t))^+\|_{L^2(\Omega)} \leq C \left(\left(- \int_\Omega A(u_n(t))^+ (u_n(t))^+ \right)^{1/2} + 1 \right), \quad \forall t \in [0, T].$$

Hence, by (5.37), $\{u_n\}$ is bounded in $L^2(0, T; L^2(\Omega))$. Since u_n is nondecreasing in n, passing to a subsequence if necessary, we can assume that

$$u_n \rightharpoonup u \quad \text{weakly in } L^2(0, T; L^2(\Omega)) \text{ as } n \to +\infty,$$

and, by (5.35),

$$u_n \to u \quad \text{in } L^2(0, T; L^2(\Omega)) \text{ as } n \to +\infty.$$

Consequently,

$$z \in \gamma(u) \quad \text{a.e. in } \Omega \times (0, T).$$

Since also $\{Au_n\}$ is bounded in $L^2(0, T; L^2(\Omega))$, passing to the limit in (5.34) we get

$$z_t - Au = 0.$$

Step 3. Let $z_0 \in L^1(\Omega)$, $\gamma_- \leq z_0 \leq \gamma_+$,

$$\gamma_- |\Omega| < \int_\Omega z_0 < \gamma_- |\Omega| \quad \text{and} \quad \int_\Omega j_\gamma^*(z_0) < +\infty$$

such that (5.33) is not satisfied. Let $z_{0n} \in L^\infty(\Omega)$,

$$z_{0n} \searrow z_0 \quad \text{as } n \text{ goes to } +\infty,$$

such that $\int_\Omega z_{0n} > \int_\Omega z_{0n+1}$ and $z_{0n} \geq m_1(n) > \gamma_-$, $m_1(n) \in \gamma(c_1(n))$ for some $c_1(n)$. By Step 2, there exists a solution z_n of problem (5.1) with initial datum z_{0n}, which is the mild solution of (5.21) with initial datum z_{0n} and satisfies $z_n \ll z_0$. It is obvious that

$$(5.40) \quad \lim_{n \to \infty} z_n = z \quad \text{in } C([0, T] : L^1(\Omega)),$$

z being the mild solution of (5.21) with initial datum z_0. Moreover $z \ll z_0$. We shall see that z is the solution of (5.1). The proof is similar to the above step and we only need to take care of the boundedness of $\{u_n\}$ in $L^2(0, T; L^2(\Omega))$. To this end we need a formula like (5.39) for u_n^-; that is, we need to prove that there exists $C > 0$ such that

$$(5.41) \quad \|(u_n(t))^-\|_{L^2(\Omega)} \leq C \left(\left(- \int_\Omega A(u_n(t))^- (u_n(t))^- \right)^{1/2} + 1 \right)$$

for every $t \in [0, T]$.

Assume first that $\gamma_- = -\infty$, and let

$$M = \sup_{t \in [0, T]} \int_\Omega z^-(t) + 1.$$

Then there exists $n_0 \in \mathbb{N}$ such that

$$\sup_{t \in [0, T]} \int_\Omega (z_n)^-(t) < M, \quad \forall n \geq n_0.$$

In the case $\gamma_- > -\infty$, there exist $M \in \mathbb{R}$, $h > 0$ and $n_0 \in \mathbb{N}$ such that, for all $n \geq n_0$,

$$\text{(5.42)} \qquad \inf_{t \in [0,T]} \int_\Omega z_n(t) > M > \gamma_-|\Omega|$$

and

$$\text{(5.43)} \qquad \sup_{t \in [0,T]} \int_{\{x \in \Omega : z_n(t)(x) > h\}} z_n(t) < \frac{M - \gamma_-|\Omega|}{4}.$$

Formula (5.42) is straightforward and (5.43) follows from (5.40). Indeed, by (5.40), there exist $n_0 \in \mathbb{N}$, $\delta > 0$ and $h > 0$ such that, for all $n \geq n_0$ and for all $t \in [0,T]$,

$$\int_E |z_n(t)| < \frac{M - \gamma_-|\Omega|}{4}, \qquad \forall E \subset \Omega, \ |E| < \delta,$$

and we can take h satisfying

$$|\{x \in \Omega : z_n(t)(x) > h\}| < \delta.$$

Therefore, by Lemma 5.6, (5.41) is proved in both cases.

Uniqueness of solutions follows from Theorem 5.15. $\square$

5.2. Rescaling the kernel. Convergence to the local problem

We show that solutions of problem (5.1), with the kernel J rescaled in a suitable way, converge to the solution of the local problem (5.2) as the rescaling parameter goes to zero.

Consider the rescaled kernels

$$J_\varepsilon(x) := \frac{C_{J,2}}{\varepsilon^{2+N}} J\left(\frac{x}{\varepsilon}\right),$$

where

$$C_{J,2}^{-1} := \frac{1}{2} \int_{\mathbb{R}^N} J(z)|z_N|^2 \, dz$$

is a normalizing constant as mentioned before.

Associated with these rescaled kernels are the solutions z_ε of the equation in (5.1) with initial condition z_0 and J replaced by J_ε. The main result now states that these functions z_ε converge strongly in $L^1(\Omega)$ to the solution of the local problem (5.2).

THEOREM 5.19. *Let Ω be a smooth bounded domain in $\mathbb{R}^N$. Suppose $J(x) \geq J(y)$ if $|x| \leq |y|$ and γ is a continuous nondecreasing function defined in all $\mathbb{R}$ with $\mathrm{R}(\gamma) = \mathbb{R}$. Let $T > 0$, $z_0 \in L^1(\Omega)$, $\int_\Omega j_\gamma^*(z_0) < +\infty$, and let z_ε be the unique solution of (5.1) with J replaced by J_ε. Then, if z is the unique solution of (5.2),*

$$\text{(5.44)} \qquad \lim_{\varepsilon \to 0} \sup_{t \in [0,T]} \|z_\varepsilon(.,t) - z(.,t)\|_{L^1(\Omega)} = 0.$$

Before giving the proof of the above theorem, we first recall some results about the local problem, obtained in [10], [8] and [11], that will be used in the proof of the convergence of the rescaled problems.

Associated to the local problem, we can define the operator $B_l^\gamma \subset L^1(\Omega) \times L^1(\Omega)$ as $(w, \hat{w}) \in B_l^\gamma$ if and only if $w \in L^1(\Omega)$, $\hat{w} \in L^1(\Omega)$ and there exists $v \in W^{1,2}(\Omega)$, $w \in \gamma(v)$ a.e. in Ω, such that

$$\int_\Omega \nabla v \cdot \nabla \xi = \int_\Omega \hat{w}\xi \quad \text{for every } \xi \in W^{1,2}(\Omega) \cap L^\infty(\Omega).$$

In [10] it is proved that B_l^γ is a T-accretive operator in $L^1(\Omega)$ with dense domain such that its closure $\mathcal{B}_l^\gamma$ in $L^1(\Omega) \times L^1(\Omega)$ is an m-T-accretive operator in $L^1(\Omega)$. Now, in [8] and [11], it is shown that for any initial datum $z_0 \in L^1(\Omega)$ satisfying the assumptions of Theorem 5.19, the unique mild solution $e^{-t\mathcal{B}_l^\gamma} z_0$ given by Crandall-Liggett's exponential formula is the unique weak/entropy solution of problem (5.2).

By Brezis-Pazy's Theorem (Theorem A.37), to prove Theorem 5.19 it is enough to show the convergence of the resolvents. To do that we need to apply a particular case of Theorem 6.11, given in Chapter 6. For a function g defined in a set D, we define

$$\overline{g}(x) = \begin{cases} g(x) & \text{if } x \in D, \\ 0 & \text{otherwise.} \end{cases}$$

We denote B^γ by $B^{J,\gamma}$ to emphasize the dependence on the kernel J.

PROPOSITION 5.20. *Under the hypothesis of Theorem 5.19, for any $\phi \in L^\infty(\Omega)$, we have that*

$$\left(I + B^{J_\varepsilon,\gamma}\right)^{-1} \phi \to \left(I + B_l^\gamma\right)^{-1} \phi \quad \text{in } L^2(\Omega) \text{ as } \varepsilon \to 0.$$

PROOF. For $\varepsilon > 0$, let $z_\varepsilon = \left(I + B^{J_\varepsilon,\gamma}\right)^{-1} \phi$. Then there exists $u_\varepsilon \in L^2(\Omega)$, $z_\varepsilon = \gamma(u_\varepsilon)$ a.e. in Ω, such that

$$(5.45) \qquad \int_\Omega z_\varepsilon \xi - \frac{C_{J,2}}{\varepsilon^{2+N}} \int_\Omega \int_\Omega J\left(\frac{x-y}{\varepsilon}\right)(u_\varepsilon(y) - u_\varepsilon(x))\, dy\, \xi(x)\, dx = \int_\Omega \phi\xi$$

for every $\xi \in L^\infty(\Omega)$.

Changing variables, we get

$$-\frac{C_{J,2}}{\varepsilon^{2+N}} \int_\Omega \int_\Omega J\left(\frac{x-y}{\varepsilon}\right)(u_\varepsilon(y) - u_\varepsilon(x))\, dy\, \xi(x)\, dx$$

$$(5.46)$$

$$= \int_{\mathbb{R}^N} \int_\Omega \frac{C_{J,2}}{2} J(z)\chi_\Omega(x + \varepsilon z)\frac{\overline{u}_\varepsilon(x + \varepsilon z) - u_\varepsilon(x)}{\varepsilon}\frac{\overline{\xi}(x + \varepsilon z) - \xi(x)}{\varepsilon}\, dx\, dz.$$

So we can write (5.45) as

$$\int_\Omega \phi(x)\xi(x)\, dx - \int_\Omega z_\varepsilon(x)\xi(x)\, dx$$

$$(5.47)$$

$$= \int_{\mathbb{R}^N} \int_\Omega \frac{C_{J,p}}{2} J(z)\chi_\Omega(x + \varepsilon z)\frac{\overline{u}_\varepsilon(x + \varepsilon z) - u_\varepsilon(x)}{\varepsilon}\frac{\overline{\xi}(x + \varepsilon z) - \xi(x)}{\varepsilon}\, dx\, dz.$$

We shall see that there exists a sequence $\varepsilon_n \to 0$ such that $z_{\varepsilon_n} \to w$ weakly in $L^1(\Omega)$, $w \in L^1(\Omega)$ and $w = \left(I + B_l^\gamma\right)^{-1} \phi$; that is, there exists $v \in W^{1,2}(\Omega)$,

$w \in \gamma(v)$ a.e. in Ω, such that

$$\int_\Omega w\xi + \int_\Omega \nabla v \cdot \nabla \xi = \int_\Omega \phi\xi \quad \text{for every } \xi \in W^{1,2}\Omega) \cap L^\infty(\Omega).$$

Since $z_\varepsilon \ll \phi$, there exists a sequence $\varepsilon_n \to 0$ such that

(5.48) $$z_{\varepsilon_n} \rightharpoonup w, \quad \text{weakly in } L^2(\Omega), \quad w \ll \phi.$$

Observe that $\|z_{\varepsilon_n}\|_{L^\infty(\Omega)}, \|w\|_{L^\infty(\Omega)} \le \|\phi\|_{L^\infty(\Omega)}$ and also that $\{u_{\varepsilon_n}\}$ is bounded in $L^\infty(\Omega)$. Taking $\varepsilon = \varepsilon_n$ and $\xi = u_{\varepsilon_n}$ in (5.47) we get

(5.49)
$$\int_\Omega \int_\Omega \frac{1}{2} \frac{C_{J,2}}{\varepsilon_n^N} J\left(\frac{x-y}{\varepsilon_n}\right) \left| \frac{u_{\varepsilon_n}(y) - u_{\varepsilon_n}(x)}{\varepsilon_n} \right|^2 dx\, dy$$

$$= \int_{\mathbb{R}^N} \int_\Omega \frac{C_{J,2}}{2} J(z) \chi_\Omega(x + \varepsilon_n z) \left| \frac{\overline{u}_{\varepsilon_n}(x + \varepsilon_n z) - u_{\varepsilon_n}(x)}{\varepsilon_n} \right|^2 dx\, dz \le M.$$

Therefore, by Theorem 6.11, there exists a subsequence, also denoted by $\{u_{\varepsilon_n}\}$, such that

(5.50) $$u_{\varepsilon_n} \to v \quad \text{in } L^2(\Omega),$$

$v \in W^{1,2}(\Omega)$, and

(5.51)
$$\left(\frac{C_{J,2}}{2} J(z)\right)^{1/2} \chi_\Omega(x + \varepsilon_n z) \frac{\overline{u}_{\varepsilon_n}(x + \varepsilon_n z) - u_{\varepsilon_n}(x)}{\varepsilon_n}$$

$$\rightharpoonup \left(\frac{C_{J,2}}{2} J(z)\right)^{1/2} z \cdot \nabla v(x)$$

weakly in $L^2(\Omega) \times L^2(\mathbb{R}^N)$. Therefore, passing to the limit in (5.47) for $\varepsilon = \varepsilon_n$, we get

$$\int_\Omega w\xi + \int_{\mathbb{R}^N} \int_\Omega \frac{C_{J,2}}{2} J(z) z \cdot \nabla v(x)\, z \cdot \nabla \xi(x)\, dx\, dz = \int_\Omega \phi\xi$$

for every smooth ξ and by approximation for every $\xi \in W^{1,2}(\Omega)$. That is,

$$\int_\Omega w\xi + \int_\Omega \mathbf{a}(\nabla v) \cdot \nabla \xi = \int_\Omega \phi\xi \quad \text{for every } \xi \in W^{1,2}(\Omega),$$

where

$$\mathbf{a}_j(\xi) = C_{J,2} \int_{\mathbb{R}^N} \frac{1}{2} J(z) z \cdot \xi\, z_j\, dz = \xi_j;$$

see Lemma 6.16.

Finally, thanks to (5.48) and (5.50) and the hypothesis on γ, we obtain that $w = \gamma(v)$ a.e. in Ω and that

$$z_{\varepsilon_n} \to w \quad \text{in } L^2(\Omega). \qquad \square$$

PROOF OF THEOREM 5.19. Since, by Corollary 5.14, $\overline{B^{J,\gamma}}^{L^1(\Omega) \times L^1(\Omega)}$ is m-T-accretive, to get (5.44) it is enough to show that

$$\left(I + B_p^{J_{p,\varepsilon}}\right)^{-1} \phi \to (I + B_p)^{-1} \phi \quad \text{in } L^1(\Omega) \text{ as } \varepsilon \to 0$$

for any $\phi \in L^\infty(\Omega)$ (see Theorem A.37), which holds by Proposition 5.20. $\qquad \square$

5.3. Asymptotic behaviour

In this section we study the asymptotic behaviour of the solutions of (5.1). Note that since the solution preserves the total mass, it is natural to expect that solutions of the diffusion problem converge to the mean value of the initial condition as $t \to \infty$. We shall see that this is the case, for instance, when γ is a continuous function; nevertheless this fails when γ has jumps.

Let us recall the ω-limit set for a given initial condition z_0,

$$\omega(z_0) = \left\{ w \in L^1(\Omega) \, : \, \exists t_n \to \infty \text{ with } S(t_n)z_0 \to w, \text{ strongly in } L^1(\Omega) \right\},$$

and the weak ω-limit set

$$\omega_\sigma(z_0) = \left\{ w \in L^1(\Omega) \, : \, \exists t_n \to \infty \text{ with } S(t_n)z_0 \rightharpoonup w, \text{ weakly in } L^1(\Omega) \right\}.$$

Since $S(t)z_0 \ll z_0$, $\omega_\sigma(z_0) \neq \emptyset$ always. Moreover, since $S(t)$ preserves the total mass, for all $w \in \omega_\sigma(z_0)$,

$$\int_\Omega w = \int_\Omega z_0.$$

We denote by F the set of fixed points of the semigroup $(S(t))$, that is,

$$F = \left\{ w \in \overline{D(B^\gamma)}^{L^1(\Omega)} \, : \, S(t)w = w \quad \forall \, t \geq 0 \right\}.$$

It is easy to see that

$$(5.52) \qquad F = \left\{ w \in L^1(\Omega) \, : \, \exists k \in D(\gamma) \text{ such that } w \in \gamma(k) \right\}.$$

THEOREM 5.21. *Let $z_0 \in L^1(\Omega)$ such that $\gamma_- \leq z_0 \leq \gamma_+$, $\gamma_- < \frac{1}{|\Omega|} \int_\Omega z_0 < \gamma_+$ and $\int_\Omega j_\gamma^*(z_0) < +\infty$. Then $\omega_\sigma(z_0) \subset F$. Moreover, if $\omega(z_0) \neq \emptyset$, then $\omega(z_0)$ consists of a unique $w \in F$, and consequently,*

$$\lim_{t \to \infty} S(t)z_0 = w \quad \text{strongly in } L^1(\Omega).$$

PROOF. In this proof we denote by $z(t) = S(t)z_0$ the solution of problem (5.1) and by $u(t)$ the corresponding function that appears in Definition 5.1.

Multiplying the equation in (5.1) by $u(t)$, integrating and arguing as in the proof of (5.37), we deduce that

$$(5.53) \qquad -\int_0^{+\infty} \int_\Omega Au(t)\, u(t)\, dt \leq \int_\Omega j_\gamma^*(z_0).$$

Therefore, thanks to (5.3), we obtain that there exists a constant C such that

$$(5.54) \qquad \int_0^{+\infty} \int_\Omega \left| u(t) - \frac{1}{|\Omega|} \int_\Omega u(t) \right|^2 dt \leq C.$$

Let $w \in \omega_\sigma(z_0)$; then there exists a sequence $t_n \to +\infty$ such that $z(t_n) \rightharpoonup w$. By (5.54), we have

$$\alpha_n := \int_{t_n}^{+\infty} \int_\Omega \left| u(t) - \frac{1}{|\Omega|} \int_\Omega u(t) \right|^2 dt \to 0.$$

Take $s_n \to 0$ such that

$$(5.55) \qquad \lim_{n \to \infty} \frac{\alpha_n}{s_n} = 0.$$

By contradiction it is easy to see that there exists $\bar{t}_n \in [t_n, t_n + \frac{C}{s_n}]$ such that

$$\int_\Omega \left| u(\bar{t}_n) - \frac{1}{|\Omega|} \int_\Omega u(\bar{t}_n) \right|^2 \leq s_n,$$

which implies that

(5.56)
$$\int_\Omega \left| u(\bar{t}_n) - \frac{1}{|\Omega|} \int_\Omega u(\bar{t}_n) \right|^2 \to 0.$$

Let us prove that

$$\frac{1}{|\Omega|} \int_\Omega u(\bar{t}_n)$$

is bounded. In fact, suppose there exists a subsequence, also denoted by $\{u(\bar{t}_n)\}$, such that

$$\frac{1}{|\Omega|} \int_\Omega u(\bar{t}_n) \to +\infty.$$

By (5.56) we get that $u(\bar{t}_n) \to +\infty$ a.e. Since $z(\bar{t}_n) \in \gamma(u(\bar{t}_n))$, then $z(\bar{t}_n) \to \gamma_+$ a.e. Moreover, as $z(\bar{t}_n) \ll z_0$ and $\gamma^+ \geq 0$, we can deduce that $\lim_{n\to\infty} z(\bar{t}_n) = \lim_{n\to\infty} z(\bar{t}_n)^+$ weakly in $L^1(\Omega)$. Hence, applying Fatou's Lemma, we get

$$\int_\Omega z_0 = \lim_{n\to\infty} \int_\Omega z(\bar{t}_n)^+ \geq \gamma_+ |\Omega|,$$

a contradiction. A similar argument shows that $\frac{1}{|\Omega|} \int_\Omega u(\bar{t}_n)$ is bounded from below. Therefore, passing to a subsequence if necessary, we may assume that

$$\frac{1}{|\Omega|} \int_\Omega u(\bar{t}_n) \to k$$

for some constant k. Using again (5.56),

(5.57)
$$u(\bar{t}_n) \to k \quad \text{strongly in } L^2(\Omega) \text{ and a.e.}$$

As $z(\bar{t}_n) \ll z_0$, we can assume, taking a subsequence if necessary, that $z(\bar{t}_n) \rightharpoonup \hat{w}$ weakly in $L^1(\Omega)$. Then, from (5.57) it follows that $\hat{w} \in \gamma(k)$, and consequently $\hat{w} \in F$. Let us show now that $w = \hat{w}$. By (5.55), we have

$$\|z(\bar{t}_n) - z(t_n)\|_{L^1(\Omega)} = \left\| \int_{t_n}^{\bar{t}_n} z_t(s)\, ds \right\|_{L^1(\Omega)} = \left\| \int_{t_n}^{\bar{t}_n} Au(s)\, ds \right\|_{L^1(\Omega)}$$

$$\leq M(\bar{t}_n - t_n)^{1/2} \left(\int_{t_n}^{+\infty} \int_\Omega \left| u(s) - \frac{1}{|\Omega|} \int_\Omega u(s) \right|^2 ds \right)^{1/2} \leq M \left(C\, \frac{\alpha_n}{s_n} \right)^{1/2} \to 0,$$

where M is a constant depending on $|\Omega|$. Therefore, taking the limit, we get $z(\bar{t}_n) - z(t_n) \to 0$ strongly in $L^1(\Omega)$, and since it converges weakly to $\hat{w} - w$, it follows that $w = \hat{w}$, which is a fixed point. Finally, if $\omega(z_0) \neq \emptyset$, then, since $\omega(z_0) \subset \omega_\sigma(z_0) \subset F$ and $(S(t))$ is a contraction semigroup, we have that $\omega(z_0) = \{w\} \subset F$ and

$$\lim_{t\to\infty} S(t)z_0 = w \quad \text{strongly in } L^1(\Omega). \qquad \square$$

REMARK 5.22. Note that in order to prove that $\omega(z_0) \neq \emptyset$, a usual tool is to show that the resolvent of B^γ is compact. In our case this fails in general as the following example shows. Let γ be any maximal monotone graph with $\gamma(0) = [0,1]$, and $z_n \in L^\infty(\Omega)$, $0 \le z_n \le 1$ such that $\{z_n\}$ is not relatively compact in $L^1(\Omega)$. It is easy to check that $z_n = (I + B^\gamma)^{-1}(z_n)$. Hence $(I + B^\gamma)^{-1}$ is not a compact operator in $L^1(\Omega)$. On the other hand, since the nonlocal operator does not have regularizing effects, here we cannot prove regularity properties of the solutions that would help to find compactness of the orbits. Nevertheless, we shall see in the next result that when γ is a continuous function, we are able to prove that $\omega(z_0) \neq \emptyset$.

Let us now consider some cases in which $\omega(z_0) \neq \emptyset$ and

$$\lim_{t\to\infty} S(t)z_0 = \frac{1}{|\Omega|} \int_\Omega z_0 \quad \text{strongly in } L^1(\Omega).$$

Given a maximal monotone graph γ in $\mathbb{R} \times \mathbb{R}$, we set

$$\gamma(r+) := \inf \gamma((r, +\infty)), \quad \gamma(r-) := \sup \gamma((-\infty, r))$$

for $r \in \mathbb{R}$, where we use the conventions $\inf \emptyset = +\infty$ and $\sup \emptyset = -\infty$. It is easy to see that

$$\gamma(r) = [\gamma(r-), \gamma(r+)] \cap \mathbb{R} \quad \text{for} \quad r \in \mathbb{R}.$$

Moreover, $\gamma(r-) = \gamma(r+)$ except at a countable set of points, which we denote by $\mathbb{J}(\gamma)$.

COROLLARY 5.23. *Let $z_0 \in L^1(\Omega)$ such that $\gamma_- \le z_0 \le \gamma_+$, $\gamma_- < \frac{1}{|\Omega|} \int_\Omega z_0 < \gamma_+$ and $\int_\Omega j_\gamma^*(z_0) < +\infty$. The following statements hold.*

(1) *If $\frac{1}{|\Omega|} \int_\Omega z_0 \notin \gamma(\mathbb{J}(\gamma))$ or $\frac{1}{|\Omega|} \int_\Omega z_0 \in \{\gamma(k+), \gamma(k-)\}$ for some $k \in \mathbb{J}(\gamma)$, then*

$$\lim_{t\to\infty} S(t)z_0 = \frac{1}{|\Omega|} \int_\Omega z_0 \quad \text{strongly in } L^1(\Omega).$$

(2) *If γ is a continuous function, then*

$$\lim_{t\to\infty} S(t)z_0 = \frac{1}{|\Omega|} \int_\Omega z_0 \quad \text{strongly in } L^1(\Omega).$$

(3) *If $\frac{1}{|\Omega|} \int_\Omega z_0 \in]\gamma(k-), \gamma(k+)[$ for some $k \in \mathbb{J}(\gamma)$, then*

$$\omega_\sigma(z_0) \subset \left\{ w \in L^1(\Omega) : w \in [\gamma(k-), \gamma(k+)] \quad a.e., \quad \int_\Omega w = \int_\Omega z_0 \right\},$$

and consequently, for any $w \in \omega_\sigma(z_0)$, there exists a nonnull set in which $w \in (\gamma(k-), \gamma(k+))$.

PROOF. (1) Throughout this proof we denote by $z(t) = S(t)z_0$ the solution of problem (5.1) and by $u(t)$ the corresponding function that appears in Definition 5.1. First, let us assume that $\frac{1}{|\Omega|} \int_\Omega z_0 \notin \gamma(\mathbb{J}(\gamma))$ and $z_0 \in L^\infty(\Omega)$. Working as in the above theorem, we have that there exists a constat k such that

(5.58) $u(t_n) \to k \quad \text{strongly in } L^2(\Omega) \text{ and a.e.}$

Since $z(t_n) \ll z_0$, there exists a subsequence such that $z(t_n) \rightharpoonup w$ weakly in $L^1(\Omega)$. Now, from $z(t_n) \in \gamma(u(t_n))$ we deduce that $w \in \gamma(k)$ and consequently, since $\frac{1}{|\Omega|} \int_\Omega z_0 = \frac{1}{|\Omega|} \int_\Omega w$, $k \notin \mathbb{J}(\gamma)$. Then there exists $\delta > 0$ such that γ is single-valued and continuous on $]k - \delta, k + \delta[$. Hence, $w = \gamma(k)$ and $z(t_n) \to \gamma(k)$ a.e. Therefore, since $z(t_n)$ is bounded in $L^\infty(\Omega)$, $z(t_n) \to \gamma(k) = \frac{1}{|\Omega|} \int_\Omega z_0$ strongly in $L^1(\Omega)$. By the above theorem we get that

$$z(t) \to \frac{1}{|\Omega|} \int_\Omega z_0 \quad \text{as } t \to \infty.$$

The general case $z_0 \in L^1(\Omega)$ follows easily from the previous arguments using again that we deal with a contraction semigroup.

Assume now that $\frac{1}{|\Omega|} \int_\Omega z_0 \in \{\gamma(k+), \gamma(k-)\}$ for some $k \in \mathbb{J}(\gamma)$. It is easy to see that we can find $z_{0,n} \in L^1(\Omega)$, with $\gamma_- \leq z_{0,n} \leq \gamma_+$, $\gamma_- < \frac{1}{|\Omega|} \int_\Omega z_{0,n} < \gamma_+$ and $\int_\Omega j_\gamma^*(z_{0,n}) < +\infty$, such that $z_{0,n} \to z_0$ strongly in $L^1(\Omega)$ and verifies that $\frac{1}{|\Omega|} \int_\Omega z_{0,n} \notin \gamma(\mathbb{J}(\gamma))$ for all n. By the above step, we have

$$S(t)z_{0,n} \to \frac{1}{|\Omega|} \int_\Omega z_{0,n} \quad \text{strongly in } L^1(\Omega),$$

from which it follows, using again that $(S(t))$ is a contraction semigroup, that

$$S(t)z_0 \to \frac{1}{|\Omega|} \int_\Omega z_0 \quad \text{strongly in } L^1(\Omega).$$

Statement (2) is an obvious consequence of (1) since in this case $\mathbb{J}(\gamma) = \emptyset$.

Finally, we prove (3). Given $w \in \omega_\sigma(z_0)$, by Theorem 5.21, there exists $k_0 \in D(\gamma)$ such that $w \in \gamma(k_0)$. Then $k_0 = k$. In fact, if we assume, for instance, that $k_0 < k$, then

$$\gamma(k_0+) \geq \frac{1}{|\Omega|} \int_\Omega w = \frac{1}{|\Omega|} \int_\Omega z_0 > \gamma(k-) > \gamma(k_0+),$$

a contradiction. Hence, we have $w \in \gamma(k)$, and

$$\frac{1}{|\Omega|} \int_\Omega w = \frac{1}{|\Omega|} \int_\Omega z_0 \in (\gamma(k-), \gamma(k+)).$$

Thus, $w \in [\gamma(k-), \gamma(k+)]$ a.e. and, moreover, there exists a nonnull set in which $w \in (\gamma(k-), \gamma(k+))$. □

REMARK 5.24. An alternative proof of the fact that $\omega(z_0) \subset F$ is the following. Let

$$\Psi : L^1(\Omega) \to (-\infty, +\infty]$$

be the functional defined by

$$\Psi(z) := \begin{cases} \int_\Omega j_\gamma^*(z) & \text{if } j_\gamma^*(z) \in L^1(\Omega), \\ +\infty & \text{if } j_\gamma^*(z) \notin L^1(\Omega). \end{cases}$$

Since j_γ^* is continuous and convex, Ψ is lower semicontinuous ([58], p. 160). Moreover, since $S(t)z_0 \ll z_0$ for all $t \geq 0$, we have $\Psi(S(t)z_0) \leq \Psi(z_0)$ for all $t \geq 0$. Therefore, Ψ is a lower semicontinuous Liapunov functional for $(S(t))$. Then, by the Invariance Principle of Dafermos ([91]), Ψ is constant on $\omega(z_0)$. Consequently, given $w_0 \in \omega(z_0)$, if $w(t) = S(t)w_0$, we have that $\Psi(w(t))$ is constant for all $t \geq 0$.

Let $u(t)$ be such that $w(t) \in \gamma(u(t))$ and $w_t = A(u(t))$. Working as in the proof of (5.36), we get

$$0 = \frac{d}{dt}\Psi(w(t)) = \frac{d}{dt}\int_\Omega j_\gamma^*(w(t)) = \frac{d}{dt}\int_\Omega j_{\gamma^{-1}}(w(t)) = \int_\Omega Au(t)\,u(t).$$

Then, by Proposition 5.3, we obtain that

$$u(t) = \frac{1}{|\Omega|}\int_\Omega u(t).$$

Hence, $w(t) \in F$ for all $t > 0$, and consequently, $w_0 \in F$.

Bibliographical notes

Most of the results of this chapter are taken from [14], the only exception being the convergence to the local problem when the kernel is rescaled, which is an original result in this monograph. The main ingredient for the proof of this result is Theorem 6.11 (proved in Chapter 6) that is a precompactness lemma inspired by a result due to Bourgain, Brezis and Mironescu, [52].

Concerning nonlocal analogs of the porous medium equation we quote [76] and [77], where the authors introduce a nonlocal equation whose solutions have finite propagation speed and it is proved that a scaling procedure such as the one performed here converges to a solution of the porous medium equation. In [49] and [92] a fractional porous medium equation is studied.

CHAPTER 6

Nonlocal p-Laplacian evolution problems

The purpose of this chapter is to study a model type of nonlocal nonlinear diffusion problems which we call *nonlocal p-Laplacian problems*, either with homogeneous Neumann boundary conditions or nonhomogeneous Dirichlet boundary conditions. Moreover, the Cauchy problem for the nonlocal p-Laplacian is also studied.

These problems can be written in the form

$$\begin{cases} u_t(x,t) = \displaystyle\int_D J(x-y)|u(y,t) - u(x,t)|^{p-2}(u(y,t) - u(x,t))\,dy, \\ u(x,0) = u_0(x), \qquad\qquad\qquad\qquad\qquad\quad x \in \Omega,\ t > 0, \end{cases}$$

where, as for the linear model (see Chapters 1, 2 and 3), if $D = \Omega \neq \mathbb{R}^N$, we are considering homogeneous Neumann boundary conditions, if $\Omega \neq D = \mathbb{R}^N$ and $u = \psi$ in $\mathbb{R}^N \setminus \Omega$, we are dealing with nonhomogeneous Dirichlet boundary conditions, and in the case $D = \Omega = \mathbb{R}^N$, the Cauchy problem is considered.

Here, as in the previous chapters, $J : \mathbb{R}^N \to \mathbb{R}$ is a nonnegative continuous radial function with compact support, $J(0) > 0$ and $\int_{\mathbb{R}^N} J(x)dx = 1$ (this last condition is not necessary to prove the results of this chapter, it is imposed just for normalization) and p is a fixed but arbitrary number, with

$$1 < p < +\infty.$$

The cases $p = 1$ and $p = +\infty$ are different and more delicate; they will be treated in subsequent chapters.

When dealing with local evolution equations, two models of nonlinear diffusion have been extensively studied in the literature, the porous medium equation, $v_t = \Delta\left(|v|^{m-1}v\right)$, and the p-Laplacian evolution, $v_t = \text{div}\left(|\nabla v|^{p-2}\nabla v\right)$. For the first case, a nonlocal analogous equation was studied in Chapter 5. Our main objective now is to study a nonlocal analog of the p-Laplacian evolution. Together with the study of existence and uniqueness, we prove that, if the kernel J is rescaled in an appropriate way, the corresponding solutions of the nonlocal p-Laplacian evolution problems converge strongly in $L^\infty((0,T); L^p(\Omega))$ to the solution of the local p-Laplacian evolution problem. We will also study the asymptotic behaviour of some of the nonlocal evolution problems.

We finish this introduction by emphasizing some facts. The first is that, in contrast with local problems, in general, there is no regularizing effects for solutions of the nonlocal problems (see Chapters 1 and 4). The second is that when $p \neq 2$, these problems are nonlinear and hence the proofs of convergence, under rescaling, are different from the ones given in the previous chapters for the linear case $p = 2$.

Moreover, the treatment of the existence and uniqueness of solutions for these nonlocal problems differs from the one given in the linear case.

6.1. The Neumann problem

The first problem we treat is the nonlocal p-Laplacian problem with homogeneous Neumann boundary conditions

$$(6.1) \qquad \begin{cases} u_t(x,t) = \displaystyle\int_\Omega J(x-y)|u(y,t)-u(x,t)|^{p-2}(u(y,t)-u(x,t))\,dy, \\ u(x,0) = u_0(x), \hspace{5.5cm} x \in \Omega,\ t>0, \end{cases}$$

where $\Omega \subset \mathbb{R}^N$ is a bounded domain.

Solutions to (6.1) will be understood in the following sense.

DEFINITION 6.1. A *solution* of (6.1) in $[0,T]$ is a function

$$u \in W^{1,1}(0,T;L^1(\Omega))$$

that satisfies $u(x,0) = u_0(x)$ a.e. $x \in \Omega$ and

$$u_t(x,t) = \int_\Omega J(x-y)|u(y,t)-u(x,t)|^{p-2}(u(y,t)-u(x,t))\,dy$$

a.e. in $\Omega \times (0,T)$.

Let us note that, with this definition of solution, the evolution problem (6.1) is the gradient flow associated to the functional

$$J_p(u) = \frac{1}{2p}\int_\Omega\int_\Omega J(x-y)|u(y)-u(x)|^p\,dy\,dx,$$

which is the nonlocal analog of the energy functional associated to the local p-Laplacian

$$F_p(v) = \frac{1}{p}\int_\Omega |\nabla v|^p.$$

The main result on existence and uniqueness of a global solution, that is, a solution in $[0,T]$ for any $T>0$, for this problem is stated in the following theorem where we also state two contraction principles.

THEOREM 6.2. *Suppose $p>1$ and let $u_0 \in L^p(\Omega)$. Then, for any $T>0$, there exists a unique solution to (6.1).*

Moreover, if $u_{i0} \in L^1(\Omega)$, and u_i is a solution in $[0,T]$ of (6.1) with initial data u_{i0}, $i=1,2$, respectively, then

$$\int_\Omega (u_1(t)-u_2(t))^+ \le \int_\Omega (u_{10}-u_{20})^+ \quad \textit{for every } t \in [0,T].$$

If $u_{i0} \in L^p(\Omega)$, $i=1,2$, then

$$\|u_1(t)-u_2(t)\|_{L^p(\Omega)} \le \|u_{10}-u_{20}\|_{L^p(\Omega)} \quad \textit{for every } t \in [0,T].$$

6.1.1. Existence and uniqueness. We study the problem (6.1) from the point of view of Nonlinear Semigroup Theory. For this we introduce the following operator in $L^1(\Omega)$ associated with our problem.

DEFINITION 6.3. Let B_p^J be defined by

$$B_p^J u(x) = - \int_\Omega J(x-y)|u(y) - u(x)|^{p-2}(u(y) - u(x))\, dy, \qquad x \in \Omega.$$

REMARK 6.4. It is easy to see that

1. B_p^J is positively homogeneous of degree $p-1$;

2. $L^{p-1}(\Omega) \subset D(B_p^J)$ if $p > 2$;

3. For $1 < p \le 2$, $D(B_p^J) = L^1(\Omega)$ and B_p^J is closed in $L^1(\Omega) \times L^1(\Omega)$.

For this kind of operator the following integration formula, which plays the same role as the integration by parts formula for the local p-Laplacian, is straightforward.

LEMMA 6.5. *For every* $u, v \in L^p(\Omega)$,

$$-\int_\Omega \int_\Omega J(x-y)|u(y) - u(x)|^{p-2}(u(y) - u(x))dy\, v(x)\, dx$$

$$= \frac{1}{2}\int_\Omega \int_\Omega J(x-y)|u(y) - u(x)|^{p-2}(u(y) - u(x))(v(y) - v(x))\, dy\, dx.$$

From this lemma the following monotonicity result can be deduced.

LEMMA 6.6. *Let* $T \colon \mathbb{R} \to \mathbb{R}$ *be a nondecreasing function. Then*

(i) *for every* $u,\ v \in L^p(\Omega)$ *such that* $T(u - v) \in L^p(\Omega)$, *we have*

$$\int_\Omega (B_p^J u(x) - B_p^J v(x))\, T(u(x) - v(x))dx$$

(6.2)
$$= \frac{1}{2}\int_\Omega \int_\Omega J(x-y)\left(T(u(y) - v(y)) - T(u(x) - v(x))\right)$$

$$\times \left(|u(y) - u(x)|^{p-2}(u(y) - u(x)) - |v(y) - v(x)|^{p-2}(v(y) - v(x))\right)\, dy\, dx.$$

(ii) *Moreover, if* T *is bounded,* (6.2) *holds for every* $u,\ v \in D(B_p^J)$.

In the next theorem we prove that B_p^J is completely accretive (see Section A.8) and satisfies the range condition $L^p(\Omega) \subset \mathrm{R}(I + B_p^J)$. In short, this means that for any $\phi \in L^p(\Omega)$ there is a unique solution of the problem $u + B_p^J u = \phi$ and the resolvent $(I + B_p^J)^{-1}$ is a contraction in $L^q(\Omega)$ for all $1 \le q \le +\infty$.

THEOREM 6.7. *The operator* B_p^J *is completely accretive and satisfies the range condition*

(6.3)
$$L^p(\Omega) \subset \mathrm{R}(I + B_p^J).$$

PROOF. Given $u_i \in D(B_p^J)$, $i = 1, 2$, and $q \in P_0$, that is, $q \in C^\infty(\mathbb{R})$, $0 \le q' \le 1$, $\mathrm{supp}(q')$ is compact, $0 \notin \mathrm{supp}(q)$, by the monotonicity Lemma 6.6, we have

$$\int_\Omega (B_p^J u_1(x) - B_p^J u_2(x))q(u_1(x) - u_2(x))\, dx \ge 0,$$

from which it follows that B_p^J is a completely accretive operator.

To show that B_p^J satisfies the range condition we have to prove that for any $\phi \in L^p(\Omega)$ there exists $u \in D(B_p^J)$ such that $u = (I + B_p^J)^{-1}\phi$. Let us first take $\phi \in L^\infty(\Omega)$. Let $A_{n,m} : L^p(\Omega) \to L^{p'}(\Omega)$ be the continuous monotone operator defined by

$$A_{n,m}(u) := T_c(u) + B_p^J u + \frac{1}{n}|u|^{p-2}u^+ - \frac{1}{m}|u|^{p-2}u^-,$$

where $T_c(r) = c \wedge (r \vee (-c))$, $c \geq 0$, $r \in \mathbb{R}$. We have that $A_{n,m}$ is coercive in $L^p(\Omega)$. In fact,

$$\lim_{\|u\|_{L^p(\Omega)} \to +\infty} \frac{\int_\Omega A_{n,m}(u)u}{\|u\|_{L^p(\Omega)}} = +\infty.$$

Then, by [**55**, Corollary 30], there exists $u_{n,m} \in L^p(\Omega)$ such that

$$T_c(u_{n,m}) + B_p^J u_{n,m} + \frac{1}{n}|u_{n,m}|^{p-2}u_{n,m}^+ - \frac{1}{m}|u_{n,m}|^{p-2}u_{n,m}^- = \phi.$$

Using the monotonicity of $B_p^J u_{n,m} + \frac{1}{n}|u_{n,m}|^{p-2}u_{n,m}^+ - \frac{1}{m}|u_{n,m}|^{p-2}u_{n,m}^-$, we obtain that $T_c(u_{n,m}) \ll \phi$ (see A.8). Consequently, taking $c > \|\phi\|_{L^\infty(\Omega)}$, we see that $u_{n,m} \ll \phi$ and

$$u_{n,m} + B_p^J u_{n,m} + \frac{1}{n}|u_{n,m}|^{p-2}u_{n,m}^+ - \frac{1}{m}|u_{n,m}|^{p-2}u_{n,m}^- = \phi.$$

Moreover, since $u_{n,m}$ is increasing in n and decreasing in m, as $u_{n,m} \ll \phi$, we can pass to the limit as $n \to \infty$ (using the monotone convergence to handle the term $B_p^J u_{n,m}$) obtaining that u_m is a solution to

$$u_m + B_p^J u_m - \frac{1}{m}|u_m|^{p-2}u_m^- = \phi,$$

and $u_m \ll \phi$.

Since u_m is decreasing in m, we can pass again to the limit to obtain

$$u + B_p^J u = \phi.$$

Now let $\phi \in L^p(\Omega)$. Take $\phi_n \in L^\infty(\Omega)$, $\phi_n \to \phi$ in $L^p(\Omega)$. Then, by the previous step, there exists $u_n = (I + B_p^J)^{-1}\phi_n$. Since B_p^J is completely accretive, $u_n \to u$ in $L^p(\Omega)$ and also $B_p^J u_n \to B_p^J u$ in $L^{p'}(\Omega)$. We conclude that $u + B_p^J u = \phi$. $\square$

If $\mathcal{B}_p^J$ denotes the closure of B_p^J in $L^1(\Omega)$, then by Theorem 6.7 we obtain that $\mathcal{B}_p^J$ is m-completely accretive in $L^1(\Omega)$ (see A.8).

As a consequence of the above results we get, by using Nonlinear Semigroup Theory (see the Appendix), the following theorem, from which Theorem 6.2 can be derived.

THEOREM 6.8. *Let $T > 0$ and $u_0 \in L^1(\Omega)$. Then there exists a unique mild solution u of*

(6.4)
$$\begin{cases} u'(t) + B_p^J u(t) = 0, & t \in (0, T), \\ u(0) = u_0. \end{cases}$$

Moreover,

(1) *if $u_0 \in L^p(\Omega)$, the unique mild solution of (6.4) is a solution of $P_p^J(u_0)$ in the sense of Definition 6.1. If $1 < p \leq 2$, this is true for any $u_0 \in L^1(\Omega)$.*

(2) *Let $u_{i0} \in L^1(\Omega)$ and let u_i be a solution in $[0,T]$ of (6.1) with initial data u_{i0}, for $i = 1, 2$, respectively. Then*

$$\int_\Omega (u_1(t) - u_2(t))^+ \leq \int_\Omega (u_{10} - u_{20})^+ \quad \text{for every } t \in [0, T].$$

Moreover, for $q \in [1, +\infty]$, if $u_{i0} \in L^q(\Omega)$, $i = 1, 2$, then

$$\|u_1(t) - u_2(t)\|_{L^q(\Omega)} \leq \|u_{10} - u_{20}\|_{L^q(\Omega)} \quad \text{for every } t \in [0, T].$$

PROOF. As a consequence of Theorems 6.7 and A.29 we get the existence of a mild solution of (6.4). On the other hand, $u(t)$ is a solution of (6.1) if and only if $u(t)$ is a strong solution of the abstract Cauchy problem (6.4). Now, $u(t)$ is a strong solution under the hypothesis of the theorem thanks to the complete accretivity of B_p^J and the range condition (6.3) (Proposition A.35). Moreover, in the case $1 < p \leq 2$, since $D(B_p^J) = L^1(\Omega)$ and B_p^J is closed in $L^1(\Omega) \times L^1(\Omega)$, the result holds for every initial datum in $L^1(\Omega)$ (Corollary A.52). Finally, the contraction principle is a consequence of Theorem A.28. □

REMARK 6.9. The above results can be extended, with minor modifications, to obtain existence and uniqueness of

$$\begin{cases} u_t(x, t) = \int_\Omega J(x, y)|u(y, t) - u(x, t)|^{p-2}(u(y, t) - u(x, t)) \, dy, \\ u(x, 0) = u_0(x), \qquad\qquad\qquad\qquad\qquad\qquad\qquad x \in \Omega, \, t > 0, \end{cases}$$

with J symmetric, that is, $J(x, y) = J(y, x)$, bounded and nonnegative.

REMARK 6.10 (Deblurring and denoising of images). S. Kindermann, S. Osher and P. W. Jones in [129] have studied deblurring and denoising of images by nonlocal functionals, motivated by the use of neighborhood filters [61]. Such filters have originally been proposed by Yaroslavsky, [153], [154], and further generalized by C. Tomasi and R. Manduchi, [149], as bilateral filter. The main aim of [129] is to relate the neighborhood filter to an energy minimization. Now, in this case the Euler-Lagrange equations are not partial differential equations but include integrals. The functional considered in [129] takes the general form

$$(6.5) \qquad J_g(u) = \int_{\Omega \times \Omega} g\left(\frac{|u(x) - u(y)|^2}{h^2}\right) w(|x - y|) \, dx \, dy,$$

with $w \in L^\infty(\Omega)$, $g \in C^1(\mathbb{R}^+)$ and $h > 0$ as a parameter. The Fréchet derivative of J_g as a functional from $L^2(\Omega)$ into $\mathbb{R}$ is given by

$$J_g'(u)(x) = \frac{4}{h^2} \int_\Omega g'\left(\frac{|u(x) - u(y)|^2}{h^2}\right) (u(x) - u(y))w(|x - y|) \, dy.$$

Note that the nonlocal functional $-B_p^J$ is of the form (6.5) with $g(t) = \frac{1}{2p}|t|^{\frac{p}{2}}$, $w(r) = J(x)$ if $|x| = r$, and $h = 1$. Then problem (6.1) appears when one uses the steepest descent method to minimize this particular nonlocal functional.

6.1.2. A precompactness result. The following precompactness result is a variant of [**52**, Theorem 4] and will be used in the sequel.

For a function g defined in a set D, we define

(6.6)
$$\overline{g}(x) = \begin{cases} g(x) & \text{if } x \in D, \\ 0 & \text{otherwise.} \end{cases}$$

We denote by $BV(D)$ the space of functions of bounded variation; see Chapter 7 for the definition.

THEOREM 6.11. *Let $1 \leq q < +\infty$ and $D \subset \mathbb{R}^N$ open. Let $\rho : \mathbb{R}^N \to \mathbb{R}$ be a nonnegative continuous radial function with compact support, non identically zero, and $\rho_n(x) := n^N \rho(nx)$. Let $\{f_n\}$ be a sequence of functions in $L^q(D)$ such that*

(6.7)
$$\int_D \int_D |f_n(y) - f_n(x)|^q \rho_n(y-x)\, dx\, dy \leq \frac{M}{n^q}.$$

1. *If $\{f_n\}$ is weakly convergent in $L^q(D)$ to f, then*
   (i) *For $q > 1$, $f \in W^{1,q}(D)$, and moreover*

$$(\rho(z))^{1/q} \chi_D\left(x + \frac{1}{n}z\right) \frac{\overline{f_n}\left(x + \frac{1}{n}z\right) - f_n(x)}{1/n} \rightharpoonup (\rho(z))^{1/q}\, z \cdot \nabla f(x)$$

weakly in $L^q(D) \times L^q(\mathbb{R}^N)$.
   (ii) *For $q = 1$, $f \in BV(D)$, and moreover*

$$\rho(z)\chi_D\left(\cdot + \frac{1}{n}z\right) \frac{\overline{f_n}\left(\cdot + \frac{1}{n}z\right) - f_n(\cdot)}{1/n} \rightharpoonup \rho(z)z \cdot Df$$

weakly in the sense of measures.

2. *Suppose D is a smooth bounded domain in $\mathbb{R}^N$ and $\rho(x) \geq \rho(y)$ if $|x| \leq |y|$. Then $\{f_n\}$ is relatively compact in $L^q(D)$, and consequently, there exists a subsequence $\{f_{n_k}\}$ such that*
   (i) *if $q > 1$, $f_{n_k} \to f$ in $L^q(D)$ with $f \in W^{1,q}(D)$;*
   (ii) *if $q = 1$, $f_{n_k} \to f$ in $L^1(D)$ with $f \in BV(D)$.*

PROOF. Suppose $f_n \to f$ weakly in $L^q(D)$. We write (6.7) as

(6.8)
$$\int_{\mathbb{R}^N} \int_D \rho(z)\chi_D\left(x + \frac{1}{n}z\right) \left| \frac{\overline{f_n}\left(x + \frac{1}{n}z\right) - f_n(x)}{1/n} \right|^q dx\, dz$$

$$= \int_D \int_D n^N \rho(n(x-y)) \left| \frac{f_n(y) - f_n(x)}{1/n} \right|^q dx\, dy \leq M.$$

On the other hand, if $\varphi \in \mathcal{D}(D)$ and $\psi \in \mathcal{D}(\mathbb{R}^N)$, then, taking n large enough,

$$\int_{\mathbb{R}^N} (\rho(z))^{1/q} \int_D \chi_D \left(x + \frac{1}{n}z \right) \frac{\overline{f}_n \left(x + \frac{1}{n}z \right) - f_n(x)}{1/n} \varphi(x)\, dx\, \psi(z)\, dz$$

$$(6.9) \qquad = \int_{\mathbb{R}^N} (\rho(z))^{1/q} \int_{\mathrm{supp}(\varphi)} \frac{\overline{f}_n \left(x + \frac{1}{n}z \right) - f_n(x)}{1/n} \varphi(x)\, dx\, \psi(z)\, dz$$

$$= -\int_{\mathbb{R}^N} (\rho(z))^{1/q} \int_D f_n(x) \frac{\varphi(x) - \overline{\varphi} \left(x - \frac{1}{n}z \right)}{1/n}\, dx\, \psi(z)\, dz.$$

Let us start with the case 1(i). By (6.8), up to a subsequence,

$$(\rho(z))^{1/q} \chi_D \left(x + \frac{1}{n}z \right) \frac{\overline{f}_n \left(x + \frac{1}{n}z \right) - f_n(x)}{1/n} \rightharpoonup (\rho(z))^{1/q} g(x,z)$$

weakly in $L^q(D) \times L^q(\mathbb{R}^N)$. Therefore, passing to the limit in (6.9), we get

$$\int_{\mathbb{R}^N} (\rho(z))^{1/q} \int_D g(x,z)\varphi(x)\, dx\, \psi(z)\, dz$$

$$= -\int_{\mathbb{R}^N} (\rho(z))^{1/q} \int_D f(x)\, z \cdot \nabla\varphi(x)\, dx\, \psi(z)\, dz.$$

Consequently,

$$\int_D g(x,z)\varphi(x)\, dx = -\int_D f(x)\, z \cdot \nabla\varphi(x)\, dx, \qquad \forall\, z \in \mathrm{int}(\mathrm{supp}(J)).$$

From this, for s small,

$$\int_D g(x, se_i)\varphi(x)\, dx = -\int_D f(x)\, s \frac{\partial}{\partial x_i}\varphi(x)\, dx,$$

which implies $f \in W^{1,q}(D)$ and $(\rho(z))^{1/q} g(x,z) = (\rho(z))^{1/q} z \cdot \nabla f(x)$ in $D \times \mathbb{R}^N$.

Let us now prove 1(ii). By (6.8), there exists a bounded Radon measure $\mu \in \mathcal{M}(D \times \mathbb{R}^N)$ such that, up to a subsequence,

$$\rho(z)\chi_D \left(x + \frac{1}{n}z \right) \frac{\overline{f}_n \left(x + \frac{1}{n}z \right) - f_n(x)}{1/n} \rightharpoonup \mu(x,z)$$

weakly in $\mathcal{M}(D \times \mathbb{R}^N)$. Hence, passing to the limit in (6.9), we get

$$(6.10) \qquad \int_{D \times \mathbb{R}^N} \varphi(x)\psi(z)d\mu(x,z) = -\int_{D \times \mathbb{R}^N} \rho(z)\psi(z)\, z \cdot \nabla\varphi(x)f(x)\, dx\, dz.$$

Now, applying the disintegration theorem (see [4, Theorem 2.28]) to the measure μ, we get that if $\pi : D \times \mathbb{R}^N \to \mathbb{R}^N$ is the projection on the first factor and $\nu = \pi_{\#}|\mu|$, then there exists a Radon measures μ_x in $\mathbb{R}^N$ such that $x \mapsto \mu_x$ is ν-measurable,

$$|\mu_x|(\mathbb{R}^N) \leq 1 \quad \nu\text{-a.e. } x \in D$$

and, for any $h \in L^1(D \times \mathbb{R}^N, |\mu|)$,

$$h(x, \cdot) \in L^1(\mathbb{R}^N, |\mu_x|) \quad \nu\text{-a.e. } x \in D,$$

$$x \mapsto \int_{\mathbb{R}^N} h(x,z)d\mu_x(z) \in L^1(D, \nu)$$

and

$$(6.11) \qquad \int_{D \times \mathbb{R}^N} h(x, z) d\mu(x, z) = \int_D \left(\int_{\mathbb{R}^N} h(x, z) d\mu_x(z) \right) d\nu(x).$$

From (6.10) and (6.11), we get, for $\varphi \in \mathcal{D}(D)$ and $\psi \in \mathcal{D}(\mathbb{R}^N)$,

$$\int_D \left(\int_{\mathbb{R}^N} \psi(z)\, d\mu_x(z) \right) \varphi(x)\, d\nu(x) = \left(\sum_{i=1}^N \int_{\mathbb{R}^N} \rho(z) z_i \psi(z) dz \, \frac{\partial f}{\partial x_i} , \, \varphi(x) \right).$$

Hence, in the sense of measures,

$$\sum_{i=1}^N \int_{\mathbb{R}^N} \rho(z) z_i \psi(z) dz \, \frac{\partial f}{\partial x_i} = \int_{\mathbb{R}^N} \psi(z) d\mu_x(z)\, \nu.$$

Now let $\tilde{\psi} \in \mathcal{D}(\mathbb{R}^N)$ be a radial function such that $\tilde{\psi} = 1$ in supp(ρ). Taking

$$\psi(z) = \tilde{\psi}(z) z_i$$

in the above expression and having in mind that

$$\int_{\mathbb{R}^N} \rho(z) z_i z_j \tilde{\psi}(z) dz = 0 \quad \text{if } i \neq j,$$

we get

$$\int_{\mathbb{R}^N} \rho(z) z_i^{\,2} dz \, \frac{\partial f}{\partial x_i} = \int_{\mathbb{R}^N} \tilde{\psi}(z) z_i d\mu_x(z)\, \nu.$$

Since $\nu \in M_b(D)$ and $x \mapsto \int_{\mathbb{R}^N} \tilde{\psi}(z) z_i d\mu_x(z) \in L^1(D, \nu)$, we obtain that $f \in BV(D)$. Going back to (6.11),

$$\mu(x, z) = \sum_{i=1}^N \frac{\partial f}{\partial x_i}(x) \cdot \rho(z) z_i \mathcal{L}^N(z).$$

Let us now prove 2. We follow the proof of [**52**, Theorem 4], and so we may assume that $D = \mathbb{R}^N$ and supp$(f_n) \subset B$, where B is a fixed ball. Since $\rho(x) \geq \rho(y)$ if $|x| \leq |y|$, (6.7) still holds. For each n and $t > 0$, let F_n be the function

$$F_n(t) = \int_{w \in S^{N-1}} \int_{\mathbb{R}^N} |f_n(x + tw) - f_n(x)|^q \, dx \, d\sigma$$

$$= \frac{1}{t^{N-1}} \int_{|h|=t} \int_{\mathbb{R}^N} |f_n(x + h) - f_n(x)|^q \, dx \, d\sigma,$$

where S^{N-1} is the N-dimensional sphere of radius 1 centered at the origin. In terms of F_n assumption (6.7) can be expressed as

$$(6.12) \qquad \int_0^1 t^{N+q-1} \frac{F_n(t)}{t^q} \tilde{\rho}_n(t)\, dt \leq \frac{M}{n^q},$$

where $\tilde{\rho}_n(t) = \rho_n(x)$ if $|x| = t$. On the other hand, applying [**52**, Lemma 2] with $g(t) = F_n(t)/t^q$ and $h(t) = \tilde{\rho}_n(t)$, there exists a constant $K = K(N + q) > 0$ such that

$$(6.13) \qquad \delta^{-N-q} \int_0^\delta t^{N+q-1} \frac{F_n(t)}{t^q} dt \leq K \frac{\displaystyle\int_0^\delta t^{N+q-1} \frac{F_n(t)}{t^q} \tilde{\rho}_n(t)}{\displaystyle\int_{[|x| < \delta]} |x|^q \rho_n(x)\, dx}.$$

Since ρ is a function with compact support, given $\delta > 0$, we can find $n_\delta \in \mathbb{N}$ such that, for $n \geq n_\delta$,

$$\int_{[|x|<\delta]} |x|^q \rho_n(x)\,dx = \int_{[|x|<\delta]} |x|^q n^N \rho(nx)\,dx$$

$$= \int_{[|y|<n\delta]} n^{-q}|y|^q \rho(y)\,dy = \frac{1}{n^q}\int_{\mathbb{R}^N} |y|^q \rho(y)\,dy.$$

Hence, by (6.12) and (6.13),

$$(6.14) \qquad \delta^{-N}\int_0^\delta t^{N-1}F_n(t)\,dt \leq C_1\delta^q \qquad \text{for } n \geq n_\delta,$$

where C_1 is a constant independent of n and δ.

Now, as in the proof of [52, Theorem 4],

$$\int |f_n(x)|^q \leq C_2 \int_0^1 t^{N-1}F_n(t)\,dt$$

and

$$\int |f_n(x) - (f_n * \Phi_\delta)(x)|^q \leq C_2\delta^{-N}\int_0^\delta t^{N-1}F_n(t)\,dt,$$

for some constant C_2 independent of n and δ, where $\Phi_\delta = \frac{1}{|B_\delta(0)|}\chi_{B_\delta(0)}$. Therefore, by (6.14), $\|f_n\|_{L^p}$ is bounded and

$$\lim_{\delta\to 0}\limsup_{n\to\infty}\|f_n - f_n * \Phi_\delta\|_{L^q(D)} = 0,$$

which implies that $\{f_n\}$ is relatively compact in $L^q(D)$ (see, for example, [57, Corollaries IV.25 et IV.27]). $\qquad\square$

6.1.3. Rescaling the kernel. Convergence to the local p-Laplacian. Let Ω be a bounded smooth domain in $\mathbb{R}^N$. We are going to see now that the solutions of problem (6.1), with the kernel J rescaled in a suitable way, converge, as the scaling parameter goes to zero, to the solution of the classical p-Laplacian evolution problem with homogeneous Neumann boundary conditions

$$(6.15) \qquad \begin{cases} v_t = \Delta_p v & \text{in } \Omega \times (0, T), \\ |\nabla v|^{p-2}\nabla v \cdot \eta = 0 & \text{on } \partial\Omega \times (0, T), \\ v(x, 0) = u_0(x) & \text{in } \Omega, \end{cases}$$

where η is the unit outward normal on $\partial\Omega$ and $\Delta_p v = \mathrm{div}(|\nabla v|^{p-2}\nabla v)$ is the so-called p-Laplacian of v.

For fixed $p > 1$ and J we consider the rescaled kernels

$$J_{p,\varepsilon}(x) := \frac{C_{J,p}}{\varepsilon^{p+N}}J\left(\frac{x}{\varepsilon}\right),$$

where

$$C_{J,p}^{-1} := \frac{1}{2}\int_{\mathbb{R}^N} J(z)|z_N|^p\,dz$$

is a normalizing constant in order to obtain the p-Laplacian in the limit instead of a multiple of it.

Associated with these rescaled kernels we have solutions u_ε of problem (6.1) with J replaced by $J_{p,\varepsilon}$ and the same initial condition u_0. The main result now states that these functions u_ε converge strongly in $L^p(\Omega)$ to the solution of the local p-Laplacian Neumann problem (6.15).

THEOREM 6.12. *Assume that $J(x) \geq J(y)$ if $|x| \leq |y|$. Let $T > 0$, $u_0 \in L^p(\Omega)$ and let u_ε be the unique solution of* (6.1) *with J replaced by $J_{p,\varepsilon}$. Then, if v is the unique solution of* (6.15),

$$\lim_{\varepsilon \to 0} \sup_{t \in [0,T]} \|u_\varepsilon(\cdot, t) - v(\cdot, t)\|_{L^p(\Omega)} = 0.$$

Note that the above result states that (6.1) is a nonlocal analog of the Neumann p-Laplacian.

REMARK 6.13. Recall that for the linear case, $p = 2$, under additional regularity hypothesis on the involved data, the convergence of the solutions of rescaled nonlocal problems of the form (6.1) to the solution of the heat equation is proved in Chapter 3. It should be interesting to extend the previous theorem to non homogeneous Neumann problems as it is done in Chapter 3 for the linear case.

The local p-Laplacian equation. Let us give some results about the p-Laplacian equation, obtained in [**12**], [**13**] and [**8**], which will be used in the proof of the convergence of the rescaled problems.

For problem (6.15) we have the following concepts of solutions. A *weak solution* of (6.15) in the time interval $[0, T]$ is a function

$$v \in L^p(0, T; W^{1,p}(\Omega)) \cap W^{1,1}(0, T; L^1(\Omega))$$

with $v(0) = u_0$ satisfying

$$\int_\Omega v'(t)\xi \, dx + \int_\Omega |\nabla v(t)|^{p-2} \nabla v(t) \cdot \nabla \xi \, dx = 0 \qquad \text{for almost all } t \in (0, T)$$

for any $\xi \in W^{1,p}(\Omega) \cap L^\infty(\Omega)$. An *entropy solution* of (6.15) in $[0, T]$ is a function $v \in W^{1,1}(0, T; L^1(\Omega))$ with $T_k(v) \in L^p(0, T; W^{1,p}(\Omega))$ for all $k > 0$, such that $v(0) = u_0$ and

$$\int_\Omega v'(t)T_k(v(t) - \xi) \, dx + \int_\Omega |\nabla v(t)|^{p-2} \nabla v(t) \cdot \nabla T_k(v(t) - \xi) \, dx = 0$$

for almost all $t \in (0, T)$ and for any $\xi \in W^{1,p}(\Omega) \cap L^\infty(\Omega)$. Recall that $T_k(r) = k \wedge (r \vee (-k))$, $k \geq 0$, $r \in \mathbb{R}$.

In [**12**], associated to the p-Laplacian with homogeneous boundary condition, the operator $B_p \subset L^1(\Omega) \times L^1(\Omega)$ is defined as $(v, \hat{v}) \in B_p$ if and only if $\hat{v} \in L^1(\Omega)$, $v \in W^{1,p}(\Omega)$ and

$$\int_\Omega |\nabla v|^{p-2} \nabla v \cdot \nabla \xi \, dx = \int_\Omega \hat{v}\xi \, dx \quad \text{for every } \xi \in W^{1,p}(\Omega) \cap L^\infty(\Omega)$$

and it is proved that B_p is a completely accretive operator in $L^1(\Omega)$ with dense domain satisfying a range condition which implies that its closure $\mathcal{B}_p$ in $L^1(\Omega)$ is an m-completely accretive operator in $L^1(\Omega)$ with dense domain. In [**13**] (see also [**8**] for weak solutions), it is showed that for any $u_0 \in L^1(\Omega)$, the unique mild solution $e^{-t\mathcal{B}_p}u_0$ given by Crandall-Liggett's exponential formula is the unique entropy solution $v(t)$ of problem (6.15). The following result states the existence and uniqueness of solutions for (6.15).

THEOREM 6.14 ([**8**],[**13**]). *Let* $T > 0$. *For any* $u_0 \in L^1(\Omega)$ *there exists a unique entropy solution* $v(t)$ *of* (6.15). *Moreover, if* $u_0 \in L^{p'}(\Omega) \cap L^2(\Omega)$, *the entropy solution* $v(t)$ *is a weak solution.*

A formal calculation. Let us perform a formal calculation just to convince the reader that the convergence result (Theorem 6.12) is correct. Let $N = 1$. Let $u(x)$ be a smooth function and consider

$$A_\varepsilon(u)(x) = \frac{1}{\varepsilon^{p+1}} \int_{\mathbb{R}} J\left(\frac{x-y}{\varepsilon}\right) |u(y) - u(x)|^{p-2}(u(y) - u(x))\, dy.$$

Changing variables, $y = x - \varepsilon z$, we get

$$(6.16) \qquad A_\varepsilon(u)(x) = \frac{1}{\varepsilon^p} \int_{\mathbb{R}} J(z)|u(x - \varepsilon z) - u(x)|^{p-2}(u(x - \varepsilon z) - u(x))\, dz.$$

Now, we expand in powers of ε to obtain

$$|u(x - \varepsilon z) - u(x)|^{p-2} = \varepsilon^{p-2} \left| u'(x)z + \frac{u''(x)}{2}\varepsilon z^2 + O(\varepsilon^2) \right|$$

$$= \varepsilon^{p-2}|u'(x)|^{p-2}|z|^{p-2} + \varepsilon^{p-1}(p-2)|u'(x)z|^{p-4}u'(x)z\frac{u''(x)}{2}z^2 + O(\varepsilon^p),$$

and

$$u(x - \varepsilon z) - u(x) = \varepsilon u'(x)z + \frac{u''(x)}{2}\varepsilon^2 z^2 + O(\varepsilon^3).$$

Hence, (6.16) becomes

$$A_\varepsilon(u)(x) = \frac{1}{\varepsilon} \int_{\mathbb{R}} J(z)|z|^{p-2}z\, dz|u'(x)|^{p-2}u'(x)$$

$$+\frac{1}{2} \int_{\mathbb{R}} J(z)|z|^p\, dz \left((p-2)|u'(x)|^{p-2}u''(x) + |u'(x)|^{p-2}u''(x)\right) + O(\varepsilon).$$

Using that J is radially symmetric, the first integral vanishes and therefore

$$\lim_{\varepsilon \to 0} A_\varepsilon(u)(x) = C(|u'(x)|^{p-2}u'(x))',$$

where

$$C = \frac{1}{2} \int_{\mathbb{R}} J(z)|z|^p\, dz.$$

Proof of Theorem 6.12. The objective now is to make this formal calculation rigorous. For this purpose we will use Theorem 6.11.

PROPOSITION 6.15. *For any* $\phi \in L^\infty(\Omega)$, *we have that*

$$\left(I + B_p^{J_{p,\varepsilon}}\right)^{-1} \phi \rightharpoonup (I + B_p)^{-1} \phi \quad \text{weakly in } L^p(\Omega) \text{ as } \varepsilon \to 0.$$

PROOF. For $\varepsilon > 0$, let $u_\varepsilon = \left(I + B_p^{J_{p,\varepsilon}}\right)^{-1} \phi$. Then

$$
\int_\Omega u_\varepsilon \xi - \frac{C_{J,p}}{\varepsilon^{p+N}} \int_\Omega \int_\Omega J\left(\frac{x-y}{\varepsilon}\right) |u_\varepsilon(y) - u_\varepsilon(x)|^{p-2}
$$

(6.17)
$$
\times (u_\varepsilon(y) - u_\varepsilon(x))\, dy\, \xi(x)\, dx
$$

$$
= \int_\Omega \phi \xi
$$

for every $\xi \in L^\infty(\Omega)$.

Changing variables, we get

$$
-\frac{C_{J,p}}{\varepsilon^{p+N}} \int_\Omega \int_\Omega J\left(\frac{x-y}{\varepsilon}\right) |u_\varepsilon(y) - u_\varepsilon(x)|^{p-2}(u_\varepsilon(y) - u_\varepsilon(x))\, dy\, \xi(x)\, dx
$$

(6.18)
$$
= \int_{\mathbb{R}^N} \int_\Omega \frac{C_{J,p}}{2} J(z) \chi_\Omega(x + \varepsilon z) \left|\frac{\overline{u}_\varepsilon(x + \varepsilon z) - u_\varepsilon(x)}{\varepsilon}\right|^{p-2}
$$

$$
\times \frac{\overline{u}_\varepsilon(x + \varepsilon z) - u_\varepsilon(x)}{\varepsilon} \frac{\overline{\xi}(x + \varepsilon z) - \xi(x)}{\varepsilon}\, dx\, dz,
$$

where we use the notation given in (6.6).

Thus, we can write (6.17) as

$$
\int_\Omega \phi(x)\xi(x)\, dx - \int_\Omega u_\varepsilon(x)\xi(x)\, dx
$$

(6.19)
$$
= \int_{\mathbb{R}^N} \int_\Omega \frac{C_{J,p}}{2} J(z) \chi_\Omega(x + \varepsilon z) \left|\frac{\overline{u}_\varepsilon(x + \varepsilon z) - u_\varepsilon(x)}{\varepsilon}\right|^{p-2}
$$

$$
\times \frac{\overline{u}_\varepsilon(x + \varepsilon z) - u_\varepsilon(x)}{\varepsilon} \frac{\overline{\xi}(x + \varepsilon z) - \xi(x)}{\varepsilon}\, dx\, dz.
$$

We shall see that there exists a sequence $\varepsilon_n \to 0$ such that $u_{\varepsilon_n} \to v$ weakly in $L^p(\Omega)$, $v \in W^{1,p}(\Omega)$ and $v = (I + B_p)^{-1} \phi$, that is,

$$
\int_\Omega v\xi + \int_\Omega |\nabla v|^{p-2}\nabla v \cdot \nabla \xi = \int_\Omega \phi\xi \quad \text{for every } \xi \in W^{1,p}(\Omega) \cap L^\infty(\Omega).
$$

It is easy to see that $u_\varepsilon \ll \phi$; therefore, by Proposition A.44, there exists a sequence $\varepsilon_n \to 0$ such that

(6.20)
$$
u_{\varepsilon_n} \rightharpoonup v \quad \text{weakly in } L^p(\Omega) \text{ and in } L^2(\Omega), \quad v \ll \phi.
$$

Observe that $\|u_{\varepsilon_n}\|_{L^\infty(\Omega)}, \|v\|_{L^\infty(\Omega)} \le \|\phi\|_{L^\infty(\Omega)}$. Taking $\varepsilon = \varepsilon_n$ and $\xi = u_{\varepsilon_n}$ in (6.19), we get

$$
\int_\Omega \int_\Omega \frac{1}{2} \frac{C_{J,p}}{\varepsilon_n^N} J\left(\frac{x-y}{\varepsilon_n}\right) \left|\frac{u_{\varepsilon_n}(y) - u_{\varepsilon_n}(x)}{\varepsilon_n}\right|^p dx\, dy
$$

(6.21)
$$
= \int_{\mathbb{R}^N} \int_\Omega \frac{C_{J,p}}{2} J(z) \chi_\Omega(x + \varepsilon_n z) \left|\frac{\overline{u}_{\varepsilon_n}(x + \varepsilon_n z) - u_{\varepsilon_n}(x)}{\varepsilon_n}\right|^p dx\, dz \le M.
$$

Therefore, by Theorem 6.11, $v \in W^{1,p}(\Omega)$ and

(6.22)
$$\left(\frac{C_{J,p}}{2} J(z)\right)^{1/p} \chi_\Omega(x + \varepsilon_n z) \frac{\overline{u}_{\varepsilon_n}(x + \varepsilon_n z) - u_{\varepsilon_n}(x)}{\varepsilon_n}$$

$$\rightharpoonup \left(\frac{C_{J,p}}{2} J(z)\right)^{1/p} z \cdot \nabla v(x)$$

weakly in $L^p(\Omega) \times L^p(\mathbb{R}^N)$. Moreover, we can also assume that

$$(J(z))^{1/p'} \left|\frac{\overline{u}_{\varepsilon_n}(x + \varepsilon_n z) - u_{\varepsilon_n}(x)}{\varepsilon_n}\right|^{p-2} \chi_\Omega(x + \varepsilon_n z) \frac{\overline{u}_{\varepsilon_n}(x + \varepsilon_n z) - u_{\varepsilon_n}(x)}{\varepsilon_n}$$

$$\rightharpoonup (J(z))^{1/p'} \chi(x, z)$$

weakly in $L^{p'}(\Omega) \times L^{p'}(\mathbb{R}^N)$, for some function $\chi \in L^{p'}(\Omega) \times L^{p'}(\mathbb{R}^N)$. Therefore, passing to the limit in (6.19) for $\varepsilon = \varepsilon_n$, we get

(6.23)
$$\int_\Omega v\xi + \int_{\mathbb{R}^N} \int_\Omega \frac{C_{J,p}}{2} J(z)\chi(x, z) z \cdot \nabla\xi(x) \, dx \, dz = \int_\Omega \phi\xi$$

for every smooth ξ and by approximation for every $\xi \in W^{1,p}(\Omega)$.

We now show that

(6.24)
$$\int_{\mathbb{R}^N} \int_\Omega \frac{C_{J,p}}{2} J(z)\chi(x, z) z \cdot \nabla\xi(x) \, dx \, dz = \int_\Omega |\nabla v|^{p-2} \nabla v \cdot \nabla\xi.$$

In fact, taking $\xi = u_{\varepsilon_n}$ in (6.19) and taking limits, since, thanks to (6.20),

$$\int_\Omega v^2 \le \liminf_n \int_\Omega u_{\varepsilon_n}^2,$$

we get, on account of (6.23),

(6.25)
$$\limsup_n \int_{\mathbb{R}^N} \int_\Omega \frac{C_{J,p}}{2} J(z)\chi_\Omega(x + \varepsilon_n z) \left|\frac{\overline{u}_{\varepsilon_n}(x + \varepsilon_n z) - u_{\varepsilon_n}(x)}{\varepsilon_n}\right|^p \, dx \, dz$$

$$\le \int_{\mathbb{R}^N} \int_\Omega \frac{C_{J,p}}{2} J(z)\chi(x, z) z \cdot \nabla v(x) \, dx \, dz.$$

By the monotonicity Lemma 6.6, for every ρ smooth,

$$-\frac{C_{J,p}}{\varepsilon_n^{p+N}} \int_\Omega \int_\Omega J\left(\frac{x - y}{\varepsilon_n}\right) |\rho(y) - \rho(x)|^{p-2} (\rho(y) - \rho(x)) \, dy \, (u_{\varepsilon_n}(x) - \rho(x)) \, dx$$

$$\le -\frac{C_{J,p}}{\varepsilon_n^{p+N}} \int_\Omega \int_\Omega J\left(\frac{x - y}{\varepsilon_n}\right) |u_{\varepsilon_n}(y) - u_{\varepsilon_n}(x)|^{p-2}$$

$$\times (u_{\varepsilon_n}(y) - u_{\varepsilon_n}(x)) \, dy \, (u_{\varepsilon_n}(x) - \rho(x)) \, dx.$$

Using the same change of variable that we used in (6.18) and taking limits, on account of (6.22) and (6.25), we obtain, for every smooth ρ,

$$\int_{\mathbb{R}^N} \int_\Omega \frac{C_{J,p}}{2} J(z) |z \cdot \nabla\rho(x)|^{p-2} z \cdot \nabla\rho(x) \, z \cdot (\nabla v(x) - \nabla\rho(x)) \, dx \, dz$$

$$\le \int_{\mathbb{R}^N} \int_\Omega \frac{C_{J,p}}{2} J(z)\chi(x, z) z \cdot (\nabla v(x) - \nabla\rho(x)) \, dx \, dz,$$

and then for every $\rho \in W^{1,p}(\Omega)$. Taking $\rho = v \pm \lambda \xi$, $\lambda > 0$ and $\xi \in W^{1,p}(\Omega)$, and letting $\lambda \to 0$, we get

$$\int_{\mathbb{R}^N} \int_\Omega \frac{C_{J,p}}{2} J(z) \chi(x, z) z \cdot \nabla \xi(x) \, dx \, dz$$

$$= \int_{\mathbb{R}^N} \frac{C_{J,p}}{2} J(z) \int_\Omega |z \cdot \nabla v(x)|^{p-2} \left(z \cdot \nabla v(x) \right) \left(z \cdot \nabla \xi(x) \right) dx \, dz.$$

Consequently,

$$\int_{\mathbb{R}^N} \int_\Omega \frac{C_{J,p}}{2} J(z) \chi(x, z) z \cdot \nabla \xi(x) \, dx \, dz = \int_\Omega \mathbf{a}(\nabla v) \cdot \nabla \xi \quad \text{for every } \xi \in W^{1,p}(\Omega),$$

where

$$\mathbf{a}_j(\xi) = C_{J,p} \int_{\mathbb{R}^N} \frac{1}{2} J(z) |z \cdot \xi|^{p-2} \, z \cdot \xi \, z_j \, dz.$$

Then, if we prove that

(6.26) $$\mathbf{a}(\xi) = |\xi|^{p-2} \xi,$$

we obtain that (6.24) is true and $v = (I + B_p)^{-1} \phi$.

So, to finish the proof we only need to show that (6.26) holds. We state it in the following lemma. $\qquad \square$

LEMMA 6.16. *Let* $\mathbf{a} : \mathbb{R}^N \to \mathbb{R}^N$ *be defined by*

$$\mathbf{a}_j(\xi) = C_{J,p} \int_{\mathbb{R}^N} \frac{1}{2} J(z) |z \cdot \xi|^{p-2} \, z \cdot \xi \, z_j \, dz.$$

Then

$$\mathbf{a}(\xi) = |\xi|^{p-2} \xi.$$

PROOF. Observe that $\mathbf{a}$ is positively homogeneous of degree $p - 1$, that is,

$$\mathbf{a}(t\xi) = t^{p-1} \mathbf{a}(\xi) \qquad \text{for all } \xi \in \mathbb{R}^N \text{ and all } t > 0.$$

Therefore, in order to prove (6.26) it is enough to see that, for each i,

$$\mathbf{a}_i(\xi) = \xi_i \quad \text{for all } \xi \in \mathbb{R}^N, \ |\xi| = 1.$$

Fix $\xi \in \mathbb{R}^N$, $|\xi| = 1$. Let $R_{\xi,i}$ be the rotation such that $R^t_{\xi,i}(\xi) = \mathbf{e}_i$, where $\mathbf{e}_i$ is the vector with components $(\mathbf{e}_i)_i = 1$, $(\mathbf{e}_i)_j = 0$ for $j \neq i$, and $R^t_{\xi,i}$ is the transpose of $R_{\xi,i}$. Observe that

$$\xi_i = \xi \cdot \mathbf{e}_i = R^t_{\xi,i}(\xi) \cdot R^t_{\xi,i}(\mathbf{e}_i) = \mathbf{e}_i \cdot R^t_{\xi,i}(\mathbf{e}_i).$$

On the other hand, since J is a radial function, we have $C^{-1}_{J,p} = \frac{1}{2} \int_{\mathbb{R}^N} J(z) |z_i|^p \, dz$ and

$$\mathbf{a}(\mathbf{e}_i) = \mathbf{e}_i \quad \text{for any } i.$$

Then, if we make the change of variables $z = R_{\xi,i}(y)$, using again that J is radial, we obtain

$$\mathbf{a}_i(\xi) = C_{J,p} \int_{\mathbb{R}^N} \frac{1}{2} J(z)|z \cdot \xi|^{p-2} z \cdot \xi \, z \cdot \mathbf{e}_i \, dz$$

$$= C_{J,p} \int_{\mathbb{R}^N} \frac{1}{2} J(y)|y \cdot \mathbf{e}_i|^{p-2} y \cdot \mathbf{e}_i \, y \cdot R_{\xi,i}^t(\mathbf{e}_i) \, dy$$

$$= \mathbf{a}(\mathbf{e}_i) \cdot R_{\xi,i}^t(\mathbf{e}_i) = \mathbf{e}_i \cdot R_{\xi,i}^t(\mathbf{e}_i) = \xi_i,$$

and the proof is complete. $\qquad\qquad\square$

THEOREM 6.17. *Suppose $J(x) \geq J(y)$ if $|x| \leq |y|$. For any $\phi \in L^\infty(\Omega)$,*

$$(6.27) \qquad \left(I + B_p^{J_{p,\varepsilon}}\right)^{-1} \phi \to (I + B_p)^{-1} \phi \quad \text{in } L^p(\Omega) \text{ as } \varepsilon \to 0.$$

PROOF. Let $u_\varepsilon = \left(I + B_p^{J_{p,\varepsilon}}\right)^{-1} \phi$ and $v = (I + B_p)^{-1}\phi$. By the previous Proposition 6.15, we have that $u_\varepsilon \to v$ weakly in $L^p(\Omega)$ and moreover (see (6.21))

$$\int_\Omega \int_\Omega \frac{1}{2} \frac{C_{J,p}}{\varepsilon^N} J\left(\frac{x-y}{\varepsilon}\right) \left|\frac{u_\varepsilon(y) - u_\varepsilon(x)}{\varepsilon}\right|^p dx\,dy \leq M.$$

Therefore, by Theorem 6.11 the result follows. $\qquad\qquad\square$

From the above theorem, by the standard results of Nonlinear Semigroup Theory, we obtain the following result, which gives Theorem 6.12.

THEOREM 6.18. *Assume that $J(x) \geq J(y)$ if $|x| \leq |y|$. Let $T > 0$ and $u_0 \in L^q(\Omega)$, $p \leq q < +\infty$. Let u_ε be the unique solution of (6.1) with J replaced by $J_{p,\varepsilon}$ and v the unique solution of (6.15). Then*

$$(6.28) \qquad \lim_{\varepsilon \to 0} \sup_{t \in [0,T]} \|u_\varepsilon(\cdot,t) - v(\cdot,t)\|_{L^q(\Omega)} = 0.$$

Moreover, if $1 < p \leq 2$, (6.28) holds for any $u_0 \in L^q(\Omega)$, $1 \leq q < +\infty$.

PROOF. Since B_p^J is completely accretive and satisfies the range condition (6.3), to get (6.28) it is enough to see that

$$\left(I + B_p^{J_{p,\varepsilon}}\right)^{-1} \phi \to (I + B_p)^{-1} \phi \qquad \text{in } L^q(\Omega) \text{ as } \varepsilon \to 0$$

for any $\phi \in L^\infty(\Omega)$ (see Theorem A.37). Taking into account that

$$\left(I + B_p^{J_{p,\varepsilon}}\right)^{-1} \phi \ll \phi,$$

the above convergence follows by (6.27). $\qquad\qquad\square$

6.1.4. A Poincaré type inequality. The following Poincaré type inequality will be used afterwards. In the linear case, this inequality has been proved using spectral theory in Chapter 3 (see also Proposition 5.3). Now we provide a different proof.

PROPOSITION 6.19. *Given $q \geq 1$, J as above and Ω a bounded domain in $\mathbb{R}^N$, the quantity*

$$\beta_{q-1} := \beta_{q-1}(J, \Omega, q) = \inf_{u \in L^q(\Omega), \int_\Omega u = 0} \frac{\frac{1}{2} \int_\Omega \int_\Omega J(x-y)|u(y) - u(x)|^q \, dy \, dx}{\int_\Omega |u(x)|^q \, dx}$$

is strictly positive. Consequently

$$(6.29) \qquad \beta_{q-1} \int_\Omega \left| u - \frac{1}{|\Omega|} \int_\Omega u \right|^q \leq \frac{1}{2} \int_\Omega \int_\Omega J(x-y)|u(y) - u(x)|^q \, dy \, dx,$$

for every $u \in L^q(\Omega)$.

PROOF. It is enough to prove that there exists a constant c such that

$$(6.30) \qquad \|u\|_q \leq c \left(\left(\int_\Omega \int_\Omega J(x-y)|u(y) - u(x)|^q dy dx \right)^{1/q} + \left| \int_\Omega u \right| \right),$$

for every $u \in L^q(\Omega)$.

Let $r > 0$ be such that $J(z) \geq \alpha > 0$ in $B(0, r)$. Since $\overline{\Omega} \subset \bigcup_{x \in \Omega} B(x, r/2)$, there exists $\{x_i\}_{i=1}^m \subset \Omega$ such that $\Omega \subset \bigcup_{i=1}^m B(x_i, r/2)$. Take $0 < \delta < r/2$ such that $B(x_i, \delta) \subset \Omega$ for all $i = 1, \ldots, m$. Then, for any $\hat{x}_i \in B(x_i, \delta)$, $i = 1, \ldots, m$,

$$(6.31) \qquad \Omega = \bigcup_{i=1}^m (B(\hat{x}_i, r) \cap \Omega).$$

Let us argue by contradiction. Suppose that (6.30) is false. Then there exists $u_n \in L^q(\Omega)$, with $\|u_n\|_{L^q(\Omega)} = 1$, and satisfying

$$1 \geq n \left(\left(\int_\Omega \int_\Omega J(x-y)|u_n(y) - u_n(x)|^q dy dx \right)^{1/q} + \left| \int_\Omega u_n \right| \right) \qquad \forall n \in \mathbb{N}.$$

Consequently,

$$(6.32) \qquad \lim_n \int_\Omega \int_\Omega J(x-y)|u_n(y) - u_n(x)|^q \, dy \, dx = 0$$

and

$$(6.33) \qquad \lim_n \int_\Omega u_n = 0.$$

Let

$$F_n(x, y) = J(x-y)^{1/q}|u_n(y) - u_n(x)|$$

and

$$f_n(x) = \int_\Omega J(x-y)|u_n(y) - u_n(x)|^q \, dy.$$

From (6.33), it follows that

$$f_n \to 0 \quad \text{in } L^1(\Omega).$$

Passing to a subsequence if necessary, we can assume that

$$(6.34) \qquad f_n(x) \to 0 \quad \forall x \in \Omega \setminus B_1, \quad B_1 \text{ null}.$$

On the other hand, by (6.32), we also have that

$$F_n \to 0 \quad \text{in } L^q(\Omega \times \Omega).$$

So we can assume that, up to a subsequence,

(6.35) $\qquad\qquad F_n(x,y) \to 0 \quad \forall\, (x,y) \in \Omega \times \Omega \setminus C, \quad C \text{ null.}$

Suppose $B_2 \subset \Omega$ is a null set satisfying that,

(6.36) $\qquad\qquad$ for all $x \in \Omega \setminus B_2$, the section C_x of C is null.

Let $\hat{x}_1 \in B(x_1, \delta) \setminus (B_1 \cup B_2)$; then there exists a subsequence such that, in the same notation,

$$u_n(\hat{x}_1) \to \lambda_1 \in [-\infty, +\infty].$$

Consider now $\hat{x}_2 \in B(x_2, \delta) \setminus (B_1 \cup B_2)$; then, up to a subsequence, we can assume that

$$u_n(\hat{x}_2) \to \lambda_2 \in [-\infty, +\infty].$$

So, successively, for $\hat{x}_m \in B(x_m, \delta) \setminus (B_1 \cup B_2)$, there exists a subsequence, again denoted the same way, such that

$$u_n(\hat{x}_m) \to \lambda_m \in [-\infty, +\infty].$$

By (6.35) and (6.36),

$$u_n(y) \to \lambda_i \quad \forall\, y \in (B(\hat{x}_i, r) \cap \Omega) \setminus C_{\hat{x}_i}.$$

Now, by (6.31),

$$\Omega = (B(\hat{x}_1, r) \cap \Omega) \cup \left(\bigcup_{i=2}^{m} (B(\hat{x}_i, r) \cap \Omega) \right).$$

Hence, since Ω is a bounded domain, there exists $i_2 \in \{2, \dots, m\}$ such that

$$(B(\hat{x}_1, r) \cap \Omega) \cap (B(\hat{x}_{i_2}, r) \cap \Omega) \neq \emptyset.$$

Therefore, $\lambda_1 = \lambda_{i_2}$. Let us call $i_1 := 1$. Again, since

$$\Omega = ((B(\hat{x}_{i_1}, r) \cap \Omega) \cup ((B(\hat{x}_{i_1}, r) \cap \Omega)) \cup \left(\bigcup_{i \in \{1, \dots, m\} \setminus \{i_1, i_2\}} (B(\hat{x}_i, r) \cap \Omega) \right),$$

there exists $i_3 \in \{1, \dots, m\} \setminus \{i_1, i_2\}$ such that

$$((B(\hat{x}_{i_1}, r) \cap \Omega) \cup (B(\hat{x}_{i_1}, r) \cap \Omega)) \cap (B(\hat{x}_{i_3}, r) \cap \Omega) \neq \emptyset.$$

Consequently

$$\lambda_{i_1} = \lambda_{i_2} = \lambda_{i_3}.$$

Using the same argument we get

$$\lambda_1 = \lambda_2 = \cdots = \lambda_m = \lambda.$$

If $|\lambda| = +\infty$, we have shown that

$$|u_n(y)|^q \to +\infty \quad \text{for almost every } y \in \Omega,$$

which contradicts $\|u_n\|_{L^q(\Omega)} = 1$ for all $n \in \mathbb{N}$. Hence λ is finite.

On the other hand, by (6.34), $f_n(\hat{x}_i) \to 0$, $i = 1, \dots, m$. Hence,

$$F_n(\hat{x}_1, \cdot) \to 0 \quad \text{in } L^q(\Omega).$$

Since $u_n(\hat{x}_1) \to \lambda$, from the above we conclude that

$$u_n \to \lambda \quad \text{in } L^q(B(\hat{x}_i, r) \cap \Omega).$$

Using again a compactness argument we get

$$u_n \to \lambda \quad \text{in } L^q(\Omega).$$

By (6.33), $\lambda = 0$, so

$$u_n \to 0 \quad \text{in } L^q(\Omega),$$

which contradicts $\|u_n\|_{L^q(\Omega)} = 1$. □

REMARK 6.20. The above Poincaré type inequality fails to be true in general if $0 \notin \text{supp}(J)$, as the following example shows. Let $\Omega = (0, 3)$ and J be such that

$$\text{supp}(J) \subset (-3, -2) \cup (2, 3).$$

Then, if

$$u(x) = \begin{cases} 1 & \text{if } 0 < x < 1 \text{ or } 2 < x < 3, \\ 2 & \text{if } 1 \leq x \leq 2, \end{cases}$$

we have that

$$\int_0^3 \int_0^3 J(x-y)|u(y) - u(x)|^p \, dx \, dy = 0,$$

but clearly

$$u(x) - \frac{1}{3} \int_0^3 u(y) \, dy \neq 0.$$

Therefore there is no Poincaré type inequality available for this particular choice of J.

This example can be easily extended to any domain in any dimension just by considering functions u that are constant on annuli intersected with Ω.

REMARK 6.21. Let $p \geq 2$. Using the Poincaré type inequality (6.29), we can solve

(6.37) $$u + B_p^J u = \phi$$

for any $\phi \in L^\infty(\Omega)$ as follows. Let

$$\mathcal{K} := \left\{ u \in L^p(\Omega) \ : \ \int_\Omega u = 0 \right\}$$

and let $A : \mathcal{K} \to L^{p'}(\Omega)$ be the continuous monotone operator defined by $A(u) := u + B_p^J u$. In view of (6.29), we have

$$\lim_{\|u\|_{L^p(\Omega)} \to +\infty, u \in \mathcal{K}} \frac{\int_\Omega A(u)u}{\|u\|_{L^p(\Omega)}} = +\infty.$$

Then, by Corollary 30 in [55], for $\phi \in L^\infty(\Omega)$, $\int_\Omega \phi = 0$, there exists $\hat{u} \in \mathcal{K}$ such that

$$\int_\Omega \hat{u}\xi + \int_\Omega B_p^J \hat{u}\xi = \int_\Omega \phi\xi \qquad \forall \xi \in \mathcal{K}.$$

Since

$$\int_\Omega \hat{u} = 0, \quad \int_\Omega \phi = 0 \quad \text{and} \quad \int_\Omega B_p^J \hat{u} = 0,$$

we have that

$$\int_\Omega \hat{u}\xi + \int_\Omega B_p^J \hat{u}\xi = \int_\Omega \hat{u}\left(\xi - \frac{1}{|\Omega|}\int_\Omega \xi\right) + \int_\Omega B_p^J \hat{u}\left(\xi - \frac{1}{|\Omega|}\int_\Omega \xi\right)$$

$$= \int_\Omega \phi\left(\xi - \frac{1}{|\Omega|}\int_\Omega \xi\right) = \int_\Omega \phi\xi$$

for any $\xi \in L^p(\Omega)$. Finally, from the above, for $\phi \in L^\infty(\Omega)$ there exists $\tilde{u} \in \mathcal{K}$ such that

$$\int_\Omega \tilde{u}\xi + \int_\Omega B_p^J \tilde{u}\xi = \int_\Omega \left(\phi - \frac{1}{|\Omega|}\int_\Omega \phi\right)\xi$$

for any $\xi \in L^p(\Omega)$. Therefore, setting $u = \tilde{u} + \frac{1}{|\Omega|}\int_\Omega \phi$, since $B_p^J(u) = B_p^J(\tilde{u})$, (6.37) holds.

6.1.5. Asymptotic behaviour. Now we study the asymptotic behaviour as $t \to \infty$ of the solution of the nonlocal p-Laplacian problem with homogeneous Neumann boundary conditions. We show that the solution of this nonlocal problem converges to the mean value of the initial condition.

THEOREM 6.22. *Let $u_0 \in L^\infty(\Omega)$. Let u be the solution of (6.1); then*

$$\left\|u(t) - \frac{1}{|\Omega|}\int_\Omega u_0(x)\,dx\right\|_{L^p(\Omega)} \leq \left(C\frac{\|u_0\|_{L^2(\Omega)}^2}{t}\right)^{1/p} \qquad \forall t > 0,$$

where $C = C(J, \Omega, p)$.

PROOF. A simple integration of the equation in space gives that the total mass is preserved, that is,

$$\frac{1}{|\Omega|}\int_\Omega u(x,t)\,dx = \frac{1}{|\Omega|}\int_\Omega u_0(x)\,dx.$$

We set

$$w(x,t) = u(x,t) - \frac{1}{|\Omega|}\int_\Omega u_0(x)\,dx.$$

Then

$$\frac{d}{dt}\int_\Omega |w(x,t)|^p\,dx = p\int_\Omega |w(x,t)|^{p-2}w(x,t)$$

$$\times \int_\Omega J(x-y)|w(y,t) - w(x,t)|^{p-2}(w(y,t) - w(x,t))\,dydx$$

$$= -\frac{p}{2}\int_\Omega\int_\Omega J(x-y)|w(y,t) - w(x,t)|^{p-2}(w(y,t) - w(x,t))$$

$$\times(|w(y,t)|^{p-2}w(y,t) - |w(x,t)|^{p-2}w(x,t))\,dydx.$$

Therefore the $L^p(\Omega)$-norm of $w(\cdot,t)$ is decreasing with t.

Moreover, as the solution preserves the total mass, using the Poincaré type inequality (6.29), we have

$$\int_\Omega |w(x,t)|^p\,dx \leq C\int_\Omega\int_\Omega J(x-y)|u(y,t) - u(x,t)|^p\,dy\,dx.$$

Consequently,

$$t \int_\Omega |w(x,t)|^p \, dx \leq \int_0^t \int_\Omega |w(x,s)|^p \, dx \, ds$$

$$\leq C \int_0^t \int_\Omega \int_\Omega J(x-y)|u(y,s) - u(x,s)|^p \, dy \, dx \, ds.$$

On the other hand, multiplying the equation by $u(x,t)$ and integrating in space and time, we get

$$\int_\Omega |u(x,t)|^2 - \int_\Omega |u_0(x)|^2 \, dx = -\int_0^t \int_\Omega \int_\Omega J(x-y)|u(y,s) - u(x,s)|^p \, dy \, dx \, ds,$$

which implies

$$\int_0^t \int_\Omega \int_\Omega J(x-y)|u(y,s) - u(x,s)|^p \, dy \, dx \, ds \leq \|u_0\|_{L^2(\Omega)}^2,$$

and therefore we conclude that

$$\int_\Omega |w(x,t)|^p \, dx \leq C \frac{\|u_0\|_{L^2(\Omega)}^2}{t}. \qquad \qquad \square$$

6.2. The Dirichlet problem

In this section we study the nonlocal diffusion equation

$$\begin{cases} u_t(x,t) = \displaystyle\int_{\mathbb{R}^N} J(x-y)|u(y,t) - u(x,t)|^{p-2}(u(y,t) - u(x,t))dy, & x \in \Omega, t > 0, \\ u(x,t) = \psi(x), & x \in \mathbb{R}^N \setminus \overline{\Omega}, \, t > 0, \\ u(x,0) = u_0(x), & x \in \Omega, \end{cases}$$

where Ω is a bounded domain and u is prescribed in $\mathbb{R}^N \setminus \overline{\Omega}$ as $\psi(x)$. This problem can be rewriten either as

$$\begin{cases} u_t(x,t) = \displaystyle\int_{\Omega_J} J(x-y)|u(y,t) - u(x,t)|^{p-2}(u(y,t) - u(x,t)) \, dy, & x \in \Omega, t > 0, \\ u(x,t) = \psi(x), & x \in \Omega_J \setminus \overline{\Omega}, t > 0, \\ u(x,0) = u_0(x), & x \in \Omega, \end{cases}$$

where $\Omega_J = \Omega + \operatorname{supp}(J)$ and ψ is only needed to be fixed in $\Omega_J \setminus \overline{\Omega}$, or as

$$(6.38) \qquad \begin{cases} u_t(x,t) = \displaystyle\int_\Omega J(x-y)|u(y,t) - u(x,t)|^{p-2}(u(y,t) - u(x,t)) \, dy \\ \qquad\qquad + \displaystyle\int_{\Omega_J \setminus \overline{\Omega}} J(x-y)|\psi(y) - u(x,t)|^{p-2}(\psi(y) - u(x,t)) \, dy, \\ u(x,0) = u_0(x), & x \in \Omega, \, t > 0, \end{cases}$$

and we call it the *nonlocal p-Laplacian problem with Dirichlet boundary condition*. Note that we are prescribing the values of u outside $\overline{\Omega}$ taking no care on the boundary. This is due to the nonlocal character of the problem.

Closely related to this problem is the homogeneous Neumann problem treated in the previous section. The difference here is that we are now considering Dirichlet boundary conditions, including the nonhomogeneous case, and this introduces new difficulties, specially when one tries to recover the local models by rescaling. Remark that in our nonlocal formulation we are not imposing any continuity between the values of u inside Ω and outside of it, ψ. However, when dealing with local problems, usually the boundary datum is taken in the sense of traces, that is, $u|_{\partial\Omega} = \psi$. Recovering this condition as $\varepsilon \to 0$ is one of the main results of this section.

Solutions of (6.38) will be understood in the following sense.

DEFINITION 6.23. A *solution* of (6.38) in $[0,T]$ is a function
$$u \in W^{1,1}(0, T; L^1(\Omega))$$
that satisfies $u(x, 0) = u_0(x)$ a.e. $x \in \Omega$ and
$$u_t(x, t) = \int_\Omega J(x - y)|u(y, t) - u(x, t)|^{p-2}(u(y, t) - u(x, t))\, dy$$
$$+ \int_{\Omega_J \setminus \overline{\Omega}} J(x - y)|\psi(y) - u(x, t)|^{p-2}(\psi(y) - u(x, t))\, dy,$$
for a.e. $t \in (0, T)$ and a.e. $x \in \Omega$.

Our main result on the existence and uniqueness of solutions for this problem is the following theorem.

THEOREM 6.24. *Let $u_0 \in L^p(\Omega)$, $\psi \in L^p(\Omega_J \setminus \overline{\Omega})$. For any $T > 0$, there exists a unique solution of (6.38). Moreover, if $u_{i0} \in L^1(\Omega)$ and u_i is a solution in $[0, T]$ of (6.38) with initial data u_{i0}, $i = 1, 2$, respectively, then*
$$\int_\Omega (u_1(t) - u_2(t))^+ \le \int_\Omega (u_{10} - u_{20})^+ \qquad \text{for every } t \in [0, T].$$
If $u_{i0} \in L^p(\Omega)$, $i = 1, 2$, then
$$\|u_1(t) - u_2(t)\|_{L^p(\Omega)} \le \|u_{10} - u_{20}\|_{L^p(\Omega)} \qquad \text{for every } t \in [0, T].$$

Note that, as in the case of the local p-Laplacian, the Dirichlet problem can be written as a Neumann problem with a particular flux that depends on the solution itself. Indeed, the problem (6.38) can be written as
$$\begin{cases} u_t(x, t) = \displaystyle\int_\Omega J(x - y)|u(y, t) - u(x, t)|^{p-2}(u(y, t) - u(x, t))\, dy + \varphi(x, u(x, t)), \\ u(x, 0) = u_0(x), \qquad\qquad\qquad\qquad\qquad\qquad\qquad\qquad x \in \Omega,\ t > 0, \end{cases}$$
where
$$\varphi(x, u(x, t)) = \int_{\Omega_J \setminus \overline{\Omega}} J(x - y)|\psi(y) - u(x, t)|^{p-2}(\psi(y) - u(x, t))\, dy.$$

In the homogeneous case $\psi \equiv 0$ and hence $\varphi(x, u(x, t))$ becomes
$$\varphi(x, u(x, t)) = -\left(\int_{\Omega_J \setminus \overline{\Omega}} J(x - y)\, dy \right) |u(x, t)|^{p-2} u(x, t).$$

Therefore, (6.38) becomes a nonhomogeneous Neumann problem (see the previous Section 6.1) with a flux given by $\varphi(x, u(x, t))$.

Notation. From now on in this section, in order to simplify the notation, we will set

$$
u_\psi(x) := \begin{cases} u(x) & \text{if } x \in \Omega, \\[2mm] \psi(x) & \text{if } x \in \Omega_J \setminus \overline{\Omega}, \\[2mm] 0 & \text{if } x \notin \Omega_J. \end{cases}
$$

Observe that, in this way, we can rewrite (6.38) as

$$
\begin{cases} u_t(x,t) = \displaystyle\int_{\Omega_J} J(x - y)|u_\psi(y,t) - u(x,t)|^{p-2}(u_\psi(y,t) - u(x,t))\, dy, \\[3mm] u(x,0) = u_0(x), \hspace{5cm} x \in \Omega,\ t > 0. \end{cases}
$$

6.2.1. A Poincaré type inequality. We have the following Poincaré type inequality. In contrast with Proposition 6.19, for which we impose the condition $J(0) > 0$ (see Remark 6.20 for a counterexample of that result if $J = 0$ in a ball centered at the origin), here we do not need to impose such a condition. This is due to the fact that the outside values influence the inside ones.

PROPOSITION 6.25. *Given* $q \geq 1$, $J : \mathbb{R}^N \to \mathbb{R}$ *a nonnegative continuous radial function with compact support,* Ω *a bounded domain in* $\mathbb{R}^N$ *and* $\psi \in L^q(\Omega_J \setminus \overline{\Omega})$, *there exists* $\lambda = \lambda(J, \Omega, q) > 0$ *such that*

$$
(6.39) \quad \lambda \int_\Omega |u(x)|^q\, dx \leq \int_\Omega \int_{\Omega_J} J(x - y)|u_\psi(y) - u(x)|^q\, dy\, dx + \int_{\Omega_J \setminus \overline{\Omega}} |\psi(y)|^q\, dy
$$

for all $u \in L^q(\Omega)$.

PROOF. First the result is proved in the simpler case $J(0) > 0$. Take $r, \alpha > 0$ such that $J(x) \geq \alpha$ in $B(0, r)$. Let

$$
B_0 = \{x \in \Omega_J \setminus \overline{\Omega}\ :\ d(x, \Omega) \leq r/2\},
$$

$$
B_1 = \{x \in \Omega\ :\ d(x, B_0) \leq r/2\},
$$

$$
B_j = \left\{ x \in \Omega \setminus \bigcup_{k=1}^{j-1} B_k\ :\ d(x, B_{j-1}) \leq r/2 \right\}, \quad j = 2, 3, \dots.
$$

Observe that we can cover Ω by a finite number of nonnull sets $\{B_j\}_{j=1}^{l_r}$. Now

$$
\int_\Omega \int_{\Omega_J} J(x - y)|u_\psi(y) - u(x)|^q\, dy\, dx \geq \int_{B_j} \int_{B_{j-1}} J(x - y)|u_\psi(y) - u(x)|^q\, dy\, dx,
$$

for $j = 1, \ldots, l_r$, and

$$\int_{B_j} \int_{B_{j-1}} J(x-y)|u_\psi(y) - u(x)|^q \, dy \, dx$$

$$\geq \frac{1}{2^q} \int_{B_j} \int_{B_{j-1}} J(x-y)|u(x)|^q \, dy \, dx - \int_{B_j} \int_{B_{j-1}} J(x-y)|u_\psi(y)|^q \, dy \, dx$$

$$= \frac{1}{2^q} \int_{B_j} \left(\int_{B_{j-1}} J(x-y) \, dy \right) |u(x)|^q \, dx - \int_{B_{j-1}} \left(\int_{B_j} J(x-y) \, dx \right) |u_\psi(y)|^q \, dy$$

$$\geq \alpha_j \int_{B_j} |u(x)|^q \, dx - \beta \int_{B_{j-1}} |u_\psi(y)|^q \, dy,$$

where

$$\alpha_j = \frac{1}{2^q} \min_{x \in \overline{B}_j} \int_{B_{j-1}} J(x-y) \, dy > 0$$

(since $J(x) \geq \alpha$ in $B(0,r)$) and

$$\beta = \int_{\mathbb{R}^N} J(x) \, dx.$$

Hence

$$\int_\Omega \int_{\Omega_J} J(x-y)|u_\psi(y) - u(x)|^q \, dy \, dx \geq \alpha_j \int_{B_j} |u(x)|^q \, dx - \beta \int_{B_{j-1}} |u_\psi(y)|^q \, dy.$$

Therefore, since $u_\psi(y) = \psi(y)$ if $y \in B_0$, $u_\psi(y) = u(y)$ if $y \in B_j$, $j = 1, \ldots, l_r$, $B_j \cap B_i = \emptyset$, for all $i \neq j$ and $|\Omega \setminus \bigcup_{j=1}^{j_r} B_j| = 0$, it is easy to see, by cancelation, that there exists $\lambda = \lambda(J, \Omega, q) > 0$ such that

$$\lambda \int_\Omega |u|^q \leq \int_\Omega \int_{\Omega_J} J(x-y)|u_\psi(y) - u(x)|^q \, dy \, dx + \int_{B_0} |\psi|^q.$$

For the general case let $\mathbf{a} \geq 0$ and $r, \alpha > 0$ satisfy

(6.40) $J(x) \geq \alpha$ in the annulus $A(0, \mathbf{a}, r) = \{\mathbf{a} < |x| < r\}$.

We consider B_j, $j = 1, \ldots, l_r$, and

$$B_{-j} = \left\{ x \in \Omega_J \setminus \left(\Omega \cup \left(\bigcup_{k=0}^{j-1} B_{-k} \right) \right) : d(x, B_{-j+1}) \leq r/2 \right\}, \quad j = 1, 2, \ldots, m_r,$$

with m_r such that $\Omega_J \subset \bigcup_{-m_r \leq j \leq l_r} B_j$, $\Omega_J \cap B_{-m_r-1} = \emptyset$. Observe that, for each B_j, $1 \leq j \leq l_r$, there exists B_{j^e}, $-m_r \leq j^e < j$, satisfying

(6.41) $|(x + A(0, \mathbf{a}, r)) \cap B_{j^e}| > 0 \quad \forall x \in \overline{B}_j$.

In this case, for $j = 1, \ldots, l_r$,

$$\int_\Omega \int_{\Omega_J} J(x-y)|u_\psi(y) - u(x)|^q \, dy \, dx \geq \int_{B_j} \int_{B_{j^e}} J(x-y)|u_\psi(y) - u(x)|^q \, dy \, dx$$

$$\geq \alpha_j \int_{B_j} |u(x)|^q \, dx - \beta \int_{B_{j^e}} |u_\psi(y)|^q \, dy,$$

where

$$\alpha_j = \frac{1}{2^q} \min_{x \in \overline{B}_j} \int_{B_{je}} J(x-y) dy > 0$$

thanks to (6.40) and (6.41). Finally, as above, by cancelation, we find a constant $\lambda = \lambda(J, \Omega, q) > 0$ such that

$$\lambda \int_\Omega |u|^q \le \int_\Omega \int_{\Omega_J} J(x-y)|u_\psi(y) - u(x)|^q \, dy \, dx + \int_{\tilde B} |\psi|^q,$$

where

$$\tilde B = \bigcup_{-m_r \le j \le 0} B_j. \qquad \square$$

6.2.2. Existence and uniqueness of solutions. As in Subsection 6.1.1, in order to study (6.38) we introduce in $L^1(\Omega)$ the following operator.

DEFINITION 6.26. For $\psi : \Omega_J \setminus \overline{\Omega} \to \mathbb{R}$ such that $|\psi|^{p-1} \in L^1(\Omega_J \setminus \overline{\Omega})$, we define in $L^1(\Omega)$ the operator $B_{p,\psi}^J$ by

$$B_{p,\psi}^J(u)(x) \;\; = -\int_{\Omega_J} J(x-y)|u_\psi(y) - u(x)|^{p-2}(u_\psi(y) - u(x)) \, dy, \quad x \in \Omega.$$

REMARK 6.27. If $\psi = 0$, then

$$B_{p,0}^J(u)(x) = -\int_\Omega J(x-y)|u(y) - u(x)|^{p-2}(u(y) - u(x)) \, dy$$

$$+ \left(\int_{\Omega_J \setminus \overline{\Omega}} J(x-y) dy \right) |u(x)|^{p-2} u(x), \qquad x \in \Omega.$$

REMARK 6.28. It is easy to see that

(i) If $\psi = 0$, $B_{p,0}^J$ is positively homogeneous of degree $p - 1$.

(ii) $L^{p-1}(\Omega) \subset D(B_{p,\psi}^J)$ if $p > 2$.

(iii) For $1 < p \le 2$, $D(B_{p,\psi}^J) = L^1(\Omega)$ and $B_{p,\psi}^J$ is closed in $L^1(\Omega) \times L^1(\Omega)$.

We have the following monotonicity lemma, whose proof is straightforward.

LEMMA 6.29. Let $1 < p < +\infty$, $\psi : \Omega_J \setminus \overline{\Omega} \to \mathbb{R}$, $|\psi|^{p-1} \in L^1(\Omega_J \setminus \overline{\Omega})$, and $T : \mathbb{R} \to \mathbb{R}$ a nondecreasing function. Then

(i) For every $u, v \in L^p(\Omega)$ such that $T(u - v) \in L^p(\Omega)$, we have

(6.42)
$$\int_\Omega (B_{p,\psi}^J u(x) - B_{p,\psi}^J v(x)) \, T(u(x) - v(x)) dx$$

$$= \frac{1}{2} \int_{\Omega_J} \int_{\Omega_J} J(x-y) \left(T(u_\psi(y) - v_\psi(y)) - T(u_\psi(x) - v_\psi(x)) \right)$$

$$\times \Big(|u_\psi(y) - u_\psi(x)|^{p-2}(u_\psi(y) - u_\psi(x)) - |v_\psi(y) - v_\psi(x)|^{p-2}(v_\psi(y) - v_\psi(x)) \Big) dy \, dx.$$

(ii) Moreover, if T is bounded, (6.42) holds for $u, v \in D(B_{p,\psi}^J)$.

In the next result we prove that $B_{p,\psi}^J$ is a completely accretive operator (see Section A.8) and satisfies a range condition.

THEOREM 6.30. *For $\psi \in L^p(\Omega_J \setminus \overline{\Omega})$, the operator $B_{p,\psi}^J$ is completely accretive and verifies the range condition*

(6.43) $$L^p(\Omega) \subset \mathrm{R}(I + B_{p,\psi}^J).$$

PROOF. Given $u_i \in D(B_{p,\psi}^J)$, $i = 1, 2$, by the monotonicity Lemma 6.29, for any $q \in P_0$ we have that

$$\int_\Omega (B_{p,\psi}^J u_1(x) - B_{p,\psi}^J u_2(x)) q(u_1(x) - u_2(x))\, dx \geq 0,$$

from which it follows that $B_{p,\psi}^J$ is a completely accretive operator.

To show that $B_{p,\psi}^J$ satisfies the range condition we have to prove that for any $\phi \in L^p(\Omega)$ there exists $u \in D(B_{p,\psi}^J)$ such that $\phi = u + B_{p,\psi}^J u$.

Assume first that $p \geq 2$. Let $\phi \in L^p(\Omega)$ and set

$$K = \left\{ w \in L^p(\Omega_J) \ : \ w = \psi \text{ in } \Omega_J \setminus \overline{\Omega} \right\}.$$

We consider the continuous monotone operator $A : K \to L^{p'}(\Omega_J)$ defined by

$$A(w)(x) := w(x) - \int_{\Omega_J} J(x - y)|w(y) - w(x)|^{p-2}(w(y) - w(x))\, dy.$$

This operator A is coercive in $L^p(\Omega_J)$. In fact, by the Poincaré type inequality given in Proposition 6.25, for any $w \in K$,

$$\int_{\Omega_J} A(w)w$$

$$= \int_{\Omega_J} w^2 - \int_{\Omega_J} \int_{\Omega_J} J(x - y)|w(y) - w(x)|^{p-2}(w(y) - w(x))\, dy w(x) dx$$

$$\geq \frac{1}{2} \int_{\Omega_J} \int_{\Omega_J} J(x - y)|w(y) - w(x)|^p\, dy dx$$

$$\geq \frac{1}{2} \int_\Omega \int_{\Omega_J} J(x - y)|w_\psi(y) - w(x)|^p\, dy dx$$

$$\geq \frac{\lambda}{2} \|w\|_{L^p(\Omega)}^p - \frac{1}{2} \int_{\Omega_J \setminus \overline{\Omega}} |\psi|^p.$$

Therefore

$$\lim_{\|w\|_{L^p(\Omega_J)} \to +\infty, w \in K} \frac{\displaystyle\int_{\Omega_J} A(w)w}{\|w\|_{L^p(\Omega_J)}} = +\infty.$$

Since $p \geq 2$, we have the function $\phi_\psi \in L^{p'}(\Omega_J)$. Then, applying [**127**, Corollary III.1.8] to the operator $B(w) := A(w) - \phi_\psi$, we see that there exists $w \in K$ such that

$$w(x) - \int_{\Omega_J} J(x - y)|w(y) - w(x)|^{p-2}(w(y) - w(x))\, dy = \phi_\psi(x) \qquad \text{for all } x \in \Omega_J.$$

Hence, $u := w_{|\Omega}$ satisfies

$$u(x) - \int_{\Omega_J} J(x-y)|u_\psi(y) - u(x)|^{p-2}(u_\psi(y) - u(x))\,dy = \phi(x) \qquad \text{for all } x \in \Omega,$$

and consequently $\phi = u + B_{p,\psi}^J u$.

Suppose now that $1 < p < 2$. By the results in Section 6.1, the operator

$$B_p^J u(x) = -\int_\Omega J(x-y)|u(y) - u(x)|^{p-2}(u(y) - u(x))\,dy$$

is m-accretive in $L^1(\Omega)$ and satisfies what is called property (M_0), that is, for any $q \in P_0$ and $(u,v) \in B_p^J$,

$$\int_\Omega q(u)v \geq 0.$$

On the other hand,

$$\varphi(x,r) = -\int_{\Omega_J \setminus \overline{\Omega}} J(x-y)|\psi(y) - r|^{p-2}(\psi(y) - r)\,dy$$

is continuous and nondecreasing in r for almost every $x \in \Omega$, and $\varphi(\cdot, r)$ is an $L^1(\Omega)$-function for all r. Therefore, by [**19**, Theorem 3.1], since $B_{p,\psi}^J u(x) = B_p^J u(x) + \varphi(x, u(x))$, $B_{p,\psi}^J$ is m-accretive in $L^1(\Omega)$. $\qquad \square$

If $\mathcal{B}_{p,\psi}^J$ denotes the closure of $B_{p,\psi}^J$ in $L^1(\Omega)$, by Theorem 6.30, $\mathcal{B}_{p,\psi}^J$ is m-completely accretive in $L^1(\Omega)$.

As a consequence of the above results, by using the Nonlinear Semigroup Theory, we get the following theorem, from which Theorem 6.24 can be derived.

THEOREM 6.31. *Let $T > 0$, $\psi \in L^p(\Omega_J \setminus \overline{\Omega})$ and $u_0 \in L^1(\Omega)$. Then there exists a unique mild solution u of*

$$(6.44) \qquad \begin{cases} u'(t) + B_{p,\psi}^J u(t) = 0, & t \in (0,T), \\ \\ u(0) = u_0. \end{cases}$$

Moreover,

(1) *If $u_0 \in L^p(\Omega)$, the unique mild solution u of (6.44) is a solution of (6.38) in the sense of Definition 6.23. If $1 < p \leq 2$, this is true for any $u_0 \in L^1(\Omega)$ and any ψ such that $|\psi|^{p-1} \in L^1(\Omega_J \setminus \overline{\Omega})$.*

(2) *Let $u_{i0} \in L^1(\Omega)$ and u_i a solution in $[0,T]$ of (6.38) with initial data u_{i0}, $i = 1, 2$. Then*

$$\int_\Omega (u_1(t) - u_2(t))^+ \leq \int_\Omega (u_{10} - u_{20})^+ \quad \text{for every } t \in [0,T].$$

If $u_{i0} \in L^q(\Omega)$, $i = 1, 2$, $q \in [1, +\infty]$, then

$$\|u_1(t) - u_2(t)\|_{L^q(\Omega)} \leq \|u_{10} - u_{20}\|_{L^q(\Omega)} \quad \text{for every } t \in [0,T].$$

PROOF. As a consequence of Theorem 6.30 we get the existence of a mild solution of (6.44). On the other hand, $u(t)$ is a solution of (6.38) if and only if $u(t)$ is a strong solution of the abstract Cauchy problem (6.44). Now, due to the complete accretivity of $B_{p,\psi}^J$ and the range condition (6.43), $u(t)$ is a strong solution. Moreover, in the case $1 < p \leq 2$, since $D(B^J_{p,\psi}) = L^1(\Omega)$ and $B^J_{p,\psi}$ is closed in $L^1(\Omega) \times L^1(\Omega)$, the result holds for L^1-data. Finally, the contraction principle is a consequence of Theorem A.28. $\qquad\square$

6.2.3. Convergence to the local p-Laplacian. Consider the local p-Laplacian evolution equation with Dirichlet boundary condition

$$(6.45) \qquad \begin{cases} v_t = \Delta_p v & \text{in } \Omega \times (0, T), \\[2mm] v = \tilde{\psi} & \text{on } \partial\Omega \times (0, T), \\[2mm] v(\cdot, 0) = u_0 & \text{in } \Omega, \end{cases}$$

where the boundary datum $\tilde{\psi}$ is assumed to be the trace of a function defined in a larger domain. We prove that the solutions of this local problem can be approximated by solutions of a sequence of nonlocal p-*Laplacian* problems of the form (6.38). Indeed, for $p > 1$ fixed and J we consider the rescaled kernels

$$(6.46) \qquad J_{p,\varepsilon}(x) := \frac{C_{J,p}}{\varepsilon^{p+N}} J\left(\frac{x}{\varepsilon}\right),$$

where

$$C_{J,p}^{-1} := \frac{1}{2} \int_{\mathbb{R}^N} J(z)|z_N|^p \, dz$$

is a normalizing constant. Then we obtain the following result.

THEOREM 6.32. *Let Ω be a smooth bounded domain in $\mathbb{R}^N$ and $\tilde{\psi} \in L^\infty(\partial\Omega) \cap W^{1/p',p}(\partial\Omega)$. Let $\psi \in L^\infty(\Omega_J) \cap W^{1,p}(\Omega_J)$ such that $\psi|_{\partial\Omega} = \tilde{\psi}$. Suppose $J(x) \geq J(y)$ if $|x| \leq |y|$. Let $T > 0$ and $u_0 \in L^p(\Omega)$. Let u_ε be the unique solution of (6.38) with J replaced by $J_{p,\varepsilon}$ and v the unique mild solution of (6.45). Then*

$$(6.47) \qquad \lim_{\varepsilon \to 0} \sup_{t \in [0,T]} \|u_\varepsilon(\cdot, t) - v(\cdot, t)\|_{L^p(\Omega)} = 0.$$

Note that the above result says that (6.38) is a nonlocal problem analogous to the p-Laplacian with a nonhomogeneous Dirichlet boundary condition.

REMARK 6.33. Recall that for the linear case, $p = 2$, the convergence of the solutions of rescaled nonlocal problems of this type to the solution of the heat equation was proved in Chapter 3.

The local p-Laplacian equation. Let us give some results about the p-Laplacian problem (6.45) that will be used in the proof of convergence of the rescaled problems.

In the case $\tilde{\psi} \in W^{1/p',p}(\partial\Omega)$, associated to the p-Laplacian with nonhomogeneous Dirichlet boundary condition, in [7] the operator $A_{p,\tilde{\psi}} \subset L^1(\Omega) \times L^1(\Omega)$ is defined as $(v, \hat{v}) \in A_{p,\tilde{\psi}}$ if and only if $\hat{v} \in L^1(\Omega)$,

$$v \in W^{1,p}_{\tilde{\psi}}(\Omega) := \{v \in W^{1,p}(\Omega) : v|_{\partial\Omega} = \tilde{\psi} \ \mathcal{H}^{N-1}\text{-a.e. on } \partial\Omega\}$$

and

$$\int_\Omega |\nabla v|^{p-2} \nabla v \cdot \nabla(v - \xi) \leq \int_\Omega \hat{v}(v - \xi) \quad \text{for every } \xi \in W^{1,p}_{\tilde{\psi}}(\Omega) \cap L^\infty(\Omega),$$

which is equivalent to

$$\int_\Omega |\nabla v|^{p-2}\nabla v \cdot \nabla \xi = \int_\Omega \hat{v}\xi \quad \text{for every } \xi \in W_0^{1,p}(\Omega) \cap L^\infty(\Omega).$$

Moreover, for $\tilde{\psi} \in W^{1/p',p}(\partial\Omega) \cap L^\infty(\partial\Omega)$, $A_{p,\tilde{\psi}}$ is proved to be a completely accretive operator in $L^1(\Omega)$ satisfying the range condition $L^\infty(\Omega) \subset \mathrm{R}(I + A_{p,\tilde{\psi}})$. Therefore, its closure $\mathcal{A}_{p,\tilde{\psi}}$ in $L^1(\Omega) \times L^1(\Omega)$ is an m-completely accretive operator in $L^1(\Omega)$. On the other hand, it is easy to see that $\overline{D(A_{p,\tilde{\psi}})}^{L^1(\Omega)} = L^1(\Omega)$, and consequently for any $u_0 \in L^1(\Omega)$ there exists a unique mild solution $v(t) = e^{-t\mathcal{A}_{p,\tilde{\psi}}}u_0$ of the abstract Cauchy problem associated to (6.45), given by the Crandall-Liggett exponential formula. Due to the complete accretivity of the operator $\mathcal{A}_{p,\tilde{\psi}}$, in the case $u_0 \in D(\mathcal{A}_{p,\tilde{\psi}})$ this mild solution is the unique strong solution of problem (6.45). In the homogeneous case $\tilde{\psi} = 0$, due to the results in [**44**], we can say that, for any $u_0 \in L^1(\Omega)$, the mild solution $v(t) = e^{-t\mathcal{A}_{p,0}}u_0$ is the unique entropy solution of problem (6.45).

Proof of Theorem 6.32. Arguing as in the proof of Theorem 6.18, since all the solutions of Theorem 6.32 coincide with the semigroup solutions, by Theorem A.37, the proof is obtained from the following result.

PROPOSITION 6.34. *Let Ω be a smooth bounded domain in $\mathbb{R}^N$ and take $\tilde{\psi} \in L^\infty(\partial\Omega) \cap W^{1/p',p}(\partial\Omega)$. Let $\psi \in L^\infty(\Omega_J) \cap W^{1,p}(\Omega_J)$ such that $\psi|_{\partial\Omega} = \tilde{\psi}$. Suppose $J(x) \geq J(y)$ if $|x| \leq |y|$. Then, for any $\phi \in L^\infty(\Omega)$,*

$$\left(I + B_{p,\psi}^{J_{p,\varepsilon}}\right)^{-1}\phi \to \left(I + A_{p,\tilde{\psi}}\right)^{-1}\phi \quad \text{in } L^p(\Omega) \text{ as } \varepsilon \to 0.$$

PROOF. We denote

$$\Omega_\varepsilon := \Omega_{J_{p,\varepsilon}} = \Omega + \mathrm{supp}(J_{p,\varepsilon}).$$

For $\varepsilon > 0$ small, let $u_\varepsilon = \left(I + B_{p,\psi}^{J_{p,\varepsilon}}\right)^{-1}\phi$. Then

(6.48)
$$\int_\Omega u_\varepsilon \xi - \frac{C_{J,p}}{\varepsilon^{p+N}} \int_\Omega\int_{\Omega_\varepsilon} J\left(\frac{x-y}{\varepsilon}\right)|(u_\varepsilon)_\psi(y) - u_\varepsilon(x)|^{p-2}$$
$$\times ((u_\varepsilon)_\psi(y) - u_\varepsilon(x))\,dy\,\xi(x)\,dx = \int_\Omega \phi\xi$$

for every $\xi \in L^\infty(\Omega)$.

Let $M := \max\{\|\phi\|_{L^\infty(\Omega)}, \|\psi\|_{L^\infty(\Omega_J)}\}$. Taking $\xi = (u_\varepsilon - M)^+$ in (6.48), we get

$$\int_\Omega u_\varepsilon(x)(u_\varepsilon(x) - M)^+ dx$$
$$- \frac{C_{J,p}}{\varepsilon^{p+N}}\int_\Omega\int_{\Omega_\varepsilon} J\left(\frac{x-y}{\varepsilon}\right)|(u_\varepsilon)_\psi(y) - u_\varepsilon(x)|^{p-2}((u_\varepsilon)_\psi(y) - u_\varepsilon(x))\,dy$$
$$\times (u_\varepsilon(x) - M)^+\,dx$$
$$= \int_\Omega \phi(x)(u_\varepsilon(x) - M)^+\,dx.$$

Now,

$$-\frac{C_{J,p}}{\varepsilon^{p+N}} \int_\Omega \int_{\Omega_\varepsilon} J\left(\frac{x-y}{\varepsilon}\right) |(u_\varepsilon)_\psi(y) - u_\varepsilon(x)|^{p-2}((u_\varepsilon)_\psi(y) - u_\varepsilon(x))\, dy$$
$$\times (u_\varepsilon(x) - M)^+ dx$$

$$= -\frac{C_{J,p}}{\varepsilon^{p+N}} \int_{\Omega_\varepsilon} \int_{\Omega_\varepsilon} J\left(\frac{x-y}{\varepsilon}\right) |(u_\varepsilon)_\psi(y) - (u_\varepsilon)_\psi(x)|^{p-2}((u_\varepsilon)_\psi(y) - (u_\varepsilon)_\psi(x))\, dy$$
$$\times ((u_\varepsilon)_\psi(x) - M)^+ dx$$

$$= \frac{C_{J,p}}{2\varepsilon^{p+N}} \int_{\Omega_\varepsilon} \int_{\Omega_\varepsilon} J\left(\frac{x-y}{\varepsilon}\right) |(u_\varepsilon)_\psi(y) - (u_\varepsilon)_\psi(x)|^{p-2}((u_\varepsilon)_\psi(y) - (u_\varepsilon)_\psi(x))$$
$$\times (((u_\varepsilon)_\psi(y) - M)^+ - ((u_\varepsilon)_\psi(x) - M)^+)\, dy\, dx \geq 0.$$

Therefore

$$\int_\Omega u_\varepsilon(x)(u_\varepsilon(x) - M)^+ dx \leq \int_\Omega \phi(x)(u_\varepsilon(x) - M)^+ dx.$$

Consequently, we have

$$\int_\Omega (u_\varepsilon(x) - M)(u_\varepsilon(x) - M)^+ dx \leq \int_\Omega (\phi(x) - M)(u_\varepsilon(x) - M)^+ dx \leq 0,$$

and $u_\varepsilon(x) \leq M$ for almost all $x \in \Omega$. Analogously, we can obtain $-M \leq u_\varepsilon(x)$ for almost all $x \in \Omega$. Thus

$$(6.49) \qquad \|u_\varepsilon\|_{L^\infty(\Omega)} \leq M \qquad \text{for all } \epsilon > 0,$$

and therefore, there exists a sequence $\varepsilon_n \to 0$ such that

$$u_{\varepsilon_n} \rightharpoonup v \quad \text{weakly in } L^1(\Omega).$$

Taking $\xi = u_\varepsilon - \psi$ in (6.48) we get

$$(6.50) \quad \int_\Omega u_\varepsilon(u_\varepsilon - \psi) - \frac{C_{J,p}}{\varepsilon^{p+N}} \int_{\Omega_\varepsilon} \int_{\Omega_\varepsilon} J\left(\frac{x-y}{\varepsilon}\right) |(u_\varepsilon)_\psi(y) - (u_\varepsilon)_\psi(x)|^{p-2}$$
$$\times ((u_\varepsilon)_\psi(y) - (u_\varepsilon)_\psi(x))\, dy\, ((u_\varepsilon)_\psi(x) - \psi(x))\, dx = \int_\Omega \phi(u_\varepsilon - \psi).$$

By (6.50) and (6.49),

$$\frac{C_{J,p}}{2\varepsilon^N} \int_{\Omega_\varepsilon} \int_{\Omega_\varepsilon} J\left(\frac{x-y}{\varepsilon}\right) \frac{|(u_\varepsilon)_\psi(y) - (u_\varepsilon)_\psi(x)|^p}{\varepsilon^p}\, dy\, dx$$

$$\leq \frac{C_{J,p}}{2\varepsilon^N} \int_{\Omega_\varepsilon} \int_{\Omega_\varepsilon} J\left(\frac{x-y}{\varepsilon}\right) \frac{|(u_\varepsilon)_\psi(y) - (u_\varepsilon)_\psi(x)|^{p-1}}{\varepsilon^{p-1}} \frac{|\psi(y) - \psi(x)|}{\varepsilon}\, dy\, dx + M_1.$$

Since $\psi \in W^{1,p}(\Omega_J)$, using Young's inequality, we obtain

$$\frac{1}{\varepsilon^N} \int_{\Omega_\varepsilon} \int_{\Omega_\varepsilon} J\left(\frac{x-y}{\varepsilon}\right) \frac{|(u_\varepsilon)_\psi(y) - (u_\varepsilon)_\psi(x)|^p}{\varepsilon^p}\, dy\, dx \leq M_2.$$

Moreover,

$$
\int_{\Omega_J} \int_{\Omega_J} \frac{1}{\varepsilon^N} J\left(\frac{x-y}{\varepsilon}\right) \left|\frac{(u_\varepsilon)_\psi(y) - (u_\varepsilon)_\psi(x)}{\varepsilon}\right|^p \, dx \, dy
$$

$$
= \int_{\Omega_\varepsilon} \int_{\Omega_\varepsilon} \frac{1}{\varepsilon^N} J\left(\frac{x-y}{\varepsilon}\right) \left|\frac{(u_\varepsilon)_\psi(y) - (u_\varepsilon)_\psi(x)}{\varepsilon}\right|^p \, dx \, dy
$$

$$
+ 2\int_{\Omega_J \setminus \overline{\Omega}_\varepsilon} \int_{\Omega_\varepsilon} \frac{1}{\varepsilon^N} J\left(\frac{x-y}{\varepsilon}\right) \left|\frac{\psi(y) - (u_\varepsilon)_\psi(x)}{\varepsilon}\right|^p \, dx \, dy
$$

$$
+ \int_{\Omega_J \setminus \overline{\Omega}_\varepsilon} \int_{\Omega_J \setminus \overline{\Omega}_\varepsilon} \frac{1}{\varepsilon^N} J\left(\frac{x-y}{\varepsilon}\right) \left|\frac{\psi(y) - \psi(x)}{\varepsilon}\right|^p \, dx \, dy
$$

$$
= \int_{\Omega_\varepsilon} \int_{\Omega_\varepsilon} \frac{1}{\varepsilon^N} J\left(\frac{x-y}{\varepsilon}\right) \left|\frac{(u_\varepsilon)_\psi(y) - (u_\varepsilon)_\psi(x)}{\varepsilon}\right|^p \, dx \, dy
$$

$$
+ 2\int_{\Omega_J \setminus \overline{\Omega}_\varepsilon} \int_{\Omega_\varepsilon \setminus \overline{\Omega}} \frac{1}{\varepsilon^N} J\left(\frac{x-y}{\varepsilon}\right) \left|\frac{\psi(y) - \psi(x)}{\varepsilon}\right|^p \, dx \, dy
$$

$$
+ \int_{\Omega_J \setminus \overline{\Omega}_\varepsilon} \int_{\Omega_J \setminus \overline{\Omega}_\varepsilon} \frac{1}{\varepsilon^N} J\left(\frac{x-y}{\varepsilon}\right) \left|\frac{\psi(y) - \psi(x)}{\varepsilon}\right|^p \, dx \, dy \le M_3.
$$

Therefore, by Theorem 6.11, there exists a subsequence, denoted as above, and $w \in W^{1,p}(\Omega_J)$ such that

$$
(u_{\varepsilon_n})_\psi \to w \quad \text{strongly in } L^p(\Omega_J);
$$

hence $w = v$ in Ω, $v \in W^{1,p}_{\frac{1}{\psi}}(\Omega)$, and, moreover,

$$
\left(\frac{C_{J,p}}{2} J(z)\right)^{1/p} \chi_\Omega(x + \varepsilon_n z) \frac{(u_{\varepsilon_n})_\psi(x + \varepsilon_n z) - (u_{\varepsilon_n})_\psi(x)}{\varepsilon_n}
$$

$$
\rightharpoonup \left(\frac{C_{J,p}}{2} J(z)\right)^{1/p} z \cdot \nabla v(x)
$$

weakly in $L^p(\Omega) \times L^p(\mathbb{R}^N)$. We can also assume that

$$
(J(z))^{1/p'} \left|\frac{(u_{\varepsilon_n})_\psi(x + \varepsilon_n z) - (u_{\varepsilon_n})_\psi(x)}{\varepsilon_n}\right|^{p-2} \chi_{\Omega_{\varepsilon_n}}(x + \varepsilon_n z)
$$

$$
\times \frac{(u_{\varepsilon_n})_\psi(x + \varepsilon_n z) - (u_{\varepsilon_n})_\psi(x)}{\varepsilon_n}
$$

$$
\rightharpoonup (J(z))^{1/p'} \chi(x, z)
$$

weakly in $L^{p'}(\Omega_J) \times L^{p'}(\mathbb{R}^N)$, for some function $\chi \in L^{p'}(\Omega_J) \times L^{p'}(\mathbb{R}^N)$.

Passing to the limit in (6.48) for $\varepsilon = \varepsilon_n$, we get

$$
\int_\Omega v\xi + \int_{\mathbb{R}^N} \int_\Omega \frac{C_{J,p}}{2} J(z)\chi(x,z)\, z \cdot \nabla\xi(x) \, dx \, dz = \int_\Omega \phi\xi
$$

for every smooth ξ with support in Ω and by approximation for every $\xi \in W^{1,p}_0(\Omega)$.

Finally, working as in Proposition 6.15 and using Lemma 6.16,

$$\int_{\mathbb{R}^N} \int_\Omega \frac{C_{J,p}}{2} J(z)\chi(x,z)z \cdot \nabla\xi(x)\, dx\, dz = \int_\Omega |\nabla v|^{p-2}\, \nabla v \cdot \nabla\xi$$

and the proof is finished.   □

6.2.4. Asymptotic behaviour. We now study the asymptotic behaviour as $t \to \infty$ of the solutions of the nonlocal problem with homogeneous Dirichlet condition by using the Poincaré type inequality given in Proposition 6.25. This inequality permits us to show that these solutions converge to zero as in the following result.

THEOREM 6.35. *Let $p > 1$ and $u_0 \in L^\infty(\Omega)$. Let u be the solution of (6.38); then*

$$\|u(t)\|_{L^p(\Omega)} \le \left(C \frac{\|u_0\|^2_{L^2(\Omega)}}{t} \right)^{1/p} \qquad \forall\, t > 0,$$

where $C = C(J, \Omega, q)$.

PROOF. First we observe that, setting $u = 0$ in $\Omega_J \setminus \overline{\Omega}$,

$$\frac{d}{dt} \int_\Omega |u(x,t)|^p\, dx$$

$$= p \int_\Omega |u(x,t)|^{p-2} u(x,t) \int_{\Omega_J} J(x-y)|u(y,t) - u(x,t)|^{p-2}(u(y,t) - u(x,t))\, dy\, dx$$

$$= -\frac{p}{2} \int_{\Omega_J} \int_{\Omega_J} J(x-y)|u(y,t) - u(x,t)|^{p-2}(u(y,t) - u(x,t))$$

$$\times (|u(y,t)|^{p-2} u(y,t) - |u(x,t)|^{p-2} u(x,t))\, dy\, dx.$$

Therefore the $L^p(\Omega)$-norm of $u(\cdot, t)$ is decreasing with t.

Moreover, using Poincaré type inequality (6.39), we have

$$\int_\Omega |u(x,t)|^p\, dx \le C \int_\Omega \int_{\Omega_J} J(x-y)|u(y,t) - u(x,t)|^p\, dy\, dx.$$

Consequently,

$$t \int_\Omega |u(x,t)|^p\, dx \le \int_0^t \int_\Omega |u(x,s)|^p\, dx\, ds$$

$$\le C \int_0^t \int_\Omega \int_{\Omega_J} J(x-y)|u(y,s) - u(x,s)|^p\, dy\, dx\, ds.$$

On the other hand, multiplying the equation by $u(x,t)$ and integrating in space and time, we get

$$\int_\Omega |u(x,t)|^2 - \int_\Omega |u_0(x)|^2\, dx = -\int_0^t \int_{\Omega_J} \int_{\Omega_J} J(x-y)|u(y,s) - u(x,s)|^p\, dy\, dx\, ds,$$

which implies

$$\int_0^t \int_\Omega \int_{\Omega_J} J(x-y)|u(y,s) - u(x,s)|^p\, dy\, dx\, ds \le \|u_0\|^2_{L^2(\Omega)},$$

and therefore

$$\int_\Omega |u(x,t)|^p \, dx \le C\frac{\|u_0\|^2_{L^2(\Omega)}}{t}. \qquad \qquad \square$$

6.3. The Cauchy problem

Our main purpose in this section is to study the nonlocal p-Laplacian evolution problem
(6.51)
$$\begin{cases} (u_p)_t(x,t) = \displaystyle\int_{\mathbb{R}^N} J(x-y)|u_p(y,t) - u_p(x,t)|^{p-2}(u_p(y,t) - u_p(x,t))dy, \\[2mm] u_p(x,0) = u_0(x), \qquad\qquad\qquad\qquad\qquad\qquad x \in \mathbb{R}^N, \, t > 0. \end{cases}$$

In Chapter 1 this problem was treated in the linear case $p = 2$.

Let us note that the evolution problem (6.51) concerns the gradient flow associated to the functional

$$G_p^J(u) = \frac{1}{2p}\int_{\mathbb{R}^N}\int_{\mathbb{R}^N} J(x-y)|u(y) - u(x)|^p \, dy \, dx,$$

which is the nonlocal analog of the energy functional

$$F_p(v) = \begin{cases} \dfrac{1}{p}\displaystyle\int_{\mathbb{R}^N} |\nabla v(y)|^p \, dy & \text{if } v \in L^2(\mathbb{R}^N) \cap W^{1,p}(\mathbb{R}^N), \\[3mm] +\infty & \text{if } v \in L^2(\mathbb{R}^N) \setminus W^{1,p}(\mathbb{R}^N), \end{cases}$$

associated to the local p-Laplacian.

6.3.1. Existence and uniqueness. Solutions to (6.51) are understood in the following sense.

DEFINITION 6.36. Let $u_0 \in L^p(\mathbb{R}^N)$. A *solution* of (6.51) in $[0,T]$ is a function $u \in W^{1,1}(0,T;L^p(\mathbb{R}^N))$ which satisfies $u(x,0) = u_0(x)$ a.e. $x \in \mathbb{R}^N$ and

$$u_t(x,t) = \int_{\mathbb{R}^N} J(x-y)|u(y,t)-u(x,t)|^{p-2}(u(y,t)-u(x,t)) \, dy \quad \text{a.e. in } \mathbb{R}^N \times (0,T).$$

Using the ideas of Section 6.1 we can obtain the following theorem on the existence and uniqueness of global solutions for this problem.

THEOREM 6.37. *Let $u_0 \in L^p(\mathbb{R}^N)$. For any $T > 0$, there exists a unique solution of (6.51).*

Moreover, if $u_i(t)$ is a solution of (6.51) with initial data $u_{i0} \in L^p(\mathbb{R}^N)$, $i = 1,2$, then

$$\|(u_1(t) - u_2(t))^+\|_{L^p(\mathbb{R}^N)} \le \|(u_{10} - u_{20})^+\|_{L^p(\mathbb{R}^N)} \quad \textit{for every } t \in [0,T].$$

To prove this result, let us first define $Q_p^J : L^p(\mathbb{R}^N) \to L^{p'}(\mathbb{R}^N)$ by

$$Q_p^J u(x) = -\int_{\mathbb{R}^N} J(x-y)|u(y) - u(x)|^{p-2}(u(y) - u(x)) \, dy, \qquad x \in \mathbb{R}^N.$$

Observe that, for every u, $v \in L^p(\mathbb{R}^N)$ and $T : \mathbb{R} \to \mathbb{R}$ such that $T(u-v) \in L^p(\mathbb{R}^N)$,

$$
(6.52) \quad
\begin{aligned}
&\int_{\mathbb{R}^N} (Q_p^J u(x) - Q_p^J v(x)) \, T(u(x) - v(x)) dx \\
&= \frac{1}{2} \int_{\mathbb{R}^N} \int_{\mathbb{R}^N} J(x-y) \left(T(u(y) - v(y)) - T(u(x) - v(x)) \right) \\
&\quad \times \left(|u(y) - u(x)|^{p-2}(u(y) - u(x)) - |v(y) - v(x)|^{p-2}(v(y) - v(x)) \right) dy \, dx.
\end{aligned}
$$

Consider the operator

$$
\mathcal{Q}_p^J = \left\{ (u,v) \in L^p(\mathbb{R}^N) \times L^p(\mathbb{R}^N) \ : \ v = Q_p^J(u) \right\}.
$$

It is easy to see that $\overline{D(\mathcal{Q}_p^J)} = L^p(\mathbb{R}^N)$ and $\mathcal{Q}_p^J$ is positively homogeneous of degree $p - 1$, and $D(\mathcal{Q}_2^J) = L^2(\mathbb{R}^N)$. The following result holds.

PROPOSITION 6.38. $\mathcal{Q}_p^J$ is completely accretive and satisfies the range condition

$$
(6.53) \qquad\qquad L^p(\mathbb{R}^N) = \mathrm{R}(I + \mathcal{Q}_p^J).
$$

PROOF. Given $u_i \in D(\mathcal{Q}_p^J)$, $i = 1, 2$ and $q \in P_0$, by (6.52) we have

$$
\int_{\mathbb{R}^N} \left(Q_p^J u_1(x) - Q_p^J u_2(x) \right) q(u_1(x) - u_2(x)) \, dx \geq 0,
$$

from which it follows that $\mathcal{Q}_p^J$ is a completely accretive operator. To show that $\mathcal{Q}_p^J$ satisfies the range condition we have to prove that for any $\phi \in L^p(\mathbb{R}^N)$ there exists $u \in D(\mathcal{Q}_p^J)$ such that $u = (I + \mathcal{Q}_p^J)^{-1}\phi$. Let us first take $\phi \in L^1(\mathbb{R}^N) \cap L^\infty(\mathbb{R}^N)$. For every $n \in \mathbb{N}$, let $\phi_n := \phi \chi_{B(0,n)}$. By the results in Section 6.1, the operator $B_{p,n}^J$ defined by

$$
B_{p,n}^J u(x) = - \int_{B(0,n)} J(x-y) |u(y) - u(x)|^{p-2}(u(y) - u(x)) \, dy, \qquad x \in B(0,n),
$$

is m-completely accretive in $L^p(B(0,n))$. Then there exists $u_n \in L^p(B(0,n))$ such that

$$
(6.54) \qquad u_n(x) + B_{p,n}^J u_n(x) = \phi_n(x), \qquad \text{a.e. in } B(0,n).
$$

Moreover, $u_n \ll \phi_n$.

We denote by $\overline{u}_n$ and H_n the extensions

$$
\overline{u}_n(x) = \begin{cases} u_n(x) & \text{if } x \in B(0,n), \\ 0 & \text{if } x \in \mathbb{R}^N \setminus B(0,n), \end{cases}
$$

and

$$
H_n(x) = \begin{cases} B_{p,n}^J u_n(x) & \text{if } x \in B(0,n), \\ 0 & \text{if } x \in \mathbb{R}^N \setminus B(0,n). \end{cases}
$$

Since $u_n \ll \phi_n$, we have

$$
\|\overline{u}_n\|_{L^q(\mathbb{R}^N)} \leq \|\phi\|_{L^q(\mathbb{R}^N)} \qquad \text{for all } 1 \leq q \leq \infty, \forall \, n \in \mathbb{N}.
$$

Hence, we can assume that

$$
(6.55) \qquad\qquad \overline{u}_n \rightharpoonup u \quad \text{in } L^{p'}(\mathbb{R}^N)
$$

with $u \in L^p(\mathbb{R}^N)$.

On the other hand, multiplying (6.54) by u_n and integrating, we get

$$(6.56) \quad \int_{B(0,n)} \int_{B(0,n)} J(x-y)|u_n(y) - u_n(x)|^p \, dy \, dx \leq \|\phi\|_{L^2(\mathbb{R}^N)} \qquad \forall\, n \in \mathbb{N},$$

which implies, by Hölder's inequality, that $\{H_n \, : \, n \in \mathbb{N}\}$ is bounded in $L^{p'}(\mathbb{R}^N)$. Therefore, we can assume that

$$(6.57) \quad H_n \rightharpoonup H \quad \text{in } L^{p'}(\mathbb{R}^N).$$

By (6.55) and (6.57), taking limit in (6.54), we get

$$(6.58) \quad u + H = \phi \qquad \text{a.e. in } \mathbb{R}^N.$$

Let us show that

$$(6.59) \quad H(x) = -\int_{\mathbb{R}^N} J(x-y)|u(y) - u(x)|^{p-2}(u(y) - u(x)) \, dy \quad \text{a.e. in } x \in \mathbb{R}^N.$$

In fact, multiplying (6.54) by u_n and integrating, we obtain

$$\int_{B(0,n)} B_{p,n}^J u_n \, u_n = \int_{B(0,n)} (\phi - u_n) u_n$$

$$= \int_{B(0,n)} (\phi - u) u - \int_{B(0,n)} \phi(u - u_n)$$

$$+ \int_{B(0,n)} 2u(u - u_n) - \int_{B(0,n)} (u - u_n)(u - u_n).$$

By (6.58), we have

$$(6.60) \quad \limsup \int_{B(0,n)} B_{p,n}^J u_n \, u_n \leq \int_{\mathbb{R}^N} (\phi - u) u = \int_{\mathbb{R}^N} H \, u.$$

Since

$$0 \leq \int_{B(0,n)} \left(B_{p,n}^J u_n - B_{p,n}^J \xi \right)(u_n - \xi) \quad \forall\, \xi \in L^1(\mathbb{R}^N) \cap L^\infty(\mathbb{R}^N),$$

then

$$\int_{B(0,n)} B_{p,n}^J u_n \, u_n + \int_{B(0,n)} B_{p,n}^J \xi \, \xi \geq \int_{B(0,n)} B_{p,n}^J u_n \, \xi + \int_{B(0,n)} B_{p,n}^J \xi \, u_n.$$

Therefore, by (6.60),

$$(6.61) \quad \int_{\mathbb{R}^N} H \, u + \int_{\mathbb{R}^N} Q_p^J \xi \, \xi \geq \int_{\mathbb{R}^N} H \, \xi + \int_{\mathbb{R}^N} Q_p^J \xi \, u.$$

Taking now $\xi = u \pm \lambda w$, $\lambda > 0$, $w \in L^1(\mathbb{R}^N) \cap L^\infty(\mathbb{R}^N)$, and letting $\lambda \to 0$, we get

$$\int_{\mathbb{R}^N} H \, w = \int_{\mathbb{R}^N} Q_p^J u \, w,$$

and consequently (6.59) is proved. Therefore, by (6.58), the range condition is satisfied for $\phi \in L^1(\mathbb{R}^N) \cap L^\infty(\mathbb{R}^N)$.

Now let $\phi \in L^p(\mathbb{R}^N)$. Take $\phi_n \in L^1(\mathbb{R}^N) \cap L^\infty(\mathbb{R}^N)$, $\phi_n \to \phi$ in $L^p(\mathbb{R}^N)$. Then, by the previous step, there exists $u_n = (I + Q_p^J)^{-1}\phi_n$. Since Q_p^J is completely accretive, $u_n \to u$ in $L^p(\mathbb{R}^N)$; also $Q_p^J u_n \to Q_p^J u$ in $L^{p'}(\mathbb{R}^N)$ and we conclude that $u + Q_p^J u = \phi$. $\qquad \square$

PROOF OF THEOREM 6.37. Since $\mathcal{Q}_p^J$ is an m-accretive operator in $L^p(\mathbb{R}^N)$, by Theorem A.29, we get the existence of a mild solution $u(t)$ of the abstract Cauchy problem

(6.62)
$$\begin{cases} u'(t) + \mathcal{Q}_p^J u(t) = 0, & t \in (0, T), \\[2mm] u(0) = u_0. \end{cases}$$

By the complete accretivity of $\mathcal{Q}_p^J$, since $D(\mathcal{Q}_2^J) = L^2(\mathbb{R}^N)$ and the operator $\mathcal{Q}_p^J$ is homogeneous of degree $p - 1$ for $p \neq 2$, for any $u_0 \in L^p(\mathbb{R}^N)$ the mild solution is a strong solution of (6.62), that is, a solution of (6.51) in the sense of Definition 6.36 (Theorems A.36 and A.53). Finally, the contraction principle follows from Theorem A.28. $\qquad\square$

6.3.2. Convergence to the Cauchy problem for the local p-Laplacian.
Our next step is to rescale the kernel J appropriately and to take limits as the scaling parameter goes to zero.

For $p > 1$ fixed and J we consider the rescaled kernels

$$J_{p,\varepsilon}(x) := \frac{C_{J,p}}{\varepsilon^{p+N}} J\left(\frac{x}{\varepsilon}\right), \quad \text{where} \quad C_{J,p}^{-1} := \frac{1}{2} \int_{\mathbb{R}^N} J(z)|z_N|^p \, dz.$$

Consider the local problem

(6.63)
$$\begin{cases} v_t - \Delta_p v = 0 & \text{in } \mathbb{R}^N \times (0, T), \\[2mm] v(x, 0) = u_0(x) & \text{in } \mathbb{R}^N, \end{cases}$$

and assume that $p > N$. Associated with $-\Delta_p$ is the operator $\mathcal{Q}_p$ defined in $L^p(\mathbb{R}^N) \times L^p(\mathbb{R}^N)$ by $\hat{v} \in \mathcal{Q}_p(v)$ if $v \in W^{1,p}(\mathbb{R}^N) \cap L^p(\mathbb{R}^N)$, $\hat{v} \in L^p(\mathbb{R}^N)$ and

$$\int_{\mathbb{R}^N} |\nabla v|^{p-2} \nabla v \cdot \nabla \xi = \int_{\mathbb{R}^N} \hat{v} \xi \quad \text{for every } \xi \in W^{1,p}(\mathbb{R}^N) \cap L^\infty(\mathbb{R}^N).$$

It is well known that $\mathcal{Q}_p$ is m-completely accretive in $L^p(\Omega)$ with dense domain and that (6.63) has a unique strong solution for any $u_0 \in L^p(\mathbb{R}^N)$.

THEOREM 6.39. Let $p > N$ and $J(x) \geq J(y)$ if $|x| \leq |y|$. Let $T > 0$, $u_0 \in L^p(\mathbb{R}^N)$ and let u_ε be the unique solution of (6.51) with J replaced by $J_{p,\varepsilon}$. Then, if v is the unique solution of (6.63),

(6.64)
$$\lim_{\varepsilon \to 0} \sup_{t \in [0,T]} \|u_\varepsilon(\cdot, t) - v(\cdot, t)\|_{L^p(\mathbb{R}^N)} = 0.$$

PROOF. In order to get (6.64), as we did before when dealing with the Dirichlet and Neumann problems, by Theorem A.37, it is enough to prove that

$$\left(I + \mathcal{Q}_p^{J_{p,\varepsilon}}\right)^{-1} \phi \to \left(I + \mathcal{Q}_p\right)^{-1} \phi \quad \text{in } L^p(\mathbb{R}^N) \text{ as } \varepsilon \to 0$$

for any $\phi \in C_c(\mathbb{R}^N)$.

Let $\phi \in C_c(\mathbb{R}^N)$ and $u_\varepsilon := \left(I + Q_p^{J_{p,\varepsilon}}\right)^{-1} \phi$. Then

$$
\int_{\mathbb{R}^N} u_\varepsilon \xi - \frac{C_{J,p}}{\varepsilon^{p+N}} \int_{\mathbb{R}^N} \int_{\mathbb{R}^N} J\left(\frac{x-y}{\varepsilon}\right) |u_\varepsilon(y) - u_\varepsilon(x)|^{p-2}
$$

(6.65)
$$
\times (u_\varepsilon(y) - u_\varepsilon(x)) \, dy \, \xi(x) \, dx
$$

$$
= \int_{\mathbb{R}^N} \phi \xi
$$

for every $\xi \in L^1(\mathbb{R}^N) \cap L^\infty(\mathbb{R}^N)$.

Changing variables,

$$
-\frac{C_{J,p}}{\varepsilon^{p+N}} \int_{\mathbb{R}^N} \int_{\mathbb{R}^N} J\left(\frac{x-y}{\varepsilon}\right) |u_\varepsilon(y) - u_\varepsilon(x)|^{p-2} (u_\varepsilon(y) - u_\varepsilon(x)) \, dy \, \xi(x) \, dx
$$

$$
= \int_{\mathbb{R}^N} \int_{\mathbb{R}^N} \frac{C_{J,p}}{2} J(z) \left|\frac{u_\varepsilon(x+\varepsilon z) - u_\varepsilon(x)}{\varepsilon}\right|^{p-2} \frac{u_\varepsilon(x+\varepsilon z) - u_\varepsilon(x)}{\varepsilon}
$$

$$
\times \frac{\xi(x+\varepsilon z) - \xi(x)}{\varepsilon} \, dx \, dz.
$$

So we can rewrite (6.65) as

$$
\int_{\mathbb{R}^N} \phi(x)\xi(x) \, dx - \int_{\mathbb{R}^N} u_\varepsilon(x)\xi(x) \, dx
$$

(6.66)
$$
= \int_{\mathbb{R}^N} \int_{\mathbb{R}^N} \frac{C_{J,p}}{2} J(z) \left|\frac{u_\varepsilon(x+\varepsilon z) - u_\varepsilon(x)}{\varepsilon}\right|^{p-2} \frac{u_\varepsilon(x+\varepsilon z) - u_\varepsilon(x)}{\varepsilon}
$$

$$
\times \frac{\xi(x+\varepsilon z) - \xi(x)}{\varepsilon} \, dx \, dz.
$$

We shall see that there exists a sequence $\varepsilon_n \to 0$ such that $u_{\varepsilon_n} \to v$ in $L^p(\mathbb{R}^N)$, $v \in W^{1,p}(\mathbb{R}^N)$ and $v = (I + Q_p)^{-1} \phi$, that is,

$$
\int_{\mathbb{R}^N} v\xi + \int_{\mathbb{R}^N} |\nabla v|^{p-2} \nabla v \cdot \nabla \xi = \int_{\mathbb{R}^N} \phi \xi \quad \text{for every } \xi \in W^{1,p}(\mathbb{R}^N) \cap L^\infty(\mathbb{R}^N).
$$

Since $u_\varepsilon \ll \phi$, by Proposition A.44, there exists a sequence $\varepsilon_n \to 0$ such that

$$
u_{\varepsilon_n} \rightharpoonup v \quad \text{weakly in } L^p(\mathbb{R}^N), \quad v \ll \phi.
$$

Observe that $\|u_{\varepsilon_n}\|_{L^\infty(\mathbb{R}^N)}, \|v\|_{L^\infty(\mathbb{R}^N)} \leq \|\phi\|_{L^\infty(\mathbb{R}^N)}$. Taking $\varepsilon = \varepsilon_n$ and $\xi = u_{\varepsilon_n}$ in (6.66) and applying Young's inequality, we get

$$
\int_{\mathbb{R}^N} \int_{\mathbb{R}^N} \frac{1}{2} \frac{C_{J,p}}{\varepsilon_n^N} J\left(\frac{x-y}{\varepsilon_n}\right) \left|\frac{u_{\varepsilon_n}(y) - u_{\varepsilon_n}(x)}{\varepsilon_n}\right|^p \, dx \, dy
$$

(6.67)
$$
= \int_{\mathbb{R}^N} \int_{\mathbb{R}^N} \frac{C_{J,p}}{2} J(z) \left|\frac{u_{\varepsilon_n}(x+\varepsilon_n z) - u_{\varepsilon_n}(x)}{\varepsilon_n}\right|^p \, dx \, dz
$$

$$
\leq M := \frac{1}{2} \int_{\mathbb{R}^N} |\phi(x)|^2 \, dx.
$$

Therefore, by Theorem 6.11,

$$v \in W^{1,p}(\mathbb{R}^N),$$

$$u_{\varepsilon_n} \to v \text{ in } L^p_{\text{loc}}(\mathbb{R}^N)$$

and

(6.68) $\qquad \left(\dfrac{C_{J,p}}{2} J(z)\right)^{1/p} \dfrac{u_{\varepsilon_n}(x + \varepsilon_n z) - u_{\varepsilon_n}(x)}{\varepsilon_n} \rightharpoonup \left(\dfrac{C_{J,p}}{2} J(z)\right)^{1/p} z \cdot \nabla v(x)$

weakly in $L^p(\mathbb{R}^N) \times L^p(\mathbb{R}^N)$. We now prove the tightness of $\{u_{\varepsilon_n}\}$, which is to say that no mass moves to infinity as $n \to +\infty$.

To this end, assume $\text{supp}(\phi) \subset B(0, R)$ and fix $S > 2R$. Select a smooth function $\varphi \in C^\infty(\mathbb{R}^N)$ such that $0 \le \varphi \le 1$, $\varphi \equiv 0$ on $B(0, R)$, $\varphi \equiv 1$ on $\mathbb{R}^N \setminus B(0, S)$ and $|\nabla \varphi| \le \frac{2}{S}$. Taking $\varepsilon = \varepsilon_n$ and $\xi = \varphi |u_{\varepsilon_n}|^{p-2} u_{\varepsilon_n}$ in (6.66) we have

$$-\int_{\mathbb{R}^N} u_{\varepsilon_n} |u_{\varepsilon_n}|^{p-2} u_{\varepsilon_n}$$

$$= \int_{\mathbb{R}^N} \phi \varphi |u_{\varepsilon_n}|^{p-2} u_{\varepsilon_n} - \int_{\mathbb{R}^N} u_{\varepsilon_n} |u_{\varepsilon_n}|^{p-2} u_{\varepsilon_n}$$

$$= \int_{\mathbb{R}^N} \int_{\mathbb{R}^N} \frac{C_{J,p}}{2} J(z) \left| \frac{u_{\varepsilon_n}(x + \varepsilon_n z) - u_{\varepsilon_n}(x)}{\varepsilon_n} \right|^{p-2} \frac{u_{\varepsilon_n}(x + \varepsilon_n z) - u_{\varepsilon_n}(x)}{\varepsilon_n}$$

$$\times \frac{|u_{\varepsilon_n}(x + \varepsilon_n z)|^{p-2} u_{\varepsilon_n}(x + \varepsilon_n z) \varphi(x + \varepsilon_n z) - |u_{\varepsilon_n}(x)|^{p-2} u_{\varepsilon_n}(x) \varphi(x)}{\varepsilon_n} \, dx \, dz$$

$$= \int_{\mathbb{R}^N} \int_{\mathbb{R}^N} \frac{C_{J,p}}{2} J(z) \left| \frac{u_{\varepsilon_n}(x + \varepsilon_n z) - u_{\varepsilon_n}(x)}{\varepsilon_n} \right|^{p-2} \frac{u_{\varepsilon_n}(x + \varepsilon_n z) - u_{\varepsilon_n}(x)}{\varepsilon_n}$$

$$\times \frac{|u_{\varepsilon_n}(x + \varepsilon_n z)|^{p-2} u_{\varepsilon_n}(x + \varepsilon_n z) \left(\varphi(x + \varepsilon_n z) - \varphi(x)\right)}{\varepsilon_n} \, dx \, dz$$

$$+ \int_{\mathbb{R}^N} \int_{\mathbb{R}^N} \frac{C_{J,p}}{2} J(z) \left| \frac{u_{\varepsilon_n}(x + \varepsilon_n z) - u_{\varepsilon_n}(x)}{\varepsilon_n} \right|^{p-2} \frac{u_{\varepsilon_n}(x + \varepsilon_n z) - u_{\varepsilon_n}(x)}{\varepsilon_n}$$

$$\times \frac{\left(|u_{\varepsilon_n}(x + \varepsilon_n z)|^{p-2} u_{\varepsilon_n}(x + \varepsilon_n z) - |u_{\varepsilon_n}(x)|^{p-2} u_{\varepsilon_n}(x)\right) \varphi(x)}{\varepsilon_n} \, dx \, dz.$$

Then, since the last integral is nonnegative and having in mind that

$$\|u_{\varepsilon_n}\|_{L^\infty(\mathbb{R}^N)} \le \|\phi\|_{L^\infty(\mathbb{R}^N)},$$

we get

$$\int_{\mathbb{R}^N} |u_{\varepsilon_n}|^p(x)\varphi(x)\,dx$$

$$\leq \frac{C_{J,p}}{2\varepsilon_n^p} \int_{\mathbb{R}^N} \int_{B(0,1)} J(z)|u_{\varepsilon_n}(x+\varepsilon_n z) - u_{\varepsilon_n}(x)|^{p-1}|u_{\varepsilon_n}(x+\varepsilon_n z)|^{p-1}$$
$$\times |\varphi(x+\varepsilon_n z) - \varphi(x)|\,dz\,dx$$

$$\leq \frac{C_{J,p}\|\phi\|_{L^\infty}^{p-1}}{S} \int_{\{|x|\leq S+1\}} \int_{B(0,1)} J(z) \left| \frac{u_{\varepsilon_n}(x+\varepsilon_n z) - u_{\varepsilon_n}(x)}{\varepsilon_n} \right|^{p-1} dy\,dx$$

$$\leq \frac{C_{J,p}\|\phi\|_{L^\infty}^{p-1}}{S} \left(\int_{\{|x|\leq S+1\}} \int_{B(0,1)} J(z) \left| \frac{u_{\varepsilon_n}(x+\varepsilon_n z) - u_{\varepsilon_n}(x)}{\varepsilon_n} \right|^{p} dy \right)^{\frac{1}{p'}}$$
$$\times \left(\int_{\{|x|\leq S+1\}} \int_{B(0,1)} J(z)\,dz \right)^{\frac{1}{p}} dx$$

$$\leq C\,S^{-1+\frac{N}{p}},$$

the last equality being true by (6.67) and since

$$\left(\int_{\{|x|\leq S+1\}} \int_{B(0,1)} J(z)\,dz \right)^{\frac{1}{p}} dx \leq C(S+1)^{\frac{N}{p}}.$$

Consequently,

$$\int_{\{|x|\geq S\}} |u_{\varepsilon_n}|^p(x)\,dx \leq C\,S^{-1+\frac{N}{p}}$$

uniformly in ε_n. Therefore,

$$u_{\varepsilon_n} \to v \text{ in } L^p(\mathbb{R}^N).$$

Moreover, from (6.67), we can also assume that

$$\left| \frac{u_{\varepsilon_n}(x+\varepsilon_n z) - u_{\varepsilon_n}(x)}{\varepsilon_n} \right|^{p-2} \frac{u_{\varepsilon_n}(x+\varepsilon_n z) - u_{\varepsilon_n}(x)}{\varepsilon_n} \rightharpoonup \chi(x,z)$$

weakly in $L^{p'}(\mathbb{R}^N) \times L^{p'}(\mathbb{R}^N)$. Passing to the limit in (6.66) for $\varepsilon = \varepsilon_n$, we get

$$\int_{\mathbb{R}^N} v\xi + \int_{\mathbb{R}^N} \int_{\mathbb{R}^N} \frac{C_{J,p}}{2} J(z)\chi(x,z)\, z\cdot\nabla\xi(x)\,dx\,dz = \int_{\mathbb{R}^N} \phi\xi$$

for every smooth ξ and by approximation for every $\xi \in W^{1,p}(\mathbb{R}^N) \cap L^\infty(\mathbb{R}^N)$.

Finally, working as in Proposition 6.15 and using Lemma 6.16, we obtain

$$\int_{\mathbb{R}^N} \int_{\mathbb{R}^N} \frac{C_{J,p}}{2} J(z)\chi(x,z)z\cdot\nabla\xi(x)\,dx\,dz = \int_{\mathbb{R}^N} |\nabla v|^{p-2}\nabla v\cdot\nabla\xi. \qquad \square$$

6.4. Nonhomogeneous problems

Nonlinear Semigroup Theory allows us to handle easily nonhomogeneous problems. To illustrate this we state the following results for the Cauchy problem that are direct consequences of this general theory. The proofs run as before, hence we omit the details.

THEOREM 6.40. *If $f \in BV(0,T; L^p(\mathbb{R}^N))$ and $u_0 \in D(\mathcal{Q}_p^J)$, then there exists a unique solution of*

(6.69)
$$\begin{cases} u_t(x,t) = \int_{\mathbb{R}^N} J(x-y)|u(y,t) - u(x,t)|^{p-2}(u(y,t) - u(x,t))dy + f(x,t), \\ u(x,0) = u_0(x), \qquad\qquad\qquad\qquad\qquad x \in \mathbb{R}^N, \ t \in (0,T). \end{cases}$$

Moreover, if $u_i(t)$ is a solution of (6.69) with initial data $u_{i0} \in L^p(\mathbb{R}^N)$ and source $f_i \in L^1(0,T; L^p(\mathbb{R}^N))$, $i = 1,2$, then, for every $t \in [0,T]$,

$$\|(u_1(t) - u_2(t))^+\|_{L^p(\mathbb{R}^N)} \leq \|(u_{10} - u_{20})^+\|_{L^p(\mathbb{R}^N)} + \int_0^t \|f_1(s) - f_2(s)\|_{L^p(\mathbb{R}^N)}ds.$$

THEOREM 6.41. *Suppose $p > N$ and $J(x) \geq J(y)$ if $|x| \leq |y|$. Let $T > 0$, $f \in L^1(0,T; L^p(\mathbb{R}^N))$ and $u_0 \in L^p(\mathbb{R}^N)$. If u_ε is the unique solution of (6.69) with J replaced by $J_{p,\varepsilon}$ and v is the unique solution of*

$$\begin{cases} v_t - \Delta_p v = f & in \ \mathbb{R}^N \times (0,T), \\ v(x,0) = u_0(x) & in \ \mathbb{R}^N, \end{cases}$$

then

$$\lim_{\varepsilon \to 0} \sup_{t \in [0,T]} \|u_\varepsilon(\cdot,t) - v(\cdot,t)\|_{L^p(\mathbb{R}^N)} = 0.$$

Bibliographical notes

This chapter is based on [15], [16] and [17]. For viscosity solutions for nonlocal p-Laplacian equations with a singular kernel see [125]. Finally, let us mention that in [18] a nonlocal p-Laplacian with a diffusion coefficient (that is assumed to be nonnegative but is allowed to vanish in a subset of positive measure) is considered.

CHAPTER 7

The nonlocal total variation flow

Motivated by problems in image processing, the Neumann and Dirichlet problems for the total variation flow are studied in [5] and [6] (see also [7]). More precisely, we refer to the problems

(7.1)
$$
\begin{cases}
v_t = \operatorname{div}\left(\dfrac{Dv}{|Dv|}\right) & \text{in } \Omega \times (0, +\infty), \\[2mm]
\dfrac{Dv}{|Dv|} \cdot \eta = 0 & \text{on } \partial\Omega \times (0, +\infty), \\[2mm]
v(\cdot, 0) = u_0 & \text{in } \Omega,
\end{cases}
$$

and

(7.2)
$$
\begin{cases}
v_t = \operatorname{div}\left(\dfrac{Dv}{|Dv|}\right) & \text{in } \Omega \times (0, +\infty), \\[2mm]
v = \tilde{\psi} & \text{on } \partial\Omega \times (0, +\infty), \\[2mm]
v(\cdot, 0) = u_0 & \text{in } \Omega,
\end{cases}
$$

respectively, where $\tilde{\psi} \in L^\infty(\partial\Omega)$. These problems are related with the minimization of the energy functionals

$$
\Phi(v) = \begin{cases}
\displaystyle\int_\Omega |Dv| & \text{if } v \in BV(\Omega) \cap L^2(\Omega), \\[3mm]
+\infty & \text{if } v \in L^2(\Omega) \setminus BV(\Omega),
\end{cases}
$$

and

$$
\Psi(v) = \begin{cases}
\displaystyle\int_\Omega |Dv| + \int_{\partial\Omega} |\tilde{\psi} - v| & \text{if } v \in BV(\Omega) \cap L^2(\Omega), \\[3mm]
+\infty & \text{if } v \in L^2(\Omega) \setminus BV(\Omega),
\end{cases}
$$

respectively. In the literature the operator

$$
\operatorname{div}\left(\frac{Dv}{|Dv|}\right)
$$

is also called the 1-Laplacian and it is denoted by $\Delta_1 v$.

Problem (7.1) appears when one uses the steepest descent method to minimize the total variation, a method introduced by L. Rudin, S. Osher and E. Fatemi [141] in the context of image denoising and reconstruction. Then solving (7.1) amounts to regularizing or, in other words, filtering the initial datum u_0. This filtering process has less destructive effect on the edges than filtering with a Gaussian, i.e., than solving the heat equation with initial condition u_0. In this context the given *image*

u_0 is a function defined on a bounded smooth or piecewise smooth open subset Ω of $\mathbb{R}^N$; typically, Ω will be a rectangle in $\mathbb{R}^2$.

One of the motivations for studying problem (7.2) comes from a numerical approach introduced in [**26**] to extend a function defined in $\mathbb{R}^2 \setminus \Omega$ inside Ω along the integral curves of a vector field $\theta^\perp$ which is the counterclockwise rotation of a vector field $\theta : \mathbb{R}^2 \to \mathbb{R}^2$ satisfying $|\theta| \leq 1$ and $\operatorname{div}(\theta) \in L^p(\Omega)$, $p \geq 1$. This interpolation has an application to filling-in problems.

The aim of this chapter is to study the nonlocal version of problems (7.1) and (7.2), which can be written formally as

$$
\begin{cases}
u_t(x,t) = \displaystyle\int_\Omega J(x-y) \frac{u(y,t) - u(x,t)}{|u(y,t) - u(x,t)|} \, dy, & x \in \Omega, \ t > 0, \\
u(x,0) = u_0(x), & x \in \Omega,
\end{cases}
$$

and

$$
\begin{cases}
u_t(x,t) = \displaystyle\int_\Omega J(x-y) \frac{u(t,y) - u(x,t)}{|u(y,t) - u(x,t)|} \, dy + \int_{\Omega_J \setminus \overline{\Omega}} J(x-y) \frac{\psi(y) - u(x,t)}{|\psi(y) - u(x,t)|} \, dy, \\
u(x,0) = u_0(x), & x \in \Omega, \quad t > 0,
\end{cases}
$$

respectively, where $\psi \in W^{1,1}(\Omega_J \setminus \overline{\Omega}) \cap L^\infty(\Omega_J \setminus \overline{\Omega})$ and $\psi|_{\partial\Omega} = \tilde{\psi}$. As in previous chapters, here $J : \mathbb{R}^N \to \mathbb{R}$ is a nonnegative continuous radial function with compact support, $J(0) > 0$ and $\int_{\mathbb{R}^N} J(x) dx = 1$.

7.1. Notation and preliminaries

Let us start by collecting some notation and results that will be used in the sequel.

Due to the linear growth of the energy functionals Φ and Ψ associated with problems (7.1) and (7.2), the natural energy space to study these problems is the space of functions of bounded variation.

Recall that a function $v \in L^1(\Omega)$ whose partial derivatives in the sense of distributions are measures with finite total variation in Ω is called a function of bounded variation. The class of such functions will be denoted by $BV(\Omega)$. Thus $v \in BV(\Omega)$ if and only if there are Radon measures $\mu_1, \dots, \mu_N$ defined in Ω with finite total mass in Ω and

$$
(7.3) \qquad\qquad \int_\Omega v D_i \varphi \, dx = - \int_\Omega \varphi \, d\mu_i
$$

for all $\varphi \in C_0^\infty(\Omega)$, $i = 1, \dots, N$, and the gradient of v is a vector-valued measure with finite total variation

$$
|Dv|(\Omega) = \sup \left\{ \int_\Omega v \operatorname{div}(\varphi) \, dx \ : \ \varphi \in C_0^\infty(\Omega, \mathbb{R}^N), \ |\varphi(x)| \leq 1 \text{ for } x \in \Omega \right\}.
$$

The space $BV(\Omega)$ is a Banach space endowed with the norm

$$
(7.4) \qquad\qquad \|v\|_{BV} = \|v\|_{L^1(\Omega)} + |Dv|(\Omega).
$$

For further information concerning functions of bounded variation we refer to [**4**], [**101**] and [**156**].

Following [20] (see also [7]), let

(7.5) $$X(\Omega) = \left\{ \zeta \in L^\infty(\Omega, \mathbb{R}^n) \; : \; \mathrm{div}(\zeta) \in L^1(\Omega) \right\},$$

and define, for $\zeta \in X(\Omega)$ and $w \in BV(\Omega) \cap L^\infty(\Omega)$, the functional $(\zeta, Dw) :$ $C_0^\infty(\Omega) \to \mathbb{R}$ by the formula

(7.6) $$\langle (\zeta, Dw), \varphi \rangle = - \int_\Omega w \, \varphi \, \mathrm{div}(\zeta) \, dx - \int_\Omega w \, \zeta \cdot \nabla \varphi \, dx.$$

Then (ζ, Dw) is a Radon measure in Ω,

(7.7) $$\int_\Omega (\zeta, Dw) = \int_\Omega \zeta \cdot \nabla w \, dx \qquad \text{for all } w \in W^{1,1}(\Omega) \cap L^\infty(\Omega),$$

and

(7.8) $$\left| \int_B (\zeta, Dw) \right| \leq \int_B |(\zeta, Dw)| \leq \|\zeta\|_\infty \int_B \|Dw\|$$

for any Borel set $B \subseteq \Omega$.

In [20] a weak trace on $\partial\Omega$ of the normal component of $\zeta \in X(\Omega)$ is defined. Concretely, it is proved that there exists a linear operator $\gamma : X(\Omega) \to L^\infty(\partial\Omega)$ such that

$$\|\gamma(\zeta)\|_\infty \leq \|\zeta\|_\infty,$$

and

$$\gamma(\zeta)(x) = \zeta(x) \cdot \nu(x) \quad \text{for all } x \in \partial\Omega \text{ if } \zeta \in C^1(\overline{\Omega}, \mathbb{R}^N).$$

Let us denote $\gamma(\zeta)(x)$ by $[\zeta, \nu](x)$. The following *Green's formula*, relating the function $[\zeta, \nu]$ and the measure (ζ, Dw), for $\zeta \in X(\Omega)$ and $w \in BV(\Omega) \cap L^\infty(\Omega)$, is established:

(7.9) $$\int_\Omega w \, \mathrm{div}(\zeta) \, dx + \int_\Omega (\zeta, Dw) = \int_{\partial\Omega} [\zeta, \nu] w \, d\mathcal{H}^{N-1}.$$

Throughout this chapter the following multivalued function will be used:

$$\mathrm{sgn}(r) = \begin{cases} 1 & \text{if } r > 0, \\ [-1, 1] & \text{if } r = 0, \\ -1 & \text{if } r < 0. \end{cases}$$

Remember also that

$$\mathrm{sgn}_0(r) = \begin{cases} 1 & \text{if } r > 0, \\ 0 & \text{if } r = 0, \\ -1 & \text{if } r < 0. \end{cases}$$

7.2. The Neumann problem

In this section we study the nonlocal total variation flow with homogeneous Neumann boundary conditions,

(7.10) $$\begin{cases} u_t(x, t) = \displaystyle\int_\Omega J(x - y) \frac{u(y, t) - u(x, t)}{|u(y, t) - u(x, t)|} \, dy, & x \in \Omega, \ t > 0, \\ u(x, 0) = u_0(x), & x \in \Omega. \end{cases}$$

The formal evolution problem

$$u_t(x,t) = \int_\Omega J(x-y) \frac{u(y,t) - u(x,t)}{|u(y,t) - u(x,t)|}\, dy$$

is the gradient flow associated to the functional

$$J_1(u) = \frac{1}{2} \int_\Omega \int_\Omega J(x-y)|u(y) - u(x)|\, dy\, dx,$$

which is the nonlocal analog of the energy functional associated to the total variation

$$F_1(v) = \int_\Omega |Dv|.$$

7.2.1. Existence and uniqueness. Solutions of (7.10) will be understood as follows.

DEFINITION 7.1. A *solution* of (7.10) in $[0,T]$ is a function

$$u \in W^{1,1}(0,T;L^1(\Omega))$$

which satisfies $u(x,0) = u_0(x)$ a.e. $x \in \Omega$ and

$$u_t(x,t) = \int_\Omega J(x-y)g(x,y,t)\, dy \quad \text{a.e. in } \Omega \times (0,T),$$

for some $g \in L^\infty(\Omega \times \Omega \times (0,T))$ with $\|g\|_\infty \leq 1$ such that $g(x,y,t) = -g(y,x,t)$ and

$$J(x-y)g(x,y,t) \in J(x-y)\text{sgn}(u(y,t) - u(x,t)).$$

To prove the existence and uniqueness of this kind of solutions, the idea is to take the limit as $p \searrow 1$ of the solutions of P_p^J with $p > 1$ that were studied in Chapter 6.

THEOREM 7.2. *Let $u_0 \in L^1(\Omega)$. Then there exists a unique solution of (7.10).*

Moreover, if u_i is a solution in $[0,T]$ of (7.10) with initial data $u_{i0} \in L^1(\Omega)$, $i = 1,2$, then

$$\int_\Omega (u_1(t) - u_2(t))^+ \leq \int_\Omega (u_{10} - u_{20})^+ \quad \text{for every } t \in [0,T].$$

Observe that the formal expression $\frac{u(y,t)-u(x,t)}{|u(y,t)-u(x,t)|}$ in the evolution equation has to be interpreted as an L^∞ function $g(x,y,t)$ antisymmetric in the space variables and such that it is related to the above expression by using the multivalued function sgn.

To prove the existence and uniqueness of solutions of (7.10) we introduce the following operator in $L^1(\Omega)$.

DEFINITION 7.3. We define the operator B_1^J in $L^1(\Omega) \times L^1(\Omega)$ by $\hat{u} \in B_1^J u$ if and only if $u, \hat{u} \in L^1(\Omega)$, there exists $g \in L^\infty(\Omega \times \Omega)$, $g(x,y) = -g(y,x)$ for almost all $(x,y) \in \Omega \times \Omega$, $\|g\|_\infty \leq 1$,

$$\hat{u}(x) = -\int_\Omega J(x-y)g(x,y)\, dy \quad \text{a.e. } x \in \Omega,$$

and

(7.11) $J(x-y)g(x,y) \in J(x-y)\,\text{sgn}(u(y) - u(x)) \quad \text{a.e. } (x,y) \in \Omega \times \Omega.$

REMARK 7.4. (i) It is not difficult to see that (7.11) is equivalent to

$$-\int_\Omega \int_\Omega J(x-y)g(x,y)\,dy\,u(x)\,dx = \frac{1}{2}\int_\Omega \int_\Omega J(x-y)|u(y)-u(x)|\,dy\,dx.$$

(ii) $L^1(\Omega) = D(B_1^J)$ and B_1^J is closed in $L^1(\Omega) \times L^1(\Omega)$.

(iii) B_1^J is positively homogeneous of degree zero; that is, if $\hat{u} \in B_1^J u$ and $\lambda > 0$ then $\hat{u} \in B_1^J(\lambda u)$.

THEOREM 7.5. *The operator B_1^J is completely accretive and satisfies the range condition*

$$L^\infty(\Omega) \subset \mathrm{R}(I + B_1^J).$$

PROOF. Let $\hat{u}_i \in B_1^J u_i$, $i = 1, 2$. Then there exists $g_i \in L^\infty(\Omega \times \Omega)$, $\|g_i\|_\infty \le 1$, $g_i(x,y) = -g_i(y,x)$, $J(x-y)g_i(x,y) \in J(x-y)\mathrm{sgn}(u_i(y)-u_i(x))$ for almost all $(x,y) \in \Omega \times \Omega$, such that

$$\hat{u}_i(x) = -\int_\Omega J(x-y)g_i(x,y)\,dy \quad \text{a.e. } x \in \Omega, \quad i = 1, 2.$$

Then, given $q \in P_0$, we have

$$\int_\Omega (\hat{u}_1(x) - \hat{u}_2(x))q(u_1(x) - u_2(x))\,dx$$

$$= \frac{1}{2}\int_\Omega \int_\Omega J(x-y)(g_1(x,y) - g_2(x,y))$$
$$\times (q(u_1(y) - u_2(y)) - q(u_1(x) - u_2(x)))\,dx\,dy$$

$$= \frac{1}{2}\int \int_{\{(x,y):u_1(y)\neq u_1(x),u_2(y)=u_2(x)\}} J(x-y)(g_1(x,y) - g_2(x,y))$$
$$\times (q(u_1(y) - u_2(y)) - q(u_1(x) - u_2(x)))\,dx\,dy$$

$$+ \frac{1}{2}\int \int_{\{(x,y):u_1(y)=u_1(x),u_2(y)\neq u_2(x)\}} J(x-y)(g_1(x,y) - g_2(x,y))$$
$$\times (q(u_1(y) - u_2(y)) - q(u_1(x) - u_2(x)))\,dx\,dy$$

$$+ \frac{1}{2}\int \int_{\{(x,y):u_1(y)\neq u_1(x),u_2(y)\neq u_2(x)\}} J(x-y)(g_1(x,y) - g_2(x,y))$$
$$\times (q(u_1(y) - u_2(y)) - q(u_1(x) - u_2(x)))\,dx\,dy,$$

the last three integrals being nonnegative. Hence

$$\int_\Omega (\hat{u}_1(x) - \hat{u}_2(x))q(u_1(x) - u_2(x))\,dx \ge 0,$$

and by Corollary A.43 it follows that B_1^J is a completely accretive operator.

To show that B_1^J satisfies the range condition, let us see that for any $\phi \in L^\infty(\Omega)$,

$$\lim_{p\to 1+} (I + B_p^J)^{-1}\phi = (I + B_1^J)^{-1}\phi \quad \text{weakly in } L^1(\Omega).$$

Let $\phi \in L^\infty(\Omega)$, and write $u_p = \left(I + B_p^J\right)^{-1} \phi$ for $1 < p < +\infty$. Then

$$u_p(x) - \int_\Omega J\left(x - y\right) |u_p(y) - u_p(x)|^{p-2}(u_p(y) - u_p(x))\, dy = \phi(x) \quad \text{a.e. } x \in \Omega.$$

Thus, for every $\xi \in L^\infty(\Omega)$, we can write

$$\int_\Omega u_p \xi - \int_\Omega \int_\Omega J\left(x - y\right) |u_p(y) - u_p(x)|^{p-2}(u_p(y) - u_p(x))\, dy\, \xi(x)\, dx$$

(7.12)

$$= \int_\Omega \phi \xi.$$

Since $u_p \ll \phi$, by Proposition A.44, we have that there exists a sequence $p_n \to 1$ such that

$$u_{p_n} \rightharpoonup u \quad \text{weakly in } L^1(\Omega), \quad u \ll \phi.$$

Observe that $\|u_{p_n}\|_{L^\infty(\Omega)}, \|u\|_{L^\infty(\Omega)} \leq \|\phi\|_{L^\infty(\Omega)}$.

Now, since

$$\left| |u_{p_n}(y) - u_{p_n}(x)|^{p_n-2} \left(u_{p_n}(y) - u_{p_n}(x)\right) \right| \leq (2\|\phi\|_\infty)^{p_n-1},$$

there exists $g(x, y)$ such that

$$|u_{p_n}(y) - u_{p_n}(x)|^{p_n-2} \left(u_{p_n}(y) - u_{p_n}(x)\right) \rightharpoonup g(x, y)$$

weakly in $L^1(\Omega \times \Omega)$, $g(x, y) = -g(y, x)$ for almost all $(x, y) \in \Omega \times \Omega$, and $\|g\|_\infty \leq 1$.

Passing to the limit in (7.12) for $p = p_n$, we get

(7.13)
$$\int_\Omega u\xi - \int_\Omega \int_\Omega J(x - y)g(x, y)\, dy\, \xi(x)\, dx = \int_\Omega \phi\xi$$

for every $\xi \in L^\infty(\Omega)$, and consequently

$$u(x) - \int_\Omega J(x - y)g(x, y)\, dy = \phi(x) \quad \text{a.e. } x \in \Omega.$$

Then, to finish the proof we have to show that

(7.14) $$-\int_\Omega \int_\Omega J(x - y)g(x, y)\, dy\, u(x)\, dx = \frac{1}{2} \int_\Omega \int_\Omega J(x - y)|u(y) - u(x)|\, dy\, dx.$$

In fact, by (7.12) with $p = p_n$, $\xi = u_{p_n}$, and (7.13) with $\xi = u$,

$$\frac{1}{2} \int_\Omega \int_\Omega J(x - y) |u_{p_n}(y) - u_{p_n}(x)|^{p_n}\, dy\, dx$$

$$= \int_\Omega \phi u_{p_n} - \int_\Omega u_{p_n} u_{p_n} = \int_\Omega \phi u - \int_\Omega uu - \int_\Omega \phi(u - u_{p_n})$$

$$+ \int_\Omega 2u(u - u_{p_n}) - \int_\Omega (u - u_{p_n})(u - u_{p_n})$$

$$\leq -\int_\Omega \int_\Omega J(x - y)g(x, y)\, dy\, u(x)\, dx - \int_\Omega \phi(u - u_{p_n}) + \int_\Omega 2u(u - u_{p_n}),$$

and so,

$$\limsup_{n \to +\infty} \frac{1}{2} \int_\Omega \int_\Omega J(x-y) \left| u_{p_n}(y) - u_{p_n}(x) \right|^{p_n} dy \, dx$$

$$\leq - \int_\Omega \int_\Omega J(x-y) g(x,y) \, dy \, u(x) \, dx.$$

By the monotonicity Lemma 6.6,

$$-\int_\Omega \int_\Omega J(x-y) |\rho(y) - \rho(x)|^{p_n-2} (\rho(y) - \rho(x)) \, dy \, (u_{p_n}(x) - \rho(x)) \, dx$$

$$\leq - \int_\Omega \int_\Omega J(x-y) |u_{p_n}(y) - u_{p_n}(x)|^{p_n-2} (u_{p_n}(y) - u_{p_n}(x)) \, dy \, (u_{p_n}(x) - \rho(x)) dx.$$

Therefore, taking limits in n,

$$-\int_\Omega \int_\Omega J(x-y) \, \mathrm{sgn}_0(\rho(y) - \rho(x)) \, dy \, (u(x) - \rho(x)) \, dx$$

$$\leq - \int_\Omega \int_\Omega J(x-y) g(x,y) \, dy \, (u(x) - \rho(x)) \, dx.$$

Taking $\rho = u \pm \lambda u$, $\lambda > 0$, and letting $\lambda \to 0$, we get (7.14), and the proof is finished. □

PROOF OF THEOREM 7.2. As a consequence of the above results, by Theorem A.29, we have that the abstract Cauchy problem

$$(7.15) \qquad \begin{cases} u'(t) + B_1^J u(t) \ni 0, & t \in (0,T), \\ u(0) = u_0, \end{cases}$$

has a unique mild solution u for every initial datum $u_0 \in L^1(\Omega)$ and $T > 0$. Moreover, due to the complete accretivity of the operator $B_{1,\psi}^J$, the mild solution of (7.15) is a strong solution (Corollary A.52). Consequently, since the concept of solution of (7.10) in the sense of Definition 7.1 coincides with the strong solution of problem (7.15), the proof is concluded. □

7.2.2. Rescaling the kernel. Convergence to the total variation flow.

Let Ω be a smooth bounded domain in $\mathbb{R}^N$. We will see that the solutions of problem (7.10), with the kernel J rescaled in a suitable way, converge, as the scaling parameter goes to zero, to the solutions of the Neumann problem for the total variation flow

$$(7.16) \qquad \begin{cases} v_t = \mathrm{div}\left(\dfrac{Dv}{|Dv|} \right) & \text{in } \Omega \times (0,T), \\[2mm] \dfrac{Dv}{|Dv|} \cdot \eta = 0 & \text{on } \partial\Omega \times (0,T), \\[2mm] v(\cdot,0) = u_0 & \text{in } \Omega. \end{cases}$$

As was mentioned in the introduction, motivated by problems in image processing, problem (7.16) was studied in [5] (see also [7]). Let us recall some of the results given in [5], which will be used in the proof of convergence of the rescaled problems. The concept of solution for problem (7.16) is the following.

DEFINITION 7.6. A measurable function $v : \Omega \times (0, T) \to \mathbb{R}$ is a *weak solution* of (7.16) in $\Omega \times (0, T)$ if $v \in C([0, T]; L^1(\Omega)) \cap W^{1,1}_{loc}(0, T; L^1(\Omega))$, $T_k(v) \in L^1_w(0, T; BV(\Omega))$ for all $k > 0$ and there exists $\zeta \in L^\infty(\Omega \times (0, T))$ with $\|\zeta\|_\infty \le 1$, $v_t = \mathrm{div}(\zeta)$ in $\mathcal{D}'(\Omega \times (0, T))$, such that

$$\int_\Omega (T_k(v(t)) - \xi) v_t(t) \, dx \le \int_\Omega \zeta(t) \cdot \nabla \xi \, dx - |DT_k(v(t))|(\Omega)$$

for every $\xi \in W^{1,1}(\Omega) \cap L^\infty(\Omega)$ and a.e. on $[0, T]$.

The main result of [5] is the following.

THEOREM 7.7. *Let $u_0 \in L^1(\Omega)$. Then there exists a unique weak solution of (7.16) in $\Omega \times (0, T)$ for every $T > 0$ such that $u(0) = u_0$. Moreover, if $v(t), \hat{v}(t)$ are weak solutions corresponding to initial data $u_0, \hat{u}_0$, respectively, then*

$$\|(v(t) - \hat{v}(t))^+\|_1 \le \|(u_0 - \hat{u}_0)^+\|_1 \quad and \quad \|v(t) - \hat{v}(t)\|_1 \le \|u_0 - \hat{u}_0\|_1$$

for all $t \ge 0$.

Theorem 7.7 was proved using the techniques of completely accretive operators and the Crandall-Liggett semigroup generation theorem (Theorem A.31). To this end, the following operator B_1 in $L^1(\Omega)$ was defined in [5] by the following rule:

$$(v, \hat{v}) \in B_1 \quad \text{if and only if} \quad v, \hat{v} \in L^1(\Omega), \ T_k(v) \in BV(\Omega) \text{ for all } k > 0 \text{ and}$$

there exists $\zeta \in L^\infty(\Omega, \mathbb{R}^N)$ with $\|\zeta\|_\infty \le 1$, $\hat{v} = -\mathrm{div}(\zeta)$ in $\mathcal{D}'(\Omega)$, such that

$$\int_\Omega (\xi - T_k(v)) \hat{v} \, dx \le \int_\Omega \zeta \cdot \nabla \xi \, dx - |DT_k(v)|(\Omega),$$

$\forall \xi \in W^{1,1}(\Omega) \cap L^\infty(\Omega), \forall k > 0$.

Theorem 7.7 is obtained from the following result given in [5].

THEOREM 7.8. *The operator B_1 is m-completely accretive in $L^1(\Omega)$ with dense domain. For any $u_0 \in L^1(\Omega)$ the semigroup solution $v(t) = e^{-tB_1} u_0$ is a strong solution of*

$$\begin{cases} \dfrac{dv}{dt} + B_1 v \ni 0, \\[2mm] v(0) = u_0. \end{cases}$$

Now we return to the analysis of the nonlocal problem and set

$$J_{1,\varepsilon}(x) := \frac{C_{J,1}}{\varepsilon^{1+N}} J\left(\frac{x}{\varepsilon}\right),$$

with $C_{J,1}^{-1} := \frac{1}{2} \int_{\mathbb{R}^N} J(z)|z_N| \, dz$ a normalizing constant in order to obtain the 1-Laplacian in the limit instead of a multiple of it.

Associated with these rescaled kernels are the solutions u_ε of the equation in (7.10) with J replaced by $J_{1,\varepsilon}$ and the same initial condition u_0. The main result now states that these functions u_ε converge strongly in $L^1(\Omega)$ to the solution of the local problem (7.16).

THEOREM 7.9. *Suppose $J(x) \geq J(y)$ if $|x| \leq |y|$. Let $T > 0$ and $u_0 \in L^1(\Omega)$. Let u_ε be the unique solution in $[0, T]$ of (7.10) with J replaced by $J_{1,\varepsilon}$ and v the unique weak solution of (7.16). Then*

$$\lim_{\varepsilon \to 0} \sup_{t \in [0,T]} \|u_\varepsilon(\cdot, t) - v(\cdot, t)\|_{L^1(\Omega)} = 0.$$

Arguing as in the proof of Theorem 6.18, since the solutions of the above theorem coincide with the semigroup solutions, by Theorem A.37, to prove Theorem 7.9 it is enough to obtain the following result.

THEOREM 7.10. *Suppose $J(x) \geq J(y)$ if $|x| \leq |y|$. Then, for any $\phi \in L^\infty(\Omega)$,*

$$\left(I + B_1^{J_{1,\varepsilon}}\right)^{-1} \phi \to (I + B_1)^{-1} \phi \quad in \ L^1(\Omega) \ as \ \varepsilon \to 0.$$

PROOF. Given $\varepsilon > 0$, we set $u_\varepsilon = \left(I + B_1^{J_{1,\varepsilon}}\right)^{-1} \phi$. Then there exists $g_\varepsilon \in L^\infty(\Omega \times \Omega)$, $g_\varepsilon(x, y) = -g_\varepsilon(y, x)$ for almost all $x, y \in \Omega$, $\|g_\varepsilon\|_\infty \leq 1$,

$$J\left(\frac{x-y}{\varepsilon}\right) g_\varepsilon(x, y) \in J\left(\frac{x-y}{\varepsilon}\right) \operatorname{sgn}(u_\varepsilon(y) - u_\varepsilon(x)) \quad a.e. \ x, y \in \Omega$$

and

(7.17) $$-\frac{C_{J,1}}{\varepsilon^{1+N}} \int_\Omega J\left(\frac{x-y}{\varepsilon}\right) g_\varepsilon(x, y) dy = \phi(x) - u_\varepsilon(x) \quad a.e. \ x \in \Omega.$$

Moreover $u_\varepsilon \ll \phi$.

Observe that

(7.18) $$-\frac{C_{J,1}}{\varepsilon^{1+N}} \int_\Omega \int_\Omega J\left(\frac{x-y}{\varepsilon}\right) g_\varepsilon(x, y) dy \, u_\varepsilon(x) \, dx$$

$$= \frac{C_{J,1}}{\varepsilon^{1+N}} \frac{1}{2} \int_\Omega \int_\Omega J\left(\frac{x-y}{\varepsilon}\right) |u_\varepsilon(y) - u_\varepsilon(x)| \, dy \, dx.$$

By (7.17), we can write

$$\frac{C_{J,1}}{2\varepsilon^{1+N}} \int_\Omega \int_\Omega J\left(\frac{x-y}{\varepsilon}\right) g_\varepsilon(x, y)(\xi(y) - \xi(x)) \, dx \, dy$$

(7.19) $$= -\frac{C_{J,1}}{\varepsilon^{1+N}} \int_\Omega \int_\Omega J\left(\frac{x-y}{\varepsilon}\right) g_\varepsilon(x, y) \, dy \, \xi(x) \, dx$$

$$= \int_\Omega (\phi(x) - u_\varepsilon(x))\xi(x) \, dx, \qquad \forall \xi \in L^\infty(\Omega).$$

Since $u_\varepsilon \ll \phi$, by Proposition A.44, there exists a sequence $\varepsilon_n \to 0$ such that

$$u_{\varepsilon_n} \rightharpoonup v \quad weakly \ in \ L^1(\Omega), \quad u \ll \phi.$$

Note that $\|u_{\varepsilon_n}\|_{L^\infty(\Omega)}, \|v\|_{L^\infty(\Omega)} \le \|\phi\|_{L^\infty(\Omega)}$. Hence taking $\varepsilon = \varepsilon_n$ and $\xi = u_{\varepsilon_n}$ in (7.19), changing variables and having in mind (7.18), we get

$$\int_{\mathbb{R}^N} \int_\Omega \frac{C_{J,1}}{2} J(z) \chi_\Omega(x + \varepsilon_n z) \left| \frac{\overline{u}_{\varepsilon_n}(x + \varepsilon_n z) - u_{\varepsilon_n}(x)}{\varepsilon_n} \right| dx\, dz$$

$$= \int_\Omega \int_\Omega \frac{1}{2} \frac{C_{J,1}}{\varepsilon_n^N} J\left(\frac{x - y}{\varepsilon_n} \right) \left| \frac{u_{\varepsilon_n}(y) - u_{\varepsilon_n}(x)}{\varepsilon_n} \right| dx\, dy$$

$$= \int_\Omega (\phi(x) - u_{\varepsilon_n}(x)) u_{\varepsilon_n}(x)\, dx \le M, \qquad\qquad \forall\, n \in \mathbb{N},$$

where we use the notation given in (6.6). Therefore, by Theorem 6.11, $v \in BV(\Omega)$,

$$(7.20) \qquad \frac{C_{J,1}}{2} J(z) \chi_\Omega(x + \varepsilon_n z) \frac{\overline{u}_{\varepsilon_n}(x + \varepsilon_n z) - u_{\varepsilon_n}(x)}{\varepsilon_n} \rightharpoonup \frac{C_{J,1}}{2} J(z) z \cdot Dv$$

weakly as measures and

$$u_{\varepsilon_n} \to v \qquad \text{strongly in } L^1(\Omega).$$

Moreover, we also can assume that

$$(7.21) \qquad J(z) \chi_\Omega(x + \varepsilon_n z) \overline{g}_{\varepsilon_n}(x, x + \varepsilon_n z) \rightharpoonup \Lambda(x, z)$$

weakly* in $L^\infty(\Omega) \times L^\infty(\mathbb{R}^N)$, with $\Lambda(x, z) \le J(z)$ almost everywhere in $\Omega \times \mathbb{R}^N$. Changing variables and having in mind (7.19), we can write

$$\frac{C_{J,1}}{2} \int_{\mathbb{R}^N} \int_\Omega J(z) \chi_\Omega(x + \varepsilon_n z) \overline{g}_{\varepsilon_n}(x, x + \varepsilon_n z)\, dz \frac{\overline{\xi}(x + \varepsilon_n z) - \xi(x)}{\varepsilon_n}\, dx$$

$$(7.22) \qquad = -\frac{C_{J,1}}{\varepsilon_n} \int_{\mathbb{R}^N} \int_\Omega J(z) \chi_\Omega(x + \varepsilon_n z) \overline{g}_{\varepsilon_n}(x, x + \varepsilon_n z)\, dz\, \xi(x)\, dx$$

$$= \int_\Omega (\phi(x) - u_{\varepsilon_n}(x)) \xi(x)\, dx, \qquad\qquad \forall\, \xi \in L^\infty(\Omega).$$

By (7.21), passing to the limit in (7.22), we get

$$(7.23) \qquad \frac{C_{J,1}}{2} \int_{\mathbb{R}^N} \int_\Omega \Lambda(x, z) z \cdot \nabla \xi(x)\, dx\, dz = \int_\Omega (\phi(x) - v(x)) \xi(x)\, dx$$

for all smooth ξ and, by approximation, for all $\xi \in L^\infty(\Omega) \cap W^{1,1}(\Omega)$. We denote by $\zeta = (\zeta_1, \ldots, \zeta_N)$ the vector field defined by

$$\zeta_i(x) := \frac{C_{J,1}}{2} \int_{\mathbb{R}^N} \Lambda(x, z) z_i\, dz, \qquad i = 1, \ldots, N.$$

Then $\zeta \in L^\infty(\Omega, \mathbb{R}^N)$, and from (7.23),

$$-\mathrm{div}(\zeta) = \phi - v \qquad \text{in } \mathcal{D}'(\Omega).$$

Let us show that

$$(7.24) \qquad\qquad\qquad \|\zeta\|_\infty \le 1.$$

Given $\xi \in \mathbb{R}^N \setminus \{0\}$, consider R_ξ the rotation such that $R_\xi^t(\xi) = \mathbf{e}_1|\xi|$. Then, if we make the change of variables $z = R_\xi(y)$, we obtain

$$\zeta(x) \cdot \xi = \frac{C_{J,1}}{2} \int_{\mathbb{R}^N} \Lambda(x,z)z \cdot \xi \, dz = \frac{C_{J,1}}{2} \int_{\mathbb{R}^N} \Lambda(x, R_\xi(y)) R_\xi(y) \cdot \xi \, dy$$

$$= \frac{C_{J,1}}{2} \int_{\mathbb{R}^N} \Lambda(x, R_\xi(y)) y_1 |\xi| \, dy.$$

On the other hand, since J is a radial function and $\Lambda(x,z) \le J(z)$ almost everywhere,

$$C_{J,1}^{-1} = \frac{1}{2} \int_{\mathbb{R}^N} J(z)|z_1| \, dz$$

and

$$|\zeta(x) \cdot \xi| \le \frac{C_{J,1}}{2} \int_{\mathbb{R}^N} J(y)|y_1| \, dy |\xi| = |\xi| \quad \text{a.e. } x \in \Omega.$$

Therefore, (7.24) holds.

Since $v \in L^\infty(\Omega)$, to finish the proof we only need to show that

$$(7.25) \quad \int_\Omega (\xi - v)(\phi - v) \, dx \le \int_\Omega \zeta \cdot \nabla \xi \, dx - |Dv|(\Omega) \quad \forall \xi \in W^{1,1}(\Omega) \cap L^\infty(\Omega).$$

Given w smooth, taking $\xi = w - u_{\varepsilon_n}$ in (7.22), we get

$$\int_\Omega (\phi(x) - u_{\varepsilon_n}(x))(w(x) - u_{\varepsilon_n}(x)) \, dx$$

$$= \frac{C_{J,1}}{2} \int_{\mathbb{R}^N} \int_\Omega J(z) \chi_\Omega(x + \varepsilon_n z) \overline{g}_{\varepsilon_n}(x, x + \varepsilon_n z) \, dz$$

$$(7.26) \qquad \times \left(\frac{\overline{w}(x + \varepsilon_n z) - w(x)}{\varepsilon_n} - \frac{\overline{u}_{\varepsilon_n}(x + \varepsilon_n z) - u_{\varepsilon_n}(x)}{\varepsilon_n} \right) dx$$

$$= \frac{C_{J,1}}{2} \int_{\mathbb{R}^N} \int_\Omega J(z) \chi_\Omega(x + \varepsilon_n z) \overline{g}_{\varepsilon_n}(x, x + \varepsilon_n z) \, dz \frac{\overline{w}(x + \varepsilon_n z) - w(x)}{\varepsilon_n} \, dx$$

$$- \frac{C_{J,1}}{2} \int_{\mathbb{R}^N} \int_\Omega J(z) \chi_\Omega(x + \varepsilon_n z) \left| \frac{\overline{u}_{\varepsilon_n}(x + \varepsilon_n z) - u_{\varepsilon_n}(x)}{\varepsilon_n} \right| dx.$$

Having in mind (7.20) and (7.21) and taking limit in (7.26) as $n \to \infty$, we have that

$$\int_\Omega (w - v)(\phi - v) \, dx$$

$$\le \frac{C_{J,1}}{2} \int_\Omega \int_{\mathbb{R}^N} \Lambda(x,z) z \cdot \nabla w(x) \, dx \, dz - \frac{C_{J,1}}{2} \int_\Omega \int_{\mathbb{R}^N} |J(z) z \cdot Dv|$$

$$= \int_\Omega \zeta \cdot \nabla w \, dx - \frac{C_{J,1}}{2} \int_\Omega \int_{\mathbb{R}^N} |J(z) z \cdot Dv|,$$

for all smooth w and, by approximation, for all $w \in W^{1,1}(\Omega) \cap L^\infty(\Omega)$. Now, for every $x \in \Omega$ such that the Radon-Nikodym derivative $\frac{Dv}{|Dv|}(x) \ne 0$, let R_x be the rotation such that $R_x^t\left[\frac{Dv}{|Dv|}(x)\right] = \mathbf{e}_1 |\frac{Dv}{|Dv|}(x)|$. Then, since J is a radial function and $|\frac{Dv}{|Dv|}(x)| = 1$ $|Dv|$-a.e. in Ω, if we make the change of variables $y = R_x(z)$, we obtain

$$\frac{C_{J,1}}{2} \int_\Omega \int_{\mathbb{R}^N} |J(z)z \cdot Dv| = \frac{C_{J,1}}{2} \int_\Omega \int_{\mathbb{R}^N} J(z) \left| z \cdot \frac{Dv}{|Dv|}(x) \right| \, dz \, d|Dv|(x)$$

$$= \frac{C_{J,1}}{2} \int_\Omega \int_{\mathbb{R}^N} J(y) |y_1| \, dy \, d|Dv|(x) = \int_\Omega |Dv|.$$

Consequently, (7.25) holds and the proof is complete. $\qquad\square$

7.2.3. Asymptotic behaviour. Now we study the asymptotic behaviour as $t \to \infty$ of the solution of the nonlocal total variation flow with homogeneous Neumann boundary conditions. We show that the solution of this nonlocal problem converges to the mean value of the initial condition.

THEOREM 7.11. *Let $u_0 \in L^\infty(\Omega)$. Let u be the solution of (7.10); then*

$$\left\| u(t) - \frac{1}{|\Omega|} \int_\Omega u_0(x) \, dx \right\|_{L^1(\Omega)} \le C \frac{\|u_0\|^2_{L^2(\Omega)}}{t} \qquad \forall\, t > 0,$$

where $C = C(J, \Omega, q)$.

PROOF. A simple integration of the equation in space gives that the total mass is preserved, that is,

$$\frac{1}{|\Omega|} \int_\Omega u(x,t) \, dx = \frac{1}{|\Omega|} \int_\Omega u_0(x) \, dx.$$

We set

$$w(x,t) = u(x,t) - \frac{1}{|\Omega|} \int_\Omega u_0(x) \, dx.$$

Since B_1^J is completely accretive, $\mathcal{V}(v) := \left\| v - \frac{1}{|\Omega|} \int_\Omega u_0(x) \, dx \right\|_1$ is a Lyapunov functional for the semigroup solution. Hence

$$\int_\Omega |w(x,t)| \, dx \le \int_\Omega |w(x,s)| \, dx \quad \text{if } t \ge s.$$

Therefore the $L^1(\Omega)$-norm of w is decreasing with t.

Moreover, as the solution preserves the total mass, using the Poincaré type inequality (6.29), we have,

$$\int_\Omega |w(x,s)| \, dx \le C \int_\Omega \int_\Omega J(x-y)|u(y,s) - u(x,s)| \, dy \, dx.$$

Consequently,

$$t \int_\Omega |w(x,t)| \, dx \le \int_0^t \int_\Omega |w(x,s)| \, dx \, ds$$

$$\le C \int_0^t \int_\Omega \int_\Omega J(x-y)|u(y,s) - u(x,s)| \, dy \, dx \, ds.$$

On the other hand, multiplying the equation in (7.10) by $u(x,t)$ and integrating in space and time, we get

$$\int_\Omega |u(x,t)|^2 - \int_\Omega |u_0(x)|^2 \, dx = -\int_0^t \int_\Omega \int_\Omega J(x-y)|u(y,s) - u(x,s)| \, dy \, dx \, ds,$$

which implies

$$\int_0^t \int_\Omega \int_\Omega J(x-y)|u(y,s) - u(x,s)| \, dy \, dx \, ds \le ||u_0||^2_{L^2(\Omega)},$$

and therefore

$$\int_\Omega |w(x,t)| \, dx \le C \frac{||u_0||^2_{L^2(\Omega)}}{t},$$

which concludes the proof. $\qquad\square$

7.3. The Dirichlet problem

The goal of this section is to study the Dirichlet problem for the nonlocal total variation flow, which can be written formally as

(7.27)
$$\begin{cases} u_t(x,t) = \displaystyle\int_\Omega J(x-y) \frac{u(t,y) - u(x,t)}{|u(y,t) - u(x,t)|} \, dy \\ \qquad\qquad + \displaystyle\int_{\Omega_J \setminus \overline{\Omega}} J(x-y) \frac{\psi(y) - u(x,t)}{|\psi(y) - u(x,t)|} \, dy, \\ u(x,0) = u_0(x), \qquad\qquad\qquad x \in \Omega, \ t > 0, \end{cases}$$

We give the following definition of what we understand by a solution of (7.27).

DEFINITION 7.12. A *solution* of (7.27) in $[0,T]$ is a function

$$u \in W^{1,1}(0,T; L^1(\Omega))$$

that satisfies $u(x,0) = u_0(x)$ a.e. $x \in \Omega$ and

$$u_t(x,t) = \int_{\Omega_J} J(x-y) g(x,y,t) \, dy \quad \text{a.e. in } \Omega \times (0,T),$$

for some $g \in L^\infty(\Omega_J \times \Omega_J \times (0,T))$ with $||g||_\infty \le 1$ such that, for almost every $t \in]0,T[$, $g(x,y,t) = -g(y,x,t)$ and

$$J(x-y)g(x,y,t) \in J(x-y)\text{sgn}(u(y,t) - u(x,t)), \quad (x,y) \in \Omega \times \Omega,$$

$$J(x-y)g(x,y,t) \in J(x-y)\text{sgn}(\psi(y) - u(x,t)), \quad (x,y) \in \Omega \times (\Omega_J \setminus \overline{\Omega}).$$

We obtain the existence and uniqueness of this type of solutions and also, with a convenient rescale of the kernel J, we show the convergence of the solutions of the corresponding rescaled problems to the solutions of the Dirichlet problem for the total variation flow (7.2).

7.3.1. Existence and uniqueness. This section deals with the existence and uniqueness of solutions of the Dirichlet problem for the total variational flow (7.27). The main result is the following existence and uniqueness theorem.

THEOREM 7.13. *Let $u_0 \in L^1(\Omega)$ and $\psi \in L^1(\Omega_J \setminus \overline{\Omega})$. Then there exists a unique solution of (7.27). Moreover, if $u_{i0} \in L^1(\Omega)$ and u_i are solutions in $[0, T]$ of (7.27) with initial data u_{i0}, $i = 1, 2$, respectively, then*

$$\int_\Omega (u_1(t) - u_2(t))^+ \leq \int_\Omega (u_{10} - u_{20})^+ \quad \text{for every } t \in [0, T].$$

As for the Neumann problem, to prove the above theorem we use Nonlinear Semigroup Theory, so we start by introducing the following operator in $L^1(\Omega)$.

DEFINITION 7.14. *Given $\psi \in L^1(\Omega_J \setminus \overline{\Omega})$, we define the operator $B_{1,\psi}^J$ in $L^1(\Omega) \times L^1(\Omega)$ by $\hat{u} \in B_{1,\psi}^J u$ if and only if $u, \hat{u} \in L^1(\Omega)$, there exists $g \in L^\infty(\Omega_J \times \Omega_J)$, $g(x, y) = -g(y, x)$ for almost all $(x, y) \in \Omega_J \times \Omega_J$, $\|g\|_\infty \leq 1$,*

$$(7.28) \qquad \hat{u}(x) = -\int_{\Omega_J} J(x - y) g(x, y) \, dy \quad \text{a.e. } x \in \Omega,$$

and

$$(7.29) \qquad J(x - y) g(x, y) \in J(x - y) \operatorname{sgn}(u(y) - u(x)) \quad \text{a.e. } (x, y) \in \Omega \times \Omega,$$

$$(7.30) \quad J(x - y) g(x, y) \in J(x - y) \operatorname{sgn}(\psi(y) - u(x)) \quad \text{a.e. } (x, y) \in \Omega \times (\Omega_J \setminus \Omega).$$

REMARK 7.15. Observe that

(i) We can rewrite (7.29) + (7.30) as

$$(7.31) \qquad J(x - y) g(x, y) \in J(x - y) \operatorname{sgn}(u_\psi(y) - u(x)) \quad \text{a.e. } (x, y) \in \Omega \times \Omega_J,$$

where in the rest of this section we set

$$u_\psi(x) := \begin{cases} u(x) & \text{if } x \in \Omega, \\ \psi(x) & \text{if } x \in \Omega_J \setminus \overline{\Omega}, \\ 0 & \text{if } x \notin \Omega_J. \end{cases}$$

(ii) We have $L^1(\Omega) = \operatorname{Dom}(B_{1,\psi}^J)$ and $B_{1,\psi}^J$ is closed in $L^1(\Omega) \times L^1(\Omega)$.

(iii) It is not difficult to see that, if $g \in L^\infty(\Omega_J \times \Omega_J)$, $g(x, y) = -g(y, x)$ for almost all $(x, y) \in \Omega_J \times \Omega_J$ and $\|g\|_\infty \leq 1$, then

$$J(x - y) g(x, y) \in J(x - y) \operatorname{sgn}(z(y) - z(x)) \quad \text{a.e. } (x, y) \in \Omega_J \times \Omega_J,$$

is equivalent to

$$-\int_{\Omega_J} \int_{\Omega_J} J(x - y) g(x, y) \, dy \, z(x) \, dx = \frac{1}{2} \int_{\Omega_J} \int_{\Omega_J} J(x - y) |z(y) - z(x)| \, dy \, dx.$$

THEOREM 7.16. *Let $\psi \in L^1(\Omega_J \setminus \overline{\Omega})$. The operator $B_{1,\psi}^J$ is completely accretive and satisfies the range condition*

$$L^\infty(\Omega) \subset \operatorname{R}(I + B_{1,\psi}^J).$$

PROOF. Let $\hat{u}_i \in B^J_{1,\psi} u_i$, $i = 1, 2$, and set $u_i(y) = \psi(y)$ in $\Omega_J \setminus \Omega$. Then there exist $g_i \in L^\infty(\Omega_J \times \Omega_J)$, $\|g_i\|_\infty \le 1$, $g_i(x, y) = -g_i(y, x)$, $J(x - y)g_i(x, y) \in J(x - y)\mathrm{sgn}(u_i(y) - u_i(x))$ for almost all $(x, y) \in \Omega \times \Omega_J$, such that

$$\hat{u}_i(x) = -\int_{\Omega_J} J(x - y)g_i(x, y)\, dy \qquad \text{a.e. } x \in \Omega,$$

for $i = 1, 2$. Given $q \in P_0$, we have

$$\int_\Omega (\hat{u}_1(x) - \hat{u}_2(x))q(u_1(x) - u_2(x))\, dx$$

$$= \frac{1}{2} \int_\Omega \int_\Omega J(x - y)(g_1(x, y) - g_2(x, y))$$
$$\times (q(u_1(y) - u_2(y)) - q(u_1(x) - u_2(x)))\, dx\, dy$$

$$- \int_\Omega \int_{\Omega_J \setminus \Omega} J(x - y)(g_1(x, y) - g_2(x, y))q(u_1(x) - u_2(x))dx\, dy$$

$$\ge \frac{1}{2} \int_\Omega \int_\Omega J(x - y)(g_1(x, y) - g_2(x, y))$$
$$\times (q(u_1(y) - u_2(y)) - q(u_1(x) - u_2(x)))\, dx\, dy.$$

Now, by the mean value theorem,

$$J(x - y)(g_1(x, y) - g_2(x, y))\left[q(u_1(y) - u_2(y)) - q(u_1(x) - u_2(x))\right]$$

$$= J(x - y)(g_1(x, y) - g_2(x, y))q'(\xi)\left[(u_1(y) - u_2(y)) - (u_1(x) - u_2(x))\right]$$

$$= J(x - y)q'(\xi)\left[g_1(x, y)(u_1(y) - u_1(x)) - g_1(x, y)(u_2(y) - u_2(x))\right]$$

$$- J(x - y)q'(\xi)\left[g_2(x, y)(u_1(y) - u_1(x)) - g_1(x, y)(u_2(y) - u_2(x))\right] \ge 0,$$

since

$$J(x - y)g_i(x, y)(u_i(y) - u_i(x)) = J(x - y)|u_i(y) - u_i(x)|, \quad i = 1, 2,$$

and

$$-J(x - y)g_i(x, y)(u_j(y) - u_j(x)) \ge -J(x - y)|u_j(y) - u_j(x)|, \quad i \ne j.$$

Consequently,

$$\int_\Omega (\hat{u}_1(x) - \hat{u}_2(x))q(u_1(x) - u_2(x))\, dx \ge 0,$$

from which it follows that $B^J_{1,\psi}$ is a completely accretive operator.

To show that $B^J_{1,\psi}$ satisfies the range condition, let us see that for any $\phi \in L^\infty(\Omega)$,

$$\lim_{p \to 1+} (I + B^J_{p,\psi})^{-1}\phi = (I + B^J_{1,\psi})^{-1}\phi \quad \text{weakly in } L^1(\Omega).$$

We prove this in several steps.

Step 1. Let us first suppose that $\psi \in L^\infty(\Omega_J \setminus \overline{\Omega})$. For $1 < p < +\infty$, by Theorem 6.30, there is u_p such that $u_p = (I + B^J_{p,\psi})^{-1}\phi$, that is,

$$(7.32) \qquad u_p(x) - \int_{\Omega_J} J(x-y)\,|(u_p)_\psi(y) - u_p(x)|^{p-2}((u_p)_\psi(y) - u_p(x))\,dy$$
$$= \phi(x),$$

a.e. $x \in \Omega$. It is easy to see that

$$\|u_p\|_\infty \le \sup\{\|\phi\|_\infty, \|\psi\|_\infty\}.$$

Therefore, there exists a sequence $p_n \to 1$ such that

$$u_{p_n} \rightharpoonup u \quad \text{weakly in } L^2(\Omega).$$

On the other hand, we also have

$$\big|\,|(u_{p_n})_\psi(y) - (u_{p_n})_\psi(x)|^{p_n-2}\,((u_{p_n})_\psi(y) - (u_{p_n})_\psi(x))\big|$$
$$\le (2\sup\{\|\phi\|_\infty, \|\psi\|_\infty\})^{p_n-1},$$

and hence there exists $g(x,y)$ such that

$$|(u_{p_n})_\psi(y) - (u_{p_n})_\psi(x)|^{p_n-2}\,((u_{p_n})_\psi(y) - (u_{p_n})_\psi(x)) \rightharpoonup g(x,y),$$

weakly in $L^1(\Omega_J \times \Omega_J)$, $g(x,y) = -g(y,x)$ for almost all $(x,y) \in \Omega_J \times \Omega_J$, and $\|g\|_\infty \le 1$.

Therefore, by (7.32),

$$(7.33) \qquad u(x) - \int_{\Omega_J} J(x-y)g(x,y)\,dy = \phi(x) \qquad \text{a.e. } x \in \Omega.$$

Then, to finish the proof it is enough to show that

$$(7.34) \qquad -\int_{\Omega_J}\int_{\Omega_J} J(x-y)g(x,y)\,dy\,u_\psi(x)\,dx$$
$$= \frac{1}{2}\int_{\Omega_J}\int_{\Omega_J} J(x-y)|u_\psi(y) - u_\psi(x)|\,dy\,dx.$$

In fact, by (7.32) and (7.33),

$$\frac{1}{2} \int_{\Omega_J} \int_{\Omega_J} J(x-y) \left| (u_{p_n})_\psi(y) - (u_{p_n})_\psi(x) \right|^{p_n} dy \, dx = \int_\Omega \phi u_{p_n} - \int_\Omega u_{p_n} u_{p_n}$$

$$- \int_{\Omega_J \setminus \overline{\Omega}} \int_{\Omega_J} J(x-y) \left| \psi(y) - (u_{p_n})_\psi(x) \right|^{p_n - 2} (\psi(y) - (u_{p_n})_\psi(x)) \, dy \, \psi(x) \, dx$$

$$= \int_\Omega \phi u - \int_\Omega uu - \int_\Omega \phi(u - u_{p_n}) + \int_\Omega 2u(u - u_{p_n}) - \int_\Omega (u - u_{p_n})(u - u_{p_n})$$

$$- \int_{\Omega_J \setminus \overline{\Omega}} \int_{\Omega_J} J(x-y) \left| \psi(y) - (u_{p_n})_\psi(x) \right|^{p_n - 2} (\psi(y) - (u_{p_n})_\psi(x)) \, dy \, \psi(x) \, dx$$

$$\leq - \int_{\Omega_J} \int_{\Omega_J} J(x-y) g(x,y) \, dy \, u(x) \, dx - \int_\Omega \phi(u - u_{p_n}) + \int_\Omega 2u(u - u_{p_n})$$

$$+ \int_{\Omega_J \setminus \overline{\Omega}} \int_{\Omega_J} J(x-y) g(x,y) \, dy \, \psi(x) \, dx$$

$$- \int_{\Omega_J \setminus \overline{\Omega}} \int_{\Omega_J} J(x-y) \left| \psi(y) - (u_{p_n})_\psi(x) \right|^{p_n - 2} (\psi(y) - (u_{p_n})_\psi(x)) \, dy \, \psi(x) \, dx,$$

and so,

$$\limsup_{n \to +\infty} \frac{1}{2} \int_{\Omega_J} \int_{\Omega_J} J(x-y) \left| u_{p_n}(y) - u_{p_n}(x) \right|^{p_n} dy \, dx$$

$$\leq - \int_{\Omega_J} \int_{\Omega_J} J(x-y) g(x,y) \, dy \, u(x) \, dx.$$

Now, by the monotonicity Lemma 6.29, for all $\rho \in L^\infty(\Omega)$,

$$- \int_{\Omega_J} \int_{\Omega_J} J(x-y) |\rho(y) - \rho(x)|^{p_n - 2} (\rho(y) - \rho(x)) \, dy \, (u_{p_n}(x) - \rho(x)) \, dx$$

$$\leq - \int_{\Omega_J} \int_{\Omega_J} J(x-y) |u_{p_n}(y) - u_{p_n}(x)|^{p_n - 2}$$

$$\times (u_{p_n}(y) - u_{p_n}(x)) \, dy \, (u_{p_n}(x) - \rho(x)) \, dx.$$

Taking limits as $n \to +\infty$,

$$- \int_{\Omega_J} \int_{\Omega_J} J(x-y) \operatorname{sgn}_0(\rho(y) - \rho(x)) \, dy \, (u(x) - \rho(x)) \, dx$$

$$\leq - \int_{\Omega_J} \int_{\Omega_J} J(x-y) g(x,y) \, dy \, (u(x) - \rho(x)) \, dx.$$

Taking $\rho = u \pm \lambda u$, $\lambda > 0$, and letting $\lambda \to 0$, we get (7.34), and the proof is finished for this class of data.

Step 2. Let us now suppose that ψ^- is bounded. Let $\psi_n = T_n(\psi)$, n large enough such that $\psi_n^- = \psi^-$. Then $\{\psi_n\}$ is a nondecreasing sequence that converges in

L^1 to ψ. By Step 1, there exists $u_n = (I + B^J_{1,\psi_n})^{-1}\phi$; that is, there exists $g_n \in L^\infty(\Omega_J \times \Omega_J)$, $g_n(x,y) = -g_n(y,x)$ for almost all $(x,y) \in \Omega_J \times \Omega_J$, $\|g_n\|_\infty \le 1$,

$$(7.35) \qquad u_n(x) - \int_{\Omega_J} J(x-y)g_n(x,y)\,dy = \phi(x) \qquad \text{a.e. } x \in \Omega$$

and

$$(7.36) \qquad \begin{aligned} &-\int_{\Omega_J}\int_{\Omega_J} J(x-y)g_n(x,y)\,dy\,(u_n)_{\psi_n}(x)\,dx \\ &= \frac{1}{2}\int_{\Omega_J}\int_{\Omega_J} J(x-y)|(u_n)_{\psi_n}(y) - (u_n)_{\psi_n}(x)|\,dy\,dx. \end{aligned}$$

Therefore, by monotonicity,

$$\int_{\Omega_J}\int_{\Omega_J} \left((u_n)_{\psi_n} - (u_{n+1})_{\psi_{n+1}}\right)\left((u_n)_{\psi_n} - (u_{n+1})_{\psi_{n+1}}\right)^+ \le 0,$$

which implies $u_n \le u_{n+1}$. Since $\{u_n\}$ is bounded in L^∞, we see that $\{u_n\}$ converges to a function u in L^2. On the other hand, we can assume that $J(x-y)g_n(x,y)$ converges weakly in L^2 to $J(x-y)g(x,y)$, $g(x,y) = -g(y,x)$ for almost all $(x,y) \in \Omega_J \times \Omega_J$, and $\|g\|_\infty \le 1$. Passing to the limit in (7.35) and (7.36) we obtain $u = (I + B^J_{1,\psi})^{-1}\phi$.

Step 3. For a general $\psi \in L^1(\Omega_J \setminus \overline{\Omega})$, we apply Step 2 to $\psi_n = \sup\{\psi, -n\}$ and using monotonicity in a similar way as before the proof can be finished. □

PROOF OF THEOREM 7.13. As a consequence of the above results, we have that the abstract Cauchy problem

$$(7.37) \qquad \begin{cases} u'(t) + B^J_{1,\psi}u(t) \ni 0, & t \in (0,T), \\ u(0) = u_0 \end{cases}$$

has a unique mild solution u for every initial datum $u_0 \in L^1(\Omega)$ and $T > 0$. Moreover, due to the complete accretivity of the operator $B^J_{1,\psi}$ and since $L^1(\Omega) = D(B^J_{1,\psi})$, the mild solution of (7.37) is a strong solution (Corollary A.52). Consequently, the proof is concluded. □

7.3.2. Convergence to the total variation flow. Let Ω be a bounded smooth domain in $\mathbb{R}^N$. We start by recalling some results from [6] (see also [7]) about the Dirichlet problem for the total variational flow

$$(7.38) \qquad \begin{cases} v_t = \text{div}\left(\dfrac{Dv}{|Dv|}\right) & \text{in } \Omega \times (0,T), \\ v = \tilde{\psi} & \text{on } \partial\Omega \times (0,T), \\ v(\cdot,0) = u_0 & \text{in } \Omega, \end{cases}$$

with $\tilde{\psi} \in L^1(\partial\Omega)$.

Associated to the operator $-\text{div}\left(\frac{Dv}{|Dv|}\right)$ with Dirichlet boundary conditions, the operator $\mathcal{A}_{\tilde{\psi}} \subset L^1(\Omega) \times L^1(\Omega)$ is defined in [6] as follows: $(v,\hat{v}) \in \mathcal{A}_{\tilde{\psi}}$ if and only if $v, \hat{v} \in L^1(\Omega)$, $q(v) \in BV(\Omega)$ for all $q \in \mathcal{P}$, where

$$\mathcal{P} := \{q \in W^{1,\infty}(\mathbb{R}) : q' \geq 0, \operatorname{supp}(q') \text{ is compact}\},$$

and there exists $\zeta \in X(\Omega)$, with $\|\zeta\|_\infty \leq 1$, $\hat{v} = -\operatorname{div}(\zeta)$ in $\mathcal{D}'(\Omega)$, such that

$$(7.39) \quad \int_\Omega (\xi - q(v))\hat{v} \leq \int_\Omega (\zeta, D\xi) - |Dq(v)| + \int_{\partial\Omega} |\xi - q(\tilde{\psi})| - \int_{\partial\Omega} |q(v) - q(\tilde{\psi})|$$

for every $\xi \in BV(\Omega) \cap L^\infty(\Omega)$ and every $q \in \mathcal{P}$.

In [6] it is proved that the following assertions are equivalent:

(a) $(v, \hat{v}) \in \mathcal{A}_{\tilde{\psi}}$;

(b) $v, \hat{v} \in L^1(\Omega)$, $q(v) \in BV(\Omega)$ for all $q \in \mathcal{P}$, and there exists $\zeta \in X(\Omega)$, with $\|\zeta\|_\infty \leq 1$, $\hat{v} = -\operatorname{div}(\zeta)$ in $\mathcal{D}'(\Omega)$, such that

$$(7.40) \quad \int_\Omega (\zeta, Dq(v)) = |Dq(v)| \qquad \forall q \in \mathcal{P},$$

$$(7.41) \quad [\zeta, \nu] \in \operatorname{sgn}(q(\tilde{\psi}) - q(v)) \quad \mathcal{H}^{N-1}\text{-a.e. on } \partial\Omega, \ \forall q \in \mathcal{P}.$$

It is shown that $\mathcal{A}_{\tilde{\psi}}$ is an m-completely accretive operator in $L^1(\Omega)$ with dense domain and that, for any $u_0 \in L^1(\Omega)$, the unique entropy solution $v(t)$ of problem (7.38) (see [6] for the definition) coincides with the unique mild solution $e^{-t\mathcal{A}_{\tilde{\psi}}}u_0$ given by Crandall-Liggett's exponential formula. Moreover, the following existence and uniqueness result is obtained.

THEOREM 7.17 ([6]). *Let $T > 0$ and $\tilde{\psi} \in L^1(\partial\Omega)$. For any $u_0 \in L^1(\Omega)$ there exists a unique entropy solution $v(t)$ of (7.38).*

Given J, we consider the rescaled kernels

$$J_{1,\varepsilon}(x) := \frac{C_{J,1}}{\varepsilon^{1+N}} J\left(\frac{x}{\varepsilon}\right),$$

with $C_{J,1}^{-1} := \frac{1}{2} \int_{\mathbb{R}^N} J(z)|z_N|\, dz$ a normalizing constant in order to obtain the 1-Laplacian in the limit instead of a multiple of it. We have the following convergence result.

THEOREM 7.18. *Suppose $J(x) \geq J(y)$ if $|x| \leq |y|$. Let $T > 0$, $u_0 \in L^1(\Omega)$, $\tilde{\psi} \in L^\infty(\partial\Omega)$ and $\psi \in W^{1,1}(\Omega_J \setminus \overline{\Omega}) \cap L^\infty(\Omega_J \setminus \overline{\Omega})$ such that $\psi|_{\partial\Omega} = \tilde{\psi}$. Let u_ε be the unique solution of (7.27). Then, if v is the unique solution of (7.38),*

$$\lim_{\varepsilon \to 0} \sup_{t \in [0,T]} \|u_\varepsilon(\cdot, t) - v(\cdot, t)\|_{L^1(\Omega)} = 0.$$

Since the solutions of the above theorem coincide with the semigroup solutions, by Theorem A.37, to prove Theorem 7.18 it is enough to obtain the following result. Its proof differs from the one given for Theorem 7.10 since we have to recover the Dirichlet boundary condition.

THEOREM 7.19. *Let $\tilde{\psi} \in L^\infty(\partial\Omega)$ and $\psi \in W^{1,1}(\Omega_J \setminus \overline{\Omega}) \cap L^\infty(\Omega_J \setminus \overline{\Omega})$ such that $\psi|_{\partial\Omega} = \tilde{\psi}$. Suppose $J(x) \geq J(y)$ if $|x| \leq |y|$. Then, for any $\phi \in L^\infty(\Omega)$,*

$$(7.42) \quad \left(I + B_{1,\psi}^{J_{1,\varepsilon}}\right)^{-1} \phi \to \left(I + \mathcal{A}_{\tilde{\psi}}\right)^{-1} \phi \quad \text{strongly in } L^1(\Omega) \text{ as } \varepsilon \to 0.$$

PROOF. Given $\varepsilon > 0$ small, we set $u_\varepsilon = \left(I + B_{1,\psi}^{J_{1,\varepsilon}}\right)^{-1} \phi$ and

$$\Omega_\varepsilon := \Omega_{J_{1,\varepsilon}} = \Omega + \operatorname{supp}(J_{1,\varepsilon}).$$

Then there exists $g_\varepsilon \in L^\infty(\Omega_\varepsilon \times \Omega_\varepsilon)$, $g_\varepsilon(x,y) = -g_\varepsilon(y,x)$ for almost all $(x,y) \in \Omega_\varepsilon \times \Omega_\varepsilon$, $\|g_\varepsilon\|_\infty \leq 1$, such that

$$J\left(\frac{x-y}{\varepsilon}\right) g_\varepsilon(x,y) \in J\left(\frac{x-y}{\varepsilon}\right) \operatorname{sgn}(u_\varepsilon(y) - u_\varepsilon(x)) \quad \text{a.e. } (x,y) \in \Omega \times \Omega,$$

$$J\left(\frac{x-y}{\varepsilon}\right) g_\varepsilon(x,y) \in J\left(\frac{x-y}{\varepsilon}\right) \operatorname{sgn}(\tilde{\psi}(y) - u_\varepsilon(x)) \quad \text{a.e. } (x,y) \in \Omega \times (\Omega_\varepsilon \setminus \overline{\Omega})$$

and

$$(7.43) \qquad u_\varepsilon(x) - \frac{C_{J,1}}{\varepsilon^{1+N}} \int_{\Omega_\varepsilon} J\left(\frac{x-y}{\varepsilon}\right) g_\varepsilon(x,y)\, dy = \phi(x) \quad \text{a.e. } x \in \Omega.$$

Therefore, for $\xi \in L^\infty(\Omega_J)$, we can write

$$(7.44)$$

$$\int_\Omega u_\varepsilon(x) \xi(x)\, dx - \frac{C_{J,1}}{\varepsilon^{1+N}} \int_\Omega \int_{\Omega_\varepsilon} J\left(\frac{x-y}{\varepsilon}\right) g_\varepsilon(x,y) \xi(x)\, dy\, dx = \int_\Omega \phi(x) \xi(x)\, dx.$$

Observe that we can extend g_ε to a function in $L^\infty(\Omega_J \times \Omega_J)$, $g_\varepsilon(x,y) = -g_\varepsilon(y,x)$ for almost all $(x,y) \in \Omega_J \times \Omega_J$, $\|g_\varepsilon\|_{L^\infty(\Omega_J)} \leq 1$, such that

$$J\left(\frac{x-y}{\varepsilon}\right) g_\varepsilon(x,y) \in J\left(\frac{x-y}{\varepsilon}\right) \operatorname{sgn}((u_\varepsilon)_\psi(y) - (u_\varepsilon)_\psi(x)) \quad \text{a.e. } (x,y) \in \Omega_J \times \Omega_J.$$

Let $M := \max\{\|\phi\|_{L^\infty(\Omega)}, \|\psi\|_{L^\infty(\Omega_J \setminus \overline{\Omega})}\}$. Taking $\xi = (u_\varepsilon - M)^+$ in (7.44), we get

$$\int_\Omega u_\varepsilon(x)(u_\varepsilon(x) - M)^+ dx - \frac{C_{J,1}}{\varepsilon^{1+N}} \int_\Omega \int_{\Omega_\varepsilon} J\left(\frac{x-y}{\varepsilon}\right) g_\varepsilon(x,y)(u_\varepsilon(x) - M)^+ dy\, dx$$

$$= \int_\Omega \phi(x)(u_\varepsilon(x) - M)^+ dx.$$

Now,

$$-\frac{C_{J,1}}{\varepsilon^{1+N}} \int_\Omega \int_{\Omega_\varepsilon} J\left(\frac{x-y}{\varepsilon}\right) g_\varepsilon(x,y)(u_\varepsilon(x) - M)^+ dy\, dx$$

$$= -\frac{C_{J,1}}{\varepsilon^{1+N}} \int_{\Omega_\varepsilon} \int_{\Omega_\varepsilon} J\left(\frac{x-y}{\varepsilon}\right) g_\varepsilon(x,y)((u_\varepsilon)_\psi(x) - M)^+ dy\, dx$$

$$= \frac{C_{J,1}}{2\varepsilon^{1+N}} \int_{\Omega_\varepsilon} \int_{\Omega_\varepsilon} J\left(\frac{x-y}{\varepsilon}\right) g_\varepsilon(x,y)(((u_\varepsilon)_\psi(y) - M)^+ - ((u_\varepsilon)_\psi(x) - M)^+)\, dy\, dx$$

$$\geq 0.$$

Hence, we get

$$\int_\Omega u_\varepsilon(x)(u_\varepsilon(x) - M)^+ dx \leq \int_\Omega \phi(x)(u_\varepsilon(x) - M)^+ dx.$$

Consequently

$$0 \le \int_\Omega (u_\varepsilon(x) - M)(u_\varepsilon(x) - M)^+ dx \le \int_\Omega (\phi(x) - M)(u_\varepsilon(x) - M)^+ dx \le 0,$$

and we deduce $u_\varepsilon(x) \le M$ for almost all $x \in \Omega$. Analogously, we can obtain $-M \le u_\varepsilon(x)$ for almost all $x \in \Omega$. Thus

(7.45) $$\|u_\varepsilon\|_{L^\infty(\Omega)} \le M \qquad \text{for all } \epsilon > 0;$$

in view of this, we can assume that there exists a sequence $\varepsilon_n \to 0$ such that

$$u_{\varepsilon_n} \rightharpoonup v \quad \text{weakly in } L^1(\Omega).$$

Taking $\xi = u_\varepsilon$ in (7.44), we have

(7.46)
$$\int_\Omega u_\varepsilon(x) u_\varepsilon(x) \, dx - \frac{C_{J,1}}{\varepsilon^{1+N}} \int_\Omega \int_{\Omega_\varepsilon} J\left(\frac{x-y}{\varepsilon}\right) g_\varepsilon(x,y) \, dy \, u_\varepsilon(x) \, dx$$
$$= \int_\Omega \phi(x) u_\varepsilon(x) dx.$$

Observe that,

$$-\frac{C_{J,1}}{\varepsilon^{1+N}} \int_\Omega \int_{\Omega_\varepsilon} J\left(\frac{x-y}{\varepsilon}\right) g_\varepsilon(x,y) \, dy \, u_\varepsilon(x) \, dx$$

$$= -\frac{C_{J,1}}{\varepsilon^{1+N}} \int_{\Omega_\varepsilon} \int_{\Omega_\varepsilon} J\left(\frac{x-y}{\varepsilon}\right) g_\varepsilon(x,y) \, dy \, (u_\varepsilon)_\psi(x) \, dx$$

$$+\frac{C_{J,1}}{\varepsilon^{1+N}} \int_{\Omega_\varepsilon \setminus \overline{\Omega}} \int_{\Omega_\varepsilon} J\left(\frac{x-y}{\varepsilon}\right) g_\varepsilon(x,y) \, dy \, \psi(x) \, dx.$$

Now,

$$\left| \frac{C_{J,1}}{\varepsilon^{1+N}} \int_{\Omega_\varepsilon \setminus \overline{\Omega}} \int_{\Omega_\varepsilon} J\left(\frac{x-y}{\varepsilon}\right) g_\varepsilon(x,y) \, dy \, \psi(x) \, dx \right|$$

$$\le \frac{C_{J,1}}{\varepsilon^{1+N}} \int_{\Omega_\varepsilon \setminus \overline{\Omega}} \int_{\Omega_\varepsilon} J\left(\frac{x-y}{\varepsilon}\right) \, dy |\psi(x)| \, dx$$

$$\le \frac{C_{J,1}}{\varepsilon} M \int_{\Omega_\varepsilon \setminus \overline{\Omega}} \left(\frac{1}{\varepsilon^N} \int_{\Omega_\varepsilon} J\left(\frac{x-y}{\varepsilon}\right) \, dy \right) \, dx$$

$$\le \frac{C_{J,1}}{\varepsilon} M |\Omega_\varepsilon \setminus \overline{\Omega}| \le M_1.$$

On the other hand,

$$-\frac{C_{J,1}}{\varepsilon^{1+N}} \int_{\Omega_\varepsilon} \int_{\Omega_\varepsilon} J\left(\frac{x-y}{\varepsilon}\right) g_\varepsilon(x,y) \, dy \, (u_\varepsilon)_\psi(x) \, dx$$

$$= \frac{C_{J,1}}{2\varepsilon^{1+N}} \int_{\Omega_\varepsilon} \int_{\Omega_\varepsilon} J\left(\frac{x-y}{\varepsilon}\right) |(u_\varepsilon)_\psi(y) - (u_\varepsilon)_\psi(x)| \, dy \, dx.$$

Consequently, from (7.45) and (7.46), it follows that

$$(7.47) \qquad \frac{C_{J,1}}{2\varepsilon^{1+N}} \int_{\Omega_\varepsilon} \int_{\Omega_\varepsilon} J\left(\frac{x-y}{\varepsilon}\right) |(u_\varepsilon)_\psi(y) - (u_\varepsilon)_\psi(x)| \, dy \, dx \le M_2.$$

Let us write

$$\frac{C_{J,1}}{2\varepsilon^{1+N}} \int_{\Omega_J} \int_{\Omega_J} J\left(\frac{x-y}{\varepsilon}\right) |(u_\varepsilon)_\psi(y) - (u_\varepsilon)_\psi(x)| \, dy \, dx$$

$$= \frac{C_{J,1}}{2\varepsilon^{1+N}} \int_{\Omega_\varepsilon} \int_{\Omega_\varepsilon} J\left(\frac{x-y}{\varepsilon}\right) |(u_\varepsilon)_\psi(y) - (u_\varepsilon)_\psi(x)| \, dy \, dx$$

$$+ \frac{C_{J,1}}{2\varepsilon^{1+N}} \int_{\Omega_\varepsilon} \int_{\Omega_J \setminus \overline{\Omega}_\varepsilon} J\left(\frac{x-y}{\varepsilon}\right) |(u_\varepsilon)_\psi(y) - (u_\varepsilon)_\psi(x)| \, dy \, dx$$

$$+ \frac{C_{J,1}}{2\varepsilon^{1+N}} \int_{\Omega_J \setminus \overline{\Omega}_\varepsilon} \int_{\Omega_\varepsilon} J\left(\frac{x-y}{\varepsilon}\right) |(u_\varepsilon)_\psi(y) - (u_\varepsilon)_\psi(x)| \, dy \, dx$$

$$+ \frac{C_{J,1}}{2\varepsilon^{1+N}} \int_{\Omega_J \setminus \overline{\Omega}_\varepsilon} \int_{\Omega_J \setminus \overline{\Omega}_\varepsilon} J\left(\frac{x-y}{\varepsilon}\right) |(u_\varepsilon)_\psi(y) - (u_\varepsilon)_\psi(x)| \, dy \, dx.$$

Since $\psi \in W^{1,1}(\Omega_J \setminus \overline{\Omega})$, we get

$$\frac{C_{J,1}}{2\varepsilon^{1+N}} \int_{\Omega_J \setminus \overline{\Omega}_\varepsilon} \int_{\Omega_J \setminus \overline{\Omega}_\varepsilon} J\left(\frac{x-y}{\varepsilon}\right) |(u_\varepsilon)_\psi(y) - (u_\varepsilon)_\psi(x)| \, dy \, dx$$

$$= \frac{C_{J,1}}{2\varepsilon^N} \int_{\Omega_J \setminus \overline{\Omega}_\varepsilon} \int_{\Omega_J \setminus \overline{\Omega}_\varepsilon} J\left(\frac{x-y}{\varepsilon}\right) \frac{|\psi(y) - \psi(x)|}{\varepsilon} \, dy \, dx \le M_3.$$

On the other hand, we have

$$\frac{C_{J,1}}{2\varepsilon^{1+N}} \int_{\Omega_J \setminus \overline{\Omega}_\varepsilon} \int_{\Omega_\varepsilon} J\left(\frac{x-y}{\varepsilon}\right) |(u_\varepsilon)_\psi(y) - (u_\varepsilon)_\psi(x)| \, dy \, dx$$

$$= \frac{C_{J,1}}{2\varepsilon^N} \int_{\Omega_J \setminus \overline{\Omega}_\varepsilon} \int_{\Omega_\varepsilon \setminus \overline{\Omega}} J\left(\frac{x-y}{\varepsilon}\right) \frac{|\psi(y) - \psi(x)|}{\varepsilon} \, dy \, dx$$

$$\le M_4 \frac{C_{J,1}}{2} \int_{\Omega_J \setminus \overline{\Omega}_\varepsilon} \left(\frac{1}{\varepsilon^N} \int_{\Omega_\varepsilon} J\left(\frac{x-y}{\varepsilon}\right) dy \right) dx \le M_5.$$

With similar arguments we obtain,

$$\frac{C_{J,1}}{2\varepsilon^{1+N}} \int_{\Omega_\varepsilon} \int_{\Omega_J \setminus \overline{\Omega}_\varepsilon} J\left(\frac{x-y}{\varepsilon}\right) |(u_\varepsilon)_\psi(y) - (u_\varepsilon)_\psi(x)| \, dy \, dx \le M_6.$$

Therefore,

$$\frac{C_{J,1}}{2\varepsilon^{1+N}} \int_{\Omega_J} \int_{\Omega_J} J\left(\frac{x-y}{\varepsilon}\right) |(u_\varepsilon)_\psi(y) - (u_\varepsilon)_\psi(x)| \, dy \, dx \le M_7.$$

In particular, we get

$$\int_{\Omega_J} \int_{\Omega_J} \frac{1}{2} \frac{C_{J,1}}{\varepsilon_n^N} J\left(\frac{x-y}{\varepsilon_n}\right) \left| \frac{(u_{\varepsilon_n})_\psi(y) - (u_{\varepsilon_n})_\psi(x)}{\varepsilon_n} \right| \, dx \, dy \le M_7 \qquad \forall \, n \in \mathbb{N}.$$

By Theorem 6.11, there exists a subsequence, denoted the same way, and $w \in BV(\Omega_J)$ such that

$$(u_{\varepsilon_n})_\psi \to w \qquad \text{strongly in } L^1(\Omega_J)$$

and

(7.48) $$\frac{C_{J,1}}{2} J(z)\chi_\Omega(\cdot + \varepsilon_n z)\frac{(u_{\varepsilon_n})_\psi(\cdot + \varepsilon_n z) - (u_{\varepsilon_n})_\psi(\cdot)}{\varepsilon_n} \rightharpoonup \frac{C_{J,1}}{2} J(z)z \cdot Dw$$

weakly as measures. Hence, it is easy to obtain that

$$w(x) = v_\psi(x) = \begin{cases} v(x) & \text{in } x \in \Omega, \\ \psi(x) & \text{in } x \in \Omega_J \setminus \Omega, \end{cases}$$

and $v \in BV(\Omega)$.

Moreover, we can also assume that

(7.49) $$J(z)\chi_{\Omega_J}(x + \varepsilon_n z)\overline{g}_{\varepsilon_n}(x, x + \varepsilon_n z) \rightharpoonup \Lambda(x, z)$$

weakly* in $L^\infty(\Omega_J) \times L^\infty(\mathbb{R}^N)$, for some function $\Lambda \in L^\infty(\Omega_J) \times L^\infty(\mathbb{R}^N)$, $\Lambda(x, z) \leq J(z)$ almost everywhere in $\Omega_J \times \mathbb{R}^N$ (for $\overline{g}_\varepsilon$ we are using the notation (6.6)). Taking $\xi \in \mathcal{D}(\Omega)$ in (7.44), we get, for $\varepsilon = \varepsilon_n$ small enough,

(7.50) $$\int_\Omega u_{\varepsilon_n}(x)\xi(x)dx - \frac{C_{J,1}}{\varepsilon_n^{1+N}}\int_\Omega \int_\Omega J\left(\frac{x-y}{\varepsilon_n}\right) g_{\varepsilon_n}(x, y)\xi(x)\, dy\, dx$$
$$= \int_\Omega \phi(x)\xi(x)\, dx.$$

Changing variables and taking into account (7.50), we can write

(7.51) $$\frac{C_{J,1}}{2}\int_{\mathbb{R}^N}\int_\Omega J(z)\chi_\Omega(x + \varepsilon_n z)\overline{g}_{\varepsilon_n}(x, x + \varepsilon_n z)\, dz\frac{\overline{\xi}(x + \varepsilon_n z) - \xi(x)}{\varepsilon_n}\, dx$$
$$= -\frac{C_{J,1}}{\varepsilon_n}\int_{\mathbb{R}^N}\int_\Omega J(z)\chi_\Omega(x + \varepsilon_n z)\overline{g}_{\varepsilon_n}(x, x + \varepsilon_n z)\, dz\, \xi(x)\, dx$$
$$= \int_\Omega (\phi(x) - u_{\varepsilon_n}(x))\xi(x)\, dx.$$

By (7.49), passing to the limit in (7.51), we get

(7.52) $$\frac{C_{J,1}}{2}\int_{\mathbb{R}^N}\int_\Omega \Lambda(x, z)z \cdot \nabla\xi(x)\, dx\, dz$$
$$= \int_\Omega (\phi(x) - v(x))\xi(x)\, dx \qquad \forall\, \xi \in \mathcal{D}(\Omega).$$

We denote by $\zeta = (\zeta_1, \ldots, \zeta_N)$ the vector field defined by

$$\zeta_i(x) := \frac{C_{J,1}}{2}\int_{\mathbb{R}^N} \Lambda(x, z)z_i\, dz, \quad i = 1, \ldots, N.$$

Then $\zeta \in L^\infty(\Omega_J, \mathbb{R}^N)$ and, from (7.52),

$$-\text{div}(\zeta) = \phi - v \quad \text{in } \mathcal{D}'(\Omega).$$

Arguing as in the proof of (7.24),

$$\|\zeta\|_{L^\infty(\Omega_J)} \leq 1.$$

Hence, to finish the proof, that is, to show that $v = \left(I + \mathcal{A}_{\tilde{\psi}}\right)^{-1}\phi$, since $v \in L^\infty(\Omega)$ and $\tilde{\psi} \in L^\infty(\partial\Omega)$, we only need to prove that

$$(7.53) \qquad (\zeta, Dv) = |Dv| \quad \text{as measures in } \Omega$$

and

$$(7.54) \qquad [\zeta, \nu] \in \text{sgn}(\tilde{\psi} - v) \quad \mathcal{H}^{N-1}\text{-a.e. on } \partial\Omega.$$

Given $0 \le \varphi \in \mathcal{D}(\Omega)$, taking $\varepsilon = \varepsilon_n$ and $\xi = \varphi u_{\varepsilon_n}$ in (7.44), we get

$$\int_\Omega (\phi(x) - u_{\varepsilon_n}(x))u_{\varepsilon_n}(x)\varphi(x)\,dx$$

$$(7.55) \quad = -\frac{C_{J,1}}{\varepsilon_n^{1+N}} \int_\Omega \int_\Omega J\left(\frac{x-y}{\varepsilon_n}\right) g_{\varepsilon_n}(x,y)u_{\varepsilon_n}(x)\varphi(x)\,dy\,dx$$

$$= \frac{C_{J,1}}{2\varepsilon_n^{1+N}} \int_\Omega \int_\Omega J\left(\frac{x-y}{\varepsilon_n}\right) g_{\varepsilon_n}(x,y)(u_{\varepsilon_n}(y)\varphi(y) - u_{\varepsilon_n}(x)\varphi(x))\,dy\,dx.$$

Now we decompose the double integral as follows:

$$I_n := \frac{C_{J,1}}{2\varepsilon_n^{1+N}} \int_\Omega \int_\Omega J\left(\frac{x-y}{\varepsilon_n}\right) g_{\varepsilon_n}(x,y)(u_{\varepsilon_n}(y)\varphi(y) - u_{\varepsilon_n}(x)\varphi(x))\,dy\,dx = I_n^1 + I_n^2,$$

where

$$I_n^1 := \frac{C_{J,1}}{2\varepsilon_n^{1+N}} \int_\Omega \int_\Omega J\left(\frac{x-y}{\varepsilon_n}\right) |u_{\varepsilon_n}(y) - u_{\varepsilon_n}(x)|\varphi(y)\,dy\,dx$$

$$= \frac{C_{J,1}}{2} \int_\Omega \int_\Omega J(z)\chi_\Omega(x+\varepsilon_n z)\frac{|\overline{u}_{\varepsilon_n}(x+\varepsilon_n z) - u_{\varepsilon_n}(x)|}{\varepsilon_n}\overline{\varphi}(x+\varepsilon_n z)\,dz\,dx$$

and

$$I_n^2 := \frac{C_{J,1}}{2\varepsilon_n^{1+N}} \int_\Omega \int_\Omega J\left(\frac{x-y}{\varepsilon_n}\right) g_{\varepsilon_n}(x,y)u_{\varepsilon_n}(x)(\varphi(y) - \varphi(x))\,dy\,dx$$

$$= \frac{C_{J,1}}{2} \int_\Omega \int_\Omega J(z)\chi_\Omega(x+\varepsilon_n z)\overline{g}_{\varepsilon_n}(x, x+\varepsilon_n z)u_{\varepsilon_n}(x)\frac{\overline{\varphi}(x+\varepsilon_n z) - \varphi(x)}{\varepsilon_n}\,dz\,dx.$$

Having in mind (7.48), it follows that

$$\lim_{n\to\infty} I_n^1 \ge \frac{C_{J,1}}{2} \int_\Omega \int_\Omega J(z)\varphi(x)|z \cdot Dv| = \int_\Omega \varphi\,|Dv|.$$

On the other hand, since

$$u_{\varepsilon_n} \to v \qquad \text{strongly in } L^1(\Omega),$$

by (7.49), we get

$$\lim_{n\to\infty} I_n^2 = \frac{C_{J,1}}{2} \int_\Omega \int_{\mathbb{R}^N} v(x)\Lambda(x,z)z \cdot \nabla\varphi(x)\,dz\,dx = \int_\Omega v(x)\zeta(x) \cdot \nabla\varphi(x)\,dx.$$

Therefore, letting $n \to +\infty$ in (7.55), we obtain

$$(7.56) \qquad \int_\Omega \varphi\,|Dv| + \int_\Omega v(x)\zeta(x) \cdot \nabla\varphi(x)\,dx \le \int_\Omega (\phi(x) - v(x))v(x)\varphi(x)\,dx.$$

By Green's formula,

$$\int_\Omega (\phi(x) - v(x))v(x)\varphi(x)dx = -\int_\Omega \operatorname{div}(\zeta)v\varphi\,dx = \int_\Omega (\zeta, D(\varphi v))$$

$$= \int_\Omega \varphi(\zeta, Dv) + \int_\Omega v(x)\zeta(x) \cdot \nabla\varphi(x)\,dx.$$

Since $|(\zeta, Dv)| \le |Dv|$, the last identity and (7.56) give (7.53).

Finally, we show that (7.54) holds. We take $w_m \in W^{1,1}(\Omega) \cap C(\Omega)$ such that $w_m = \tilde{\psi}$ $\mathcal{H}^{N-1}$-a.e. on $\partial\Omega$, and $w_m \to v$ in $L^1(\Omega)$. Taking $\xi = v_{m,n} := (u_{\varepsilon_n})_\psi - (w_m)_\psi$ in (7.44), we get

$$\int_\Omega (\phi(x) - u_{\varepsilon_n}(x))(u_{\varepsilon_n}(x) - w_m(x))\,dx$$

(7.57)

$$= -\frac{C_{J,1}}{\varepsilon_n^{1+N}} \int_{\Omega_J} \int_{\Omega_J} J\left(\frac{x-y}{\varepsilon_n}\right) g_{\varepsilon_n}(x,y) v_{m,n}(x)\,dy\,dx$$

$$= \frac{C_{J,1}}{2\varepsilon_n^{1+N}} \int_{\Omega_J} \int_{\Omega_J} J\left(\frac{x-y}{\varepsilon_n}\right) g_{\varepsilon_n}(x,y)(v_{m,n}(x) - v_{m,n}(x))\,dy\,dx$$

$$= H_n^1 + H_{m,n}^1,$$

where

$$H_n^1 = \frac{C_{J,1}}{2} \int_{\Omega_J} \int_{\mathbb{R}^N} J(z) \chi_{\Omega_J}(x + \varepsilon_n z) \left| \frac{(u_{\varepsilon_n})_\psi(x + \varepsilon_n z) - (u_{\varepsilon_n})_\psi(x)}{\varepsilon_n} \right| dz\,dx$$

and

$$H_{m,n}^2 = -\frac{C_{J,1}}{2} \int_{\Omega_J} \int_{\mathbb{R}^N} J(z) \chi_{\Omega_J}(x + \varepsilon_n z) \overline{g}_{\varepsilon_n}(x, x + \varepsilon_n z)$$

$$\times \frac{(w_m)_\psi(x + \varepsilon_n z) - (w_m)_\psi(x)}{\varepsilon_n}\,dz\,dx.$$

Arguing as before,

$$\lim_{n\to\infty} H_n^1 \ge \int_{\Omega_J} |Dv_\psi| = \int_\Omega |Dv| + \int_{\partial\Omega} |v - \tilde{\psi}|\,d\mathcal{H}^{N-1} + \int_{\Omega_J \setminus \overline{\Omega}} |\nabla\psi|.$$

On the other hand, since $(w_m)_\psi \in W^{1,1}(\Omega_J)$, by (7.49),

$$\lim_{n\to\infty} H_{m,n}^2 = -\frac{C_{J,1}}{2} \int_{\Omega_J} \int_{\mathbb{R}^N} \Lambda(x,z) z \cdot \nabla(w_m)_\psi(x)\,dz\,dx = -\int_{\Omega_J} \zeta(x) \cdot \nabla(w_m)_\psi(x)\,dx.$$

Consequently, letting $n \to \infty$ in (7.57), we get

(7.58)

$$\int_\Omega (\phi(x) - v(x))(v(x) - w_m(x))\,dx$$

$$\ge \int_\Omega |Dv| + \int_{\partial\Omega} |v - \tilde{\psi}|\,d\mathcal{H}^{N-1} + \int_{\Omega_J \setminus \overline{\Omega}} |\nabla\psi| - \int_{\Omega_J} \zeta \cdot \nabla(w_m)_\psi.$$

Now,

$$-\int_{\Omega_J} \zeta(x) \cdot \nabla(w_m)_\psi(x)\, dx = -\int_\Omega \zeta(x) \cdot \nabla w_m(x)\, dx - \int_{\Omega_J \setminus \overline\Omega} \zeta(x) \cdot \nabla\psi(x)\, dx$$

$$= \int_\Omega \mathrm{div}\zeta(x) w_m(x)\, dx - \int_{\partial\Omega} [\zeta, \nu]\tilde\psi\, d\mathcal{H}^{N-1} - \int_{\Omega_J \setminus \overline\Omega} \zeta(x) \cdot \nabla\psi(x)\, dx.$$

Since

$$\int_{\Omega_J \setminus \overline\Omega} |\nabla\psi(x)|\, dx - \int_{\Omega_J \setminus \overline\Omega} \zeta(x) \cdot \nabla\psi(x)\, dx \geq 0,$$

from (7.58), we have

$$\int_\Omega (\phi(x) - v(x))(v(x) - w_m(x))\, dx$$

$$\geq \int_\Omega |Dv| + \int_{\partial\Omega} |v - \tilde\psi|\, d\mathcal{H}^{N-1} + \int_\Omega \mathrm{div}\zeta(x) w_m(x)\, dx - \int_{\partial\Omega} [\zeta, \nu]\tilde\psi\, d\mathcal{H}^{N-1}.$$

Letting $m \to \infty$, and using Green's formula, we deduce

$$0 \geq \int_\Omega |Dv| + \int_{\partial\Omega} |v - \tilde\psi|\, d\mathcal{H}^{N-1} + \int_\Omega \mathrm{div}\zeta(x) v(x)\, dx - \int_{\partial\Omega} [\zeta, \nu]\tilde\psi\, d\mathcal{H}^{N-1}$$

$$= \int_\Omega |Dv| + \int_{\partial\Omega} |v - \tilde\psi|\, d\mathcal{H}^{N-1} - \int_\Omega (\zeta, Dv) + \int_{\partial\Omega} [\zeta, \nu]v\, d\mathcal{H}^{N-1}$$

$$- \int_{\partial\Omega} [\zeta, \nu]\tilde\psi\, d\mathcal{H}^{N-1}.$$

By (7.53), we obtain

$$\int_{\partial\Omega} |v - \tilde\psi|\, d\mathcal{H}^{N-1} \leq \int_{\partial\Omega} [\zeta, \nu](\tilde\psi - v)\, d\mathcal{H}^{N-1} \leq \int_{\partial\Omega} |v - \tilde\psi|\, d\mathcal{H}^{N-1}.$$

Therefore,

$$[\zeta, \nu] \in \mathrm{sgn}(\tilde\psi - v) \qquad \mathcal{H}^{N-1}\text{-a.e. on } \partial\Omega,$$

and the proof is finished. □

7.3.3. Asymptotic behaviour. We study the asymptotic behaviour as $t \to \infty$ of the solution of the nonlocal total variation flow with homogeneous Dirichlet boundary conditions. We show that the solution of this nonlocal problem converges to 0.

THEOREM 7.20. *Let $u_0 \in L^2(\Omega)$. Let u be the solution of (7.27) with $\psi = 0$. Then*

$$\|u(t)\|_{L^1(\Omega)} \leq C\frac{\|u_0\|_{L^2(\Omega)}^2}{t} \qquad \forall\, t > 0,$$

where $C = C(J, \Omega, q)$.

PROOF. Since $B_{1,0}^J$ is completely accretive, $\mathcal{V}(v) := \|v\|_1$ is a Lyapunov functional for the semigroup solution. Hence

$$\int_\Omega |u(x, t)|\, dx \leq \int_\Omega |u(x, s)|\, dx \quad \text{if } t \geq s.$$

Therefore the $L^1(\Omega)$-norm of w is decreasing with t.

Moreover, using Poincaré type inequality (Proposition 6.25), we have

$$\int_\Omega |u(x,s)|\, dx \le C \int_\Omega \int_{\Omega_J} J(x-y)|u_0(y,s) - u(x,s)|\, dy\, dx$$

$$= C \int_\Omega \int_\Omega J(x-y)|u(y,s) - u(x,s)|\, dy\, dx.$$

Consequently,

$$t \int_\Omega |u(x,t)|\, dx \le \int_0^t \int_\Omega |u(x,s)|\, dx\, ds$$

$$\le C \int_0^t \int_\Omega \int_\Omega J(x-y)|u(y,s) - u(x,s)|\, dy\, dx\, ds.$$

On the other hand, multiplying the equation in (7.27) by $u(x,t)$ and integrating in space and time, we get

$$\int_\Omega |u(x,t)|^2 - \int_\Omega |u_0(x)|^2\, dx = - \int_0^t \int_\Omega \int_\Omega J(x-y)|u(y,s) - u(x,s)|\, dy\, dx\, ds,$$

which implies

$$\int_0^t \int_\Omega \int_\Omega J(x-y)|u(y,s) - u(x,s)|\, dy\, dx\, ds \le ||u_0||_{L^2(\Omega)}^2,$$

and therefore

$$\int_\Omega |u(x,t)|\, dx \le C \frac{||u_0||_{L^2(\Omega)}^2}{t}. \qquad \square$$

Bibliographical notes

The results of this chapter are essentially taken from the papers [15] and [17].

CHAPTER 8

Nonlocal models for sandpiles

In the last years an increasing attention has been paid to the study of differential models in granular matter theory (see, e.g., [22] for an overview of different theoretical approaches and models). This field of research, which is of course of strong relevance in applications, has also been the source of many new and challenging problems in the theory of partial differential equations. In this context, the continuous models for the dynamics of a sandpile, introduced, independently, by L. Prigozhin ([139], [139]) and by G. Aronsson, L. C. Evans and Y. Wu ([23]) have been of interest. These two pile growth models, obtained using different arguments, yield to a model in the form of a variational inequality. Now, there is a difference between the two models. In the Prigozhin model the sand is dropped on a bounded table, represented by a bounded domain $\Omega \subset \mathbb{R}^2$, and the sand might fall off the side of the table; in other words, homogeneous Dirichlet boundary conditions are considered in the model. In the Aronsson-Evans-Wu model the Cauchy problem is considered, that is, $\Omega = \mathbb{R}^2$.

Our main purpose in this chapter is to study the nonlocal version of the Prigozhin and Aronsson-Evans-Wu models. These nonlocal models are obtained by means of the diffusion equation which is the limit as $p \to \infty$ of the nonlocal analog of the p-Laplacian evolution studied in the previous chapters. We also show that, by reescaling, we recover the local sandpile models from the nonlocal ones.

8.1. A nonlocal version of the Aronsson-Evans-Wu model for sandpiles

In this section we study a nonlocal version of the Aronsson-Evans-Wu model for sandpiles obtained as the limit as $p \to \infty$ in the nonlocal Cauchy problem for the p-Laplacian studied in Chapter 6. We begin with recalling the Aronsson-Evans-Wu model for sandpiles. Next, we introduce the nonlocal version of the Aronsson-Evans-Wu model. By rescaling, we also prove the convergence of the nonlocal model to the one introduced by Aronsson-Evans-Wu. The next subsection is devoted to the collapse of the initial condition for the nonlocal sandpile model. We also present concrete examples of explicit solutions that illustrate the general convergence result and we give an interpretation of the nonlocal sandpile model in terms of Monge-Kantorovich mass transport theory. The section is finished showing that results analogous to the previous ones are also valid when we consider the Neumann problem in a bounded convex domain Ω, that is, for the case in which the boundary is an impermeable wall.

8.1.1. The Aronsson-Evans-Wu model for sandpiles. First, we begin with recalling the Aronsson-Evans-Wu model for sandpiles. In [100], [23] and

[99] the authors investigated the limiting behaviour as $p \to \infty$ of solutions to the quasilinear parabolic problem

(8.1)
$$\begin{cases} (v_p)_t - \Delta_p v_p = f & \text{in } \mathbb{R}^N \times (0, T), \\ v_p(x, 0) = u_0(x) & \text{in } \mathbb{R}^N, \end{cases}$$

where f is a nonnegative function that represents a given source term, which is interpreted physically as adding material to an evolving system, within which mass particles are continually rearranged by diffusion.

Let us define for $1 < p < \infty$ the functional

$$F_p(v) = \begin{cases} \dfrac{1}{p} \displaystyle\int_{\mathbb{R}^N} |\nabla v(y)|^p \, dy & \text{if } v \in L^2(\mathbb{R}^N) \cap W^{1,p}(\mathbb{R}^N), \\ +\infty & \text{if } v \in L^2(\mathbb{R}^N) \setminus W^{1,p}(\mathbb{R}^N). \end{cases}$$

Then the PDE problem (8.1) can be written as the abstract Cauchy problem associated to the subdifferential of F_p (see Appendix A), that is,

$$\begin{cases} f(\cdot, t) - (v_p)_t(\cdot, t) = \partial F_p(v_p(\cdot, t)) & \text{a.e. } t \in (0, T), \\ v_p(x, 0) = u_0(x) & \text{in } \mathbb{R}^N. \end{cases}$$

In [23], assuming that u_0 is a Lipschitz function with compact support such that

$$\|\nabla u_0\|_\infty \leq 1,$$

and f is a smooth nonnegative function with compact support in $\mathbb{R}^N \times (0, T)$, it is proved that there exist a sequence $p_i \to +\infty$ and a limit function v_∞ such that, for each $T > 0$,

$$\begin{cases} v_{p_i} \to v_\infty & \text{in } L^2(\mathbb{R}^N \times (0, T)) \text{ and a.e.,} \\ \nabla v_{p_i} \rightharpoonup \nabla v_\infty, \ (v_{p_i})_t \rightharpoonup (v_\infty)_t & \text{weakly in } L^2(\mathbb{R}^N \times (0, T)). \end{cases}$$

Moreover, the limit function v_∞ satisfies

(8.2)
$$\begin{cases} f(\cdot, t) - (v_\infty)_t(\cdot, t) \in \partial F_\infty(v_\infty(\cdot, t)) & \text{a.e. } t \in (0, T), \\ v_\infty(x, 0) = u_0(x) & \text{in } \mathbb{R}^N \end{cases}$$

where the limit functional is given by

$$F_\infty(v) = \begin{cases} 0 & \text{if } v \in L^2(\mathbb{R}^N), \ |\nabla v| \leq 1, \\ +\infty & \text{otherwise.} \end{cases}$$

This limit problem (8.2) is interpreted in [23] to explain the movement of a sandpile ($v_\infty(x, t)$ describes the amount of the sand at the point x at time t), the main assumption being that the sandpile is stable when the slope is less than or equal to one and unstable if not.

8.1.2. Limit as $p \to \infty$ in the nonlocal p-Laplacian Cauchy problem. In Section 6.4 we considered the existence and uniqueness of solutions of the nonlocal p-Laplacian Cauchy problem

$$
(8.3) \quad
\begin{cases}
(u_p)_t(x,t) = \displaystyle\int_{\mathbb{R}^N} J(x-y)|u_p(y,t) - u_p(x,t)|^{p-2}(u_p(y,t) - u_p(x,t))dy \\
\qquad\qquad\qquad\qquad\qquad\qquad\qquad\qquad\qquad\qquad\qquad +f(x,t), \\
u_p(x,0) = u_0(x), \qquad\qquad\qquad\qquad\qquad\qquad x \in \mathbb{R}^N,\ t > 0,
\end{cases}
$$

where $J : \mathbb{R}^N \to \mathbb{R}$ is a nonnegative continuous radial function with compact support, $J(0) > 0$ and $\int_{\mathbb{R}^N} J(x)\,dx = 1$. Also we showed that if the kernel J is rescaled in an appropriate way, the solutions to the corresponding nonlocal problems converge to the solution of the p-Laplacian evolution problem.

Let us note that the evolution problem (8.3) is associated to the energy functional

$$
G_p^J(u) = \frac{1}{2p} \int_{\mathbb{R}^N} \int_{\mathbb{R}^N} J(x-y)|u(y) - u(x)|^p\,dy\,dx,
$$

which is the nonlocal analog of the functional F_p associated to the p-Laplacian.

Our aim in this section concerns the limit as $p \to \infty$ in (8.3). With a formal computation, taking limits, we arrive at the functional

$$
G_\infty^J(u) =
\begin{cases}
0 & \text{if } |u(x) - u(y)| \leq 1, \ \text{for } x - y \in \text{supp}(J), \\
+\infty & \text{otherwise.}
\end{cases}
$$

Now, if we define

$$
K_\infty^J := \left\{ u \in L^2(\mathbb{R}^N) \ : \ |u(x) - u(y)| \leq 1, \ \text{for } \ x - y \in \text{supp}(J)\right\},
$$

we have that the functional G_∞^J is given by the indicator function of K_∞^J, that is, $G_\infty^J = I_{K_\infty^J}$. Then the *nonlocal limit problem* can be written as

$$
(8.4) \quad
\begin{cases}
f(\cdot,t) - u_t(\cdot,t) \in \partial I_{K_\infty^J}(u(\cdot,t)) & \text{a.e. } t \in (0,T), \\
u(x,0) = u_0(x).
\end{cases}
$$

We now prove the following result on the convergence of the solutions of problem (8.3).

THEOREM 8.1. *Let $T > 0$, $f \in L^2(0,T; L^2(\mathbb{R}^N) \cap L^\infty(\mathbb{R}^N))$, $u_0 \in L^2(\mathbb{R}^N) \cap L^\infty(\mathbb{R}^N)$ such that $|u_0(x) - u_0(y)| \leq 1$ for $x - y \in \text{supp}(J)$, and let u_p be the unique solution of (8.3), $p \geq 2$. Then, if u_∞ is the unique solution of (8.4),*

$$
\lim_{p \to \infty} \sup_{t \in [0,T]} \|u_p(\cdot,t) - u_\infty(\cdot,t)\|_{L^2(\mathbb{R}^N)} = 0.
$$

PROOF. By Theorem A.38, to prove the result it is enough to show that the functionals

$$
G_p^J(u) = \frac{1}{2p} \int_{\mathbb{R}^N} \int_{\mathbb{R}^N} J(x-y)|u(y) - u(x)|^p\,dy\,dx
$$

converge to

$$G_\infty^J(u) = \begin{cases} 0 & \text{if } |u(x) - u(y)| \leq 1 \text{ for } x - y \in \text{supp}(J), \\ +\infty & \text{otherwise,} \end{cases}$$

as $p \to \infty$, in the sense of Mosco.

First, let us check that

$$(8.5) \qquad \text{Epi}(G_\infty^J) \subset \text{s-}\liminf_{p \to \infty} \text{Epi}(G_p^J).$$

To this end let $(u, \lambda) \in \text{Epi}(G_\infty^J)$. We can assume that $u \in K_\infty^J$ and $\lambda \geq 0$ (since $G_\infty^J(u) = 0$). Now take

$$(8.6) \qquad v_p = u\chi_{B(0,R(p))} \qquad \text{and} \qquad \lambda_p = G_p^J(u_p) + \lambda.$$

Then, as $\lambda \geq 0$, we have $(v_p, \lambda_p) \in \text{Epi}(G_p^J)$. It is obvious that if $R(p) \to \infty$ as $p \to \infty$, we have

$$v_p \to u \quad \text{in } L^2(\mathbb{R}^N),$$

and, if we choose $R(p) = p^{\frac{1}{4N}}$,

$$G_p^J(v_p) = \frac{1}{2p} \int_{\mathbb{R}^N} \int_{\mathbb{R}^N} J(x - y) |v_p(y) - v_p(x)|^p \, dy \, dx \leq C \frac{R(p)^{2N}}{p} \to 0$$

as $p \to \infty$, and (8.5) holds.

Finally, let us prove that

$$(8.7) \qquad \text{w-}\limsup_{p \to \infty} \text{Epi}(G_p^J) \subset \text{Epi}(G_\infty^J).$$

To this end, consider a sequence $(u_{p_j}, \lambda_{p_j}) \in \text{Epi}(G_{p_j}^J)$ $(p_j \to \infty)$, that is,

$$G_{p_j}^J(u_{p_j}) \leq \lambda_{p_j},$$

with

$$u_{p_j} \rightharpoonup u, \qquad \text{and} \qquad \lambda_{p_j} \to \lambda.$$

Therefore we obtain that $0 \leq \lambda$, since

$$0 \leq G_{p_j}^J(u_{p_j}) \leq \lambda_{p_j} \to \lambda.$$

On the other hand, we have that

$$\left(\int_{\mathbb{R}^N} \int_{\mathbb{R}^N} J(x - y) \left| u_{p_j}(y) - u_{p_j}(x) \right|^{p_j} \, dy \, dx \right)^{1/p_j} \leq (Cp_j)^{1/p_j}.$$

Now, fix a bounded domain $\Omega \subset \mathbb{R}^N$ and $q < p_j$. Then, by the above inequality,

$$\left(\int_\Omega \int_\Omega J(x - y) \left| u_{p_j}(y) - u_{p_j}(x) \right|^q dy \, dx \right)^{1/q}$$

$$\leq \left(\int_\Omega \int_\Omega J(x - y) \, dy \, dx \right)^{(p_j - q)/p_j q}$$

$$\times \left(\int_{\mathbb{R}^N} \int_{\mathbb{R}^N} J(x - y) \left| u_{p_j}(y) - u_{p_j}(x) \right|^{p_j} dy \, dx \right)^{1/p_j}$$

$$\leq \left(\int_\Omega \int_\Omega J(x - y) \, dy \, dx \right)^{(p_j - q)/p_j q} (C p_j)^{1/p_j}.$$

Hence, we can extract a subsequence, if necessary, and consider $p_j \to \infty$ to obtain

$$\left(\int_\Omega \int_\Omega J(x - y) |u(y) - u(x)|^q dy \, dx \right)^{1/q} \leq \left(\int_\Omega \int_\Omega J(x - y) \, dy \, dx \right)^{1/q}.$$

Now, just letting $q \to \infty$, we get

$$|u(x) - u(y)| \leq 1 \qquad \text{a.e. } (x, y) \in \Omega \times \Omega, \ x - y \in \text{supp}(J).$$

As Ω is arbitrary, we conclude that

$$u \in K_\infty^J.$$

This ends the proof. $\qquad\qquad\qquad\qquad\qquad\qquad\qquad\qquad\qquad\qquad\qquad$ $\square$

8.1.3. Rescaling the kernel. Convergence to the local problem. Along this section we assume that $\text{supp}(J) = \overline{B}(0, 1)$. For $\varepsilon > 0$, we rescale the functional G_∞^J as follows:

$$G_\infty^\varepsilon(u) = \begin{cases} 0 & \text{if } |u(x) - u(y)| \leq \varepsilon, \ \text{for } |x - y| \leq \varepsilon, \\ +\infty & \text{otherwise.} \end{cases}$$

In other words, $G_\infty^\varepsilon = I_{K_\varepsilon}$, where

$$K_\varepsilon := \{ u \in L^2(\mathbb{R}^N) \ : \ |u(x) - u(y)| \leq \varepsilon, \ \text{for } |x - y| \leq \varepsilon \}.$$

Consider the gradient flow associated to the functional G_∞^ε with a source term

(8.8)
$$\begin{cases} f(\cdot, t) - u_t(\cdot, t) \in \partial I_{K_\varepsilon}(u(\cdot, t)) & \text{a.e. } t \in (0, T), \\ u(x, 0) = u_0(x) & \text{in } \mathbb{R}^N, \end{cases}$$

and the problem

(8.9)
$$\begin{cases} f(\cdot, t) - (v_\infty)_t(\cdot, t) \in \partial I_{K_0}(v_\infty(\cdot, t)) & \text{a.e. } t \in (0, T), \\ v_\infty(x, 0) = u_0(x) & \text{in } \mathbb{R}^N, \end{cases}$$

where

$$K_0 := \{ u \in L^2(\mathbb{R}^N) \cap W^{1,\infty}(\mathbb{R}^N) \ : \ |\nabla u| \leq 1 \},$$

for which $F_\infty(v) = I_{K_0}$.

Observe that if $u \in K_0$, then $|\nabla u| \leq 1$. Hence, $|u(x) - u(y)| \leq |x - y|$, from which it follows that $u \in K_\varepsilon$, that is, $K_0 \subset K_\varepsilon$. We have the following theorem.

THEOREM 8.2. *Let $T > 0$, $f \in L^2(0, T; L^2(\mathbb{R}^N))$, $u_0 \in L^2(\mathbb{R}^N) \cap W^{1,\infty}(\mathbb{R}^N)$ such that $\|\nabla u_0\|_\infty \leq 1$ and consider $u_{\infty,\varepsilon}$ the unique solution of* (8.8). *Then, if v_∞ is the unique solution of* (8.9), *we have*

$$\lim_{\varepsilon \to 0} \sup_{t \in [0,T]} \|u_{\infty,\varepsilon}(\cdot, t) - v_\infty(\cdot, t)\|_{L^2(\mathbb{R}^N)} = 0.$$

Consequently, we are approximating the sandpile model described in Subsection 8.1.1 by a nonlocal model. In this nonlocal approximation a configuration of sand is stable when its height u satisfies $|u(x) - u(y)| \leq \varepsilon$ if $|x - y| \leq \varepsilon$. This is a sort of measure of how large is the size of irregularities of the sand; the sand can be completely irregular for sizes smaller than ε but it has to be arranged for sizes greater than ε.

PROOF OF THEOREM 8.2. We have to show that if v_∞ is the unique solution of (8.9), then

$$\lim_{\varepsilon \to 0} \sup_{t \in [0,T]} \|u_{\infty,\varepsilon}(\cdot, t) - v_\infty(\cdot, t)\|_{L^2(\mathbb{R}^N)} = 0.$$

Since $u_0 \in K_0$, $u_0 \in K_\varepsilon$ for all $\varepsilon > 0$, and consequently there exists $u_{\infty,\varepsilon}$, the unique solution of (8.8).

By Theorem A.38, to prove the result it is enough to show that I_{K_ε} converges to I_{K_0} in the sense of Mosco. It is easy to see that

(8.10) $K_{\varepsilon_1} \subset K_{\varepsilon_2}$ if $\varepsilon_1 \leq \varepsilon_2$.

Since $K_0 \subset K_\varepsilon$ for all $\varepsilon > 0$, we have

$$K_0 \subset \bigcap_{\varepsilon > 0} K_\varepsilon.$$

On the other hand, if

$$u \in \bigcap_{\varepsilon > 0} K_\varepsilon,$$

we have

$$|u(y) - u(x)| \leq |y - x|, \quad \text{a.e. } x, y \in \mathbb{R}^N,$$

from which it follows that $u \in K_0$. Therefore,

(8.11) $K_0 = \bigcap_{\varepsilon > 0} K_\varepsilon.$

Note that

(8.12) $\text{Epi}(I_{K_0}) = K_0 \times [0, \infty), \quad \text{Epi}(I_{K_\varepsilon}) = K_\varepsilon \times [0, \infty) \ \forall \varepsilon > 0.$

By (8.11) and (8.12), we obtain

(8.13) $\text{Epi}(I_{K_0}) \subset \text{s-}\liminf_{\varepsilon \to 0} \text{Epi}(I_{K_\varepsilon}).$

On the other hand, given $(u, \lambda) \in \text{w-}\limsup_{\varepsilon \to 0} \text{Epi}(I_{K_\varepsilon})$ there exists $(u_{\varepsilon_k}, \lambda_k) \in K_{\varepsilon_k} \times [0, \infty)$ such that $\varepsilon_k \to 0$ and

$$u_{\varepsilon_k} \rightharpoonup u \quad \text{in } L^2(\mathbb{R}^N), \quad \lambda_k \to \lambda \quad \text{in } \mathbb{R}.$$

By (8.10), given $\varepsilon > 0$, there exists k_0, such that $u_{\varepsilon_k} \in K_\varepsilon$ for all $k \geq k_0$. Then, since K_ε is a closed convex set, we get $u \in K_\varepsilon$, and, by (8.11), we obtain that $u \in K_0$. Consequently,

$$(8.14) \qquad \text{w-lim sup}_{n \to \infty} \text{Epi}(I_{K_\varepsilon}) \subset \text{Epi}(I_{K_0}).$$

Finally, by (8.13) and (8.14), and having in mind (A.19), we obtain that I_{K_ε} converges to I_{K_0} in the sense of Mosco. $\qquad\square$

8.1.4. Collapse of the initial condition. Evans, Feldman and Gariepy [**100**] study the collapsing initial condition phenomena for the local problem (8.1) with null source when the initial condition u_0 satisfies $\|\nabla u_0\|_\infty = L > 1$. They find that the limit of the solutions $v_p(x, t)$ of (8.1) is independent of time but does not coincide with u_0. They prove that for each time $t > 0$,

$$v_p(\cdot, t) \to v_\infty(\cdot) \qquad \text{uniformly as } p \to +\infty,$$

where v_∞ is independent of time and satisfies

$$\|\nabla v_\infty\|_\infty \leq 1.$$

They also describe the small layer in which the solution rapidly changes from being u_0 at $t = 0$ to something close to the final stationary limit for $t > 0$. They prove that $v_\infty(x) = v(x, 1)$, v solving the nonautonomous evolution equation

$$\begin{cases} \dfrac{v}{t} - v_t \in \partial I_{K_0}(v), & t \in (\tau, \infty), \\[2mm] v(x, \tau) = \tau u_0(x), \end{cases}$$

where $\tau = L^{-1}$. They interpret this fact as a crude model for the collapse of a sandpile from an initially unstable configuration. The proof of this result is based on a scaling argument, which was extended by Bénilan, Evans and Gariepy in [**47**], to cover general nonlinear evolution equations governed by homogeneous accretive operators (see Theorem A.54). We use this result here to get the nonlocal version of the collapsing problem.

Consider the nonlocal problem

$$(8.15) \quad \begin{cases} (u_p)_t(x, t) = \displaystyle\int_{\mathbb{R}^N} J(x - y)|u_p(y, t) - u_p(x, t)|^{p-2}(u_p(y, t) - u_p(x, t))dy, \\[2mm] u_p(x, 0) = u_0(x), & x \in \mathbb{R}^N, t > 0. \end{cases}$$

We want to take the limit as $p \to \infty$ of the solutions u_p of the nonlocal problem when the initial condition u_0 does not satisfy $|u_0(x) - u_0(y)| \leq 1$ for $x - y \in \text{supp}(J)$. We get that the nonlinear nature of the problem creates an initial short-time layer in which the solution changes very rapidly. We describe this layer by means of a limit evolution problem. We have the following result.

THEOREM 8.3. *Let u_p be the solution of (8.15) with initial condition $u_0 \in L^2(\mathbb{R}^N) \cap L^\infty(\mathbb{R}^N)$ such that*

$$1 < L = \sup_{|x-y| \in \text{supp}(J)} |u_0(x) - u_0(y)|.$$

Then there exists the limit

$$\lim_{p \to \infty} u_p(\cdot, t) = u_\infty \qquad in \ L^2(\mathbb{R}^N),$$

which is a function independent of t such that $|u_\infty(x) - u_\infty(y)| \leq 1$ for $x - y \in$ supp(J). *Moreover, $u_\infty(x) = w(x, 1)$, where w is the unique strong solution of the evolution equation*

$$\begin{cases} \dfrac{w}{t} - w_t \in \partial G_\infty^J(w), & t \in (\tau, \infty), \\[2mm] w(x, \tau) = \tau u_0(x), \end{cases}$$

with $\tau = L^{-1}$.

Note that when u_0 satisfies $|u_0(x) - u_0(y)| \leq 1$ for $x - y \in$ supp(J), then it is an immediate consequence of Theorem 8.1 that the limit exists and is given by

$$\lim_{p \to \infty} u_p(x, t) = u_0(x).$$

We look for the limit as $p \to \infty$ of the solutions to the nonlocal problem (8.3) when the initial datum u_0 satisfies

$$1 < L = \sup_{x - y \in \text{supp}(J)} |u_0(x) - u_0(y)|.$$

For $p > N$, we consider in the Banach space $X = L^2(\mathbb{R}^N)$ the operators ∂G_p^J. Then ∂G_p^J are m-accretive operators in $L^2(\mathbb{R}^N)$ and positively homogeneous of degree $p - 1$. Moreover, the solution u_p of the nonlocal problem (8.3) with null source coincides with the strong solution of the abstract Cauchy problem

$$\begin{cases} -u_t(\cdot, t) \in \partial G_p^J(u(\cdot, t)), & \text{a.e. } t \in (0, T), \\[2mm] u(x, 0) = u_0(x), & x \in \mathbb{R}^N. \end{cases}$$

Let

$$C := \left\{ u \in L^2(\mathbb{R}^N) \ : \ \exists (u_p, v_p) \in \partial G_p^J \text{ with } u_p \to u, \ v_p \to 0 \text{ as } p \to \infty \right\}.$$

It is easy to see that

$$C = K_\infty^J = \left\{ u \in L^2(\mathbb{R}^N) \ : \ |u(x) - u(y)| \leq 1, \text{ for } x - y \in \text{supp}(J) \right\}.$$

Then

$$X_0 := \overline{\bigcup_{\lambda > 0} \lambda C}^{L^2(\mathbb{R}^N)} = L^2(\mathbb{R}^N).$$

LEMMA 8.4. *For $f \in L^2(\mathbb{R}^N)$ and $p > N$, let $u_p := (I + \partial G_p^J)^{-1} f$. Then the set of functions $\{u_p \ : \ p > N\}$ is precompact in $L^2(\mathbb{R}^N)$.*

PROOF. First assume that f is bounded and the support of f lies in the ball $B(0, R)$. Since the operator ∂G_p^J is completely accretive, observe that

$$\partial G_p^J = \overline{\mathcal{B}_p^J \cap (L^2(\mathbb{R}^N) \times L^2(\mathbb{R}^N))}^{L^2(\mathbb{R}^N)},$$

we have the estimates

$$\|u_p\|_{L^\infty(\mathbb{R}^N)} \leq \|f\|_{L^\infty(\mathbb{R}^N)}, \quad \|u_p\|_{L^2(\mathbb{R}^N)} \leq \|f\|_{L^2(\mathbb{R}^N)}$$

and

$$\|u_p(\cdot) - u_p(\cdot + h)\|_{L^2(\mathbb{R}^N)} \leq \|f(\cdot) - f(\cdot + h)\|_{L^2(\mathbb{R}^N)}$$

for each $h \in \mathbb{R}^N$. Consequently, $\{u_p \ : \ p > N\}$ is precompact in $L^2(K)$ for each compact set $K \subset \mathbb{R}^N$. We must show that $\{u_p \ : \ p > N\}$ is tight. For this, fix

$S > 2R$ and select a smooth function $\varphi \in C^\infty(\mathbb{R}^N)$ such that $0 \leq \varphi \leq 1$, $\varphi \equiv 0$ on $B(0, R)$, $\varphi \equiv 1$ on $\mathbb{R}^N \setminus B(0, S)$ and $|\nabla \varphi| \leq \frac{2}{S}$.

We have

$$u_p(x) = \int_{\mathbb{R}^N} J(x - y)|u_p(y) - u_p(x)|^{p-2}(u_p(y) - u_p(x)) \, dy + f(x).$$

Then, multiplying by φu_p and integrating, we get

$$\int_{\mathbb{R}^N} u_p^2(x)\varphi(x) \, dx$$

$$= \int_{\mathbb{R}^N} \int_{\mathbb{R}^N} J(x - y)|u_p(y) - u_p(x)|^{p-2}(u_p(y) - u_p(x))u_p(x)\varphi(x) \, dy \, dx$$

$$= -\frac{1}{2} \int_{\mathbb{R}^N} \int_{\mathbb{R}^N} J(x - y)|u_p(y) - u_p(x)|^{p-2}$$
$$\times (u_p(y) - u_p(x))(u_p(y)\varphi(y) - u_p(x)\varphi(x)) \, dy \, dx$$

$$\leq -\frac{1}{2} \int_{\mathbb{R}^N} \int_{\mathbb{R}^N} J(x - y)|u_p(y) - u_p(x)|^{p-2}$$
$$\times (u_p(y) - u_p(x))u_p(y)(\varphi(y) - \varphi(x)) \, dy \, dx.$$

Now, since $|\nabla \varphi| \leq \frac{2}{S}$, by Hölder's inequality we obtain

$$\left| \frac{1}{2} \int_{\mathbb{R}^N} \int_{\mathbb{R}^N} J(x - y)|u_p(y) - u_p(x)|^{p-2}(u_p(y) - u_p(x))u_p(y)(\varphi(y) - \varphi(x)) \, dy \, dx \right|$$

$$\leq \frac{\|f\|_{L^\infty}}{S} \int_{\{|x| \leq S+1\}} \int_{B(x,1)} J(x - y)|u_p(y) - u_p(x)|^{p-1} \, dy \, dx$$

$$\leq \frac{\|f\|_{L^\infty}}{S} \left(\int_{\{|x| \leq S+1\}} \int_{B(x,1)} J(x - y)|u_p(y) - u_p(x)|^p \, dy \, dx \right)^{\frac{1}{p'}}$$
$$\times \left(\int_{\{|x| \leq S+1\}} \int_{B(x,1)} J(x - y) \, dy \, dx \right)^{\frac{1}{p}}$$

$$\leq M(S + 1)^{\frac{N}{p} - 1} = O(S^{-1 + \frac{N}{p}}),$$

the last inequality being true since $\iint J(x-y)|u_p(y) - u_p(x)|^p$ is uniformly bounded in p. Hence,

$$\int_{\{|x| \geq S\}} u_p^2(x) \, dx = O(S^{-1 + \frac{N}{p}})$$

uniformly in $p > N$. This proves tightness and we have established compactness in $L^2(\mathbb{R}^N)$ provided f is bounded and has compact support. The general case follows since such functions are dense in $L^2(\mathbb{R}^N)$. $\qquad \square$

PROOF OF THEOREM 8.3. By the above Lemma, given $f \in L^2(\mathbb{R}^N)$, if $u_p := (I + \partial G_p^J)^{-1} f$, there exists a sequence $p_j \to +\infty$ such that $u_{p_j} \to v$ in $L^2(\mathbb{R}^N)$ as $j \to \infty$. In the proof of Theorem 8.1 we have established that the functionals G_p^J

converge to $I_{K_\infty^J}$, as $p \to \infty$, in the sense of Mosco. Then, by Theorem A.38, we have $v = (I + I_{K_\infty^J})^{-1} f$. Hence, the limit

$$Pf := \lim_{p \to \infty} (I + \partial G_p^J)^{-1} f$$

exists in $L^2(\mathbb{R}^N)$, for all $f \in X_0 = L^2(\mathbb{R}^N)$, and $Pf = f$ if $f \in C = K_\infty^J$. Moreover,

$$P^{-1} - I = \partial I_{K_\infty^J}$$

and $u = Pf$ is the unique solution of

$$u + \partial I_{K_\infty^J} u \ni f.$$

Therefore Theorem 8.3 is obtained as a consequence of Theorem A.54. $\square$

8.1.5. Explicit solutions. In this section we give some examples of explicit solutions of the problem

(8.16)
$$\begin{cases} f(\cdot, t) - u_t(\cdot, t) \in \partial G_\infty^\varepsilon(u(\cdot, t)) & \text{a.e. } t \in (0, T), \\ u(x, 0) = u_0(x) & \text{in } \mathbb{R}^N, \end{cases}$$

where

$$G_\infty^\varepsilon(u) = \begin{cases} 0 & \text{if } u \in L^2(\mathbb{R}^N), \ |u(x) - u(y)| \le \varepsilon, \text{ for } |x - y| \le \varepsilon, \\ +\infty & \text{otherwise.} \end{cases}$$

In order to verify that a function $u(x, t)$ is a solution of (8.16) we need to check that

(8.17) $\qquad G_\infty^\varepsilon(v) \ge G_\infty^\varepsilon(u) + \langle f - u_t, \ v - u \rangle, \qquad$ for all $v \in L^2(\mathbb{R}^N)$.

To this end we can assume that $v \in K_\varepsilon$ (otherwise $G_\infty^\varepsilon(v) = +\infty$ and then (8.17) becomes trivial). Therefore, we need to check that

(8.18) $\qquad\qquad\qquad\qquad u(\cdot, t) \in K_\varepsilon$

and, by (8.17), that

(8.19) $\qquad\qquad\qquad 0 \ge \int_{\mathbb{R}^N} (f(x, t) - u_t(x, t))(v(x) - u(x, t)) \, dx$

for every $v \in K_\varepsilon$.

EXAMPLE 8.5. Let us consider as source, in one space dimension, an approximation of a delta function

$$f(x, t) = \frac{1}{\eta} \chi_{[-\frac{\eta}{2}, \frac{\eta}{2}]}(x), \quad 0 < \eta \le 2\varepsilon,$$

and as initial datum

$$u_0(x) = 0.$$

Now, let us find the solution by looking at its evolution between some critical times. First, for small times, the solution of (8.16) is given by

(8.20) $\qquad\qquad\qquad u(x, t) = \frac{t}{\eta} \chi_{[-\frac{\eta}{2}, \frac{\eta}{2}]}(x),$

for

$$t \in [0, \eta\varepsilon).$$

Note that $t_1 = \eta\varepsilon$ is the first time when $u(x,t) = \varepsilon$ and hence it is immediate that $u(\cdot, t) \in K_\varepsilon$. Moreover, as $u_t(x,t) = f(x,t)$, (8.19) holds.

For times greater than t_1 the support of the solution is greater than the support of f. Indeed the solution cannot be larger than ε in $[-\frac{\eta}{2}, \frac{\eta}{2}]$ without being larger than zero in the adjacent intervals of size ε, $[\frac{\eta}{2}, \frac{\eta}{2} + \varepsilon]$ and $[-\frac{\eta}{2} - \varepsilon, -\frac{\eta}{2}]$.

We have

$$(8.21) \qquad u(x,t) = \begin{cases} \varepsilon + k_1(t - t_1) & \text{for } x \in [-\frac{\eta}{2}, \frac{\eta}{2}], \\[2mm] k_1(t - t_1) & \text{for } x \in [-\frac{\eta}{2} - \varepsilon, \frac{\eta}{2} + \varepsilon] \setminus [-\frac{\eta}{2}, \frac{\eta}{2}], \\[2mm] 0 & \text{for } x \notin [-\frac{\eta}{2} - \varepsilon, \frac{\eta}{2} + \varepsilon], \end{cases}$$

for times t such that

$$t \in [t_1, t_2),$$

where

$$k_1 = \frac{1}{2\varepsilon + \eta} \quad \text{and} \quad t_2 = t_1 + \frac{\varepsilon}{k_1} = 2\varepsilon^2 + 2\varepsilon\eta.$$

Note that t_2 is the first time when $u(x,t) = 2\varepsilon$ for $x \in [-\frac{\eta}{2}, \frac{\eta}{2}]$. Again it is immediate to see that $u(\cdot, t) \in K_\varepsilon$, since for $|x - y| < \varepsilon$ the maximum of the difference $u(x,t) - u(y,t)$ is exactly ε. Now let us check (8.19).

Using the explicit formula for $u(x,t)$ given in (8.21), we obtain

$$\int_{\mathbb{R}} (f(x,t) - u_t(x,t))(v(x) - u(x,t))\, dx = \int_{-\frac{\eta}{2}}^{\frac{\eta}{2}} \left(\frac{1}{\eta} - u_t(x,t) \right)(v(x) - u(x,t))\, dx$$

$$+ \int_{\frac{\eta}{2}}^{\frac{\eta}{2} + \varepsilon} (-u_t(x,t))(v(x) - u(x,t))\, dx + \int_{-\frac{\eta}{2} - \varepsilon}^{-\frac{\eta}{2}} (-u_t(x,t))(v(x) - u(x,t))\, dx$$

$$= \int_{-\frac{\eta}{2}}^{\frac{\eta}{2}} \left(\frac{1}{\eta} - k_1 \right)(v(x) - (\varepsilon + k_1(t - t_1)))\, dx$$

$$+ \int_{\frac{\eta}{2}}^{\frac{\eta}{2} + \varepsilon} (-k_1)(v(x) - (k_1(t - t_1)))\, dx + \int_{-\frac{\eta}{2} - \varepsilon}^{-\frac{\eta}{2}} (-k_1)(v(x) - (k_1(t - t_1)))\, dx$$

$$= \left(-\eta \left(\frac{1}{\eta} - k_1 \right) + 2\varepsilon k_1 \right) k_1(t - t_1) - \varepsilon\eta \left(\frac{1}{\eta} - k_1 \right) + \int_{-\frac{\eta}{2}}^{\frac{\eta}{2}} \left(\frac{1}{\eta} - k_1 \right) v(x)\, dx$$

$$- \int_{\frac{\eta}{2}}^{\frac{\eta}{2} + \varepsilon} k_1 v(x)\, dx - \int_{-\frac{\eta}{2} - \varepsilon}^{-\frac{\eta}{2}} k_1 v(x)\, dx.$$

From our choice of k_1 we get

$$-\eta \left(\frac{1}{\eta} - k_1 \right) + 2\varepsilon k_1 = 0$$

and, since $v \in K_\varepsilon$, we have

$$\int_{\mathbb{R}} (f(x,t) - u_t(x,t))(v(x) - u(x,t))\, dx$$

$$= -2\varepsilon^2 k_1 + \frac{2\varepsilon k_1}{\eta} \int_{-\frac{\eta}{2}}^{\frac{\eta}{2}} v(x)\, dx - k_1 \int_{\frac{\eta}{2}}^{\frac{\eta}{2}+\varepsilon} v(x)\, dx - k_1 \int_{-\frac{\eta}{2}-\varepsilon}^{-\frac{\eta}{2}} v(x)\, dx \le 0.$$

In fact, without loss of generality we can assume that

$$\int_{-\frac{\eta}{2}}^{\frac{\eta}{2}} v(x)\, dx = 0.$$

Then

(8.22) $$\int_0^{\eta/2} (-v) = a, \qquad \int_{-\eta/2}^0 (-v) = -a.$$

Consequently,

(8.23) $$-v \le \frac{2}{\eta} a + \varepsilon \quad \text{in } [0, \varepsilon].$$

Indeed, if (8.23) does not hold, then $-v > \frac{2}{\eta} a$ in $[0, \varepsilon]$, which contradicts (8.22).

By (8.22), since $v \in K_\varepsilon$,

(8.24)
$$\int_\varepsilon^{\varepsilon+\eta/2} (-v(x))dx = \int_0^{\eta/2} (-v(y+\varepsilon))dy$$

$$= \int_0^{\eta/2} (-v(y+\varepsilon) + v(y))dy + \int_0^{\eta/2} (-v(y))dy \le \varepsilon \frac{\eta}{2} + a.$$

Therefore, by (8.23) and (8.24),

(8.25)
$$\int_{\eta/2}^{\varepsilon+\eta/2} (-v) = \int_{\eta/2}^\varepsilon (-v) + \int_\varepsilon^{\varepsilon+\eta/2} (-v)$$

$$\le \left(\frac{2}{\eta} a + \varepsilon\right)\left(\varepsilon - \frac{\eta}{2}\right) + \varepsilon \frac{\eta}{2} + a = \frac{2}{\eta} a\varepsilon + \varepsilon^2.$$

Similarly,

(8.26) $$\int_{-\varepsilon-\eta/2}^{-\eta/2} (-v) \le -\frac{2}{\eta} a\varepsilon + \varepsilon^2.$$

Consequently, by (8.25) and (8.26),

$$\int_{\frac{\eta}{2}}^{\frac{\eta}{2}+\varepsilon} (-v) + \int_{-\frac{\eta}{2}-\varepsilon}^{-\frac{\eta}{2}} (-v) \le 2\varepsilon^2.$$

Now, it is easy to generalize and verify the following general formula that describes the solution for every $t \geq 0$. For any given integer $l \geq 0$ we have
(8.27)
$$u(x,t) = \begin{cases} l\varepsilon + k_l(t - t_l), & x \in [-\frac{\eta}{2}, \frac{\eta}{2}], \\ (l-1)\varepsilon + k_l(t - t_l), & x \in [-\frac{\eta}{2} - \varepsilon, \frac{\eta}{2} + \varepsilon] \setminus [-\frac{\eta}{2}, \frac{\eta}{2}], \\ \dots \\ k_l(t - t_l), & x \in [-\frac{\eta}{2} - l\varepsilon, \frac{\eta}{2} + l\varepsilon] \setminus [-\frac{\eta}{2} - (l-1)\varepsilon, \frac{\eta}{2} + (l-1)\varepsilon], \\ 0, & x \notin [-\frac{\eta}{2} - l\varepsilon, \frac{\eta}{2} + l\varepsilon], \end{cases}$$

for
$$t \in [t_l, t_{l+1}),$$

where
$$k_l = \frac{1}{2l\varepsilon + \eta} \quad \text{and} \quad t_{l+1} = t_l + \frac{\varepsilon}{k_l}, \quad t_0 = 0.$$

From formula (8.27) we get, taking the limit as $\eta \to 0$, that the expected solution of (8.16) with $f = \delta_0$ is, for any given integer $l \geq 1$,

(8.28) $$u(x,t) = \begin{cases} (l-1)\varepsilon + k_l(t - t_l), & x \in [-\varepsilon, \varepsilon], \\ (l-2)\varepsilon + k_l(t - t_l), & x \in [-2\varepsilon, 2\varepsilon] \setminus [-\varepsilon, \varepsilon], , \\ \dots \\ k_l(t - t_l), & x \in [-l\varepsilon, l\varepsilon] \setminus [-(l-1)\varepsilon, (l-1)\varepsilon], \\ 0, & x \notin [-l\varepsilon, l\varepsilon], \end{cases}$$

for
$$t \in [t_l, t_{l+1}),$$

where
$$k_l = \frac{1}{2l\varepsilon}, \quad t_{l+1} = t_l + \frac{\varepsilon}{k_l}, \quad t_1 = 0.$$

Observe that, since the space of functions K_ε is not contained in $C(\mathbb{R})$, formula (8.19) with $f = \delta_0$ does not make sense. Hence the function $u(x,t)$ described by (8.28) is to be understood as a generalized solution of (8.16) (it is obtained as a limit of solutions of approximating problems).

Note that the function $u(x, t_l)$ is a "regular and symmetric pyramid" composed by squares of side ε.

8.1.5.1. *Recovering the local sandpile model as $\varepsilon \to 0$.* Now, to recover the local sandpile model, let us fix
$$l\varepsilon = L,$$
and take the limit as $\varepsilon \to 0$ in the previous example. We get that $u(x,t) \to v(x,t)$, where
$$v(x,t) = (L - |x|)^+ \qquad \text{for } t = L^2,$$
that is, exactly the evolution given by the local sandpile model with initial datum $u_0 = 0$ and a point source δ_0; see [23].

FIGURE 1. Letting $\varepsilon \to 0$ in the 2-dimensional case

Therefore, this concrete example illustrates the general convergence result, Theorem 8.2.

EXAMPLE 8.6. The explicit formula (8.27) can be easily generalized to the case in which the source depends on t in the form

$$f(x,t) = \varphi(t)\chi_{[-\frac{\eta}{2},\frac{\eta}{2}]}(x),$$

with φ a nonnegative integrable function and $0 < \eta \leq \varepsilon$. We arrive at the following formulas, setting

$$g(t) = \int_0^t \varphi(s)ds,$$

for any given integer $l \geq 0$:

$$u(x,t) = \begin{cases} l\varepsilon + \hat{k}_l\left(g(t) - g(t_l)\right), & x \in [-\frac{\eta}{2}, \frac{\eta}{2}], \\[2mm] (l-1)\varepsilon + \hat{k}_l\left(g(t) - g(t_l)\right), & x \in [-\frac{\eta}{2} - \varepsilon, \frac{\eta}{2} + \varepsilon] \setminus [-\frac{\eta}{2}, \frac{\eta}{2}], \\[2mm] \ldots \\[2mm] \hat{k}_l\left(g(t) - g(t_l)\right), & x \in [-\frac{\eta}{2} - l\varepsilon, \frac{\eta}{2} + l\varepsilon] \setminus [-\frac{\eta}{2} - (l-1)\varepsilon, \frac{\eta}{2} + (l-1)\varepsilon], \\[2mm] 0, & x \notin [-\frac{\eta}{2} - l\varepsilon, \frac{\eta}{2} + l\varepsilon], \end{cases}$$

for

$$t \in [t_l, t_{l+1}),$$

where

$$\hat{k}_l = \frac{\eta}{\eta + 2l\varepsilon} \quad \text{and} \quad g(t_{l+1}) - g(t_l) = \frac{\varepsilon}{\hat{k}_l}, \quad t_0 = 0.$$

Observe that t_l is the first time at which the solution reaches the level $l\varepsilon$.

We can also consider φ changing sign. In this case the solution increases if $\varphi(t)$ is positive in every interval of size ε (around the support of the source $[-\frac{\eta}{2}, \frac{\eta}{2}]$) for which $u(x) - u(y) = i\varepsilon$ with $|x - y| = i\varepsilon$ for some $x \in [-\frac{\eta}{2}, \frac{\eta}{2}]$ (here i is any integer). If $\varphi(t)$ is negative, the solution decreases in every interval of size ε for which $u(x) - u(y) = -i\varepsilon$ with $|x - y| = i\varepsilon$ for some $x \in [-\frac{\eta}{2}, \frac{\eta}{2}]$.

EXAMPLE 8.7. Observe that if $\eta > 2\varepsilon$, then $u(x,t)$ given in (8.21) does not satisfy (8.19) for any test function $v \in K_\varepsilon$ whose values in $[-\frac{\eta}{2} - \varepsilon, \frac{\eta}{2} + \varepsilon]$ are

$$v(x) = \begin{cases} -\beta\frac{\varepsilon}{2} + 2\varepsilon & \text{for } x \in [-\frac{\eta}{2} + \varepsilon, \frac{\eta}{2} - \varepsilon], \\ -\beta\frac{\varepsilon}{2} + \varepsilon & \text{for } x \in [-\frac{\eta}{2}, \frac{\eta}{2}] \setminus [-\frac{\eta}{2} + \varepsilon, \frac{\eta}{2} - \varepsilon], \\ -\beta\frac{\varepsilon}{2} & \text{for } x \in [-\frac{\eta}{2} - \varepsilon, \frac{\eta}{2} + \varepsilon] \setminus [-\frac{\eta}{2}, \frac{\eta}{2}], \end{cases}$$

for $\beta = 4(1 - \varepsilon/\eta)$, which is greater than 2.

At this point one can ask what happens in the previous situation when $\eta > 2\varepsilon$. In this case the solution begins to grow as before with constant speed in the support of f but after the first time when it reaches the level ε the situation changes. Consider, for example, that the source is given by

$$f(x,t) = \frac{1}{\varepsilon}\chi_{[-2\varepsilon,2\varepsilon]}(x).$$

In this case the solution to our nonlocal problem with $u_0(x) = 0$, $u(x,t)$, can be described as follows. First we have

$$u(x,t) = \frac{t}{\varepsilon}\chi_{[-2\varepsilon,2\varepsilon]}(x), \quad \text{for } t \in [0, \varepsilon^2).$$

Note that $t_1 = \varepsilon^2$ is the first time when $u(x,t) = \varepsilon$ and hence it is immediate that $u(\cdot, t) \in K_\varepsilon$. Moreover, as $u_t(x,t) = f(x,t)$, (8.19) holds.

For times greater that t_1 we have

$$u(x,t) = \begin{cases} \varepsilon + \frac{1}{\varepsilon}(t - t_1) & \text{for } x \in [-\varepsilon, \varepsilon], \\ \varepsilon + k_1(t - t_1) & \text{for } x \in [-2\varepsilon, -\varepsilon] \cup [\varepsilon, 2\varepsilon], \\ k_1(t - t_1) & \text{for } x \in [-3\varepsilon, -2\varepsilon] \cup [2\varepsilon, 3\varepsilon], \\ 0 & \text{for } x \notin [-3\varepsilon, 3\varepsilon], \end{cases}$$

for

$$t \in [t_1, t_2),$$

where

$$k_1 = \frac{1}{2\varepsilon} \quad \text{and} \quad t_2 = \varepsilon^2 + 2\varepsilon^2 = 3\varepsilon^2.$$

With this expression of $u(x,t)$ it is easy to see that it satisfies (8.19).

For times greater than t_2 an expression similar to (8.27) holds. We leave the details to the reader.

EXAMPLE 8.8. For two or more dimensions we can get similar formulas. Given a bounded domain $\Omega_0 \subset \mathbb{R}^N$ let us define inductively

$$\Omega_1 = \left\{ x \in \mathbb{R}^N \ : \ \exists y \in \Omega_0 \text{ with } |x - y| < \varepsilon \right\}$$

and

$$\Omega_j = \left\{ x \in \mathbb{R}^N \ : \ \exists y \in \Omega_{j-1} \text{ with } |x - y| < \varepsilon \right\}.$$

In the sequel, for simplicity, we consider the two-dimensional case $N = 2$. Let us take as the source

$$f(x,t) = \chi_{\Omega_0}(x), \quad \Omega_0 = B(0, \varepsilon/2),$$

and as the initial datum,

$$u_0(x) = 0.$$

In this case, for any integer $l \geq 0$, the solution to (8.16) is given by

$$(8.29) \quad u(x,t) = \begin{cases} l\varepsilon + \hat{k}_l(t - t_l), & x \in \Omega_0, \\[2mm] (l-1)\varepsilon + \hat{k}_l(t - t_l), & x \in \Omega_1 \setminus \Omega_0, \\[2mm] \ldots \\[2mm] \hat{k}_l(t - t_l), & x \in \Omega_l \setminus \bigcup_{j=1}^{l-1} \Omega_j, \\[2mm] 0, & x \notin \Omega_l, \end{cases}$$

for

$$t \in [t_l, t_{l+1}),$$

where

$$\hat{k}_l = \frac{|\Omega_0|}{|\Omega_l|}, \qquad t_{l+1} = t_l + \frac{\varepsilon}{\hat{k}_l}, \quad t_0 = 0.$$

Note that the solution grows in strips of width ε around the set Ω_0 where the source is localized.

FIGURE 2. Source and five time steps: t_1, t_2, t_3, t_4, t_5

As in the previous examples, the result is evident for $t \in [0, t_1)$. Let us prove it for $t \in [t_1, t_2)$; a similar argument works for later times. It is clear that $u(\cdot, t) \in K_\varepsilon$; let us check (8.19). Working as in Example 1, we must show that

$$(1 - \hat{k}_1) \int_{\Omega_0} v - \hat{k}_1 \int_{\Omega_1 \setminus \Omega_0} v \leq (1 - \hat{k}_1)\varepsilon|\Omega_0| \qquad \forall v \in K_\varepsilon,$$

where $\Omega_1 = B(0, 3\varepsilon/2)$. Since $\hat{k}_1 = |\Omega_0|/|\Omega_1|$, the last inequality is equivalent to

$$(8.30) \quad \left| \frac{1}{|\Omega_0|} \int_{\Omega_0} v - \frac{1}{|\Omega_1 \setminus \Omega_0|} \int_{\Omega_1 \setminus \Omega_0} v \right| \leq \varepsilon \qquad \forall v \in K_\varepsilon.$$

By density, it is enough to prove (8.30) for any continuous $v \in K_\varepsilon$.

Let us now subdivide

$$\Omega_0 = \left\{ r(\cos\theta, \sin\theta) \ : \ 0 \leq \theta \leq 2\pi, 0 \leq r < \frac{\varepsilon}{2} \right\}$$

and

$$\Omega_1 \setminus \Omega_0 = \left\{ r(\cos\theta, \sin\theta) \ : \ 0 \le \theta \le 2\pi, \varepsilon \le r < \frac{3}{2}\varepsilon \right\}$$

as follows. Consider the partitions

$$0 = \theta_0 < \theta_1 < \cdots < \theta_N = 2\pi,$$

with $\theta_i - \theta_{i-1} = 2\pi/N$, $N \in \mathbb{N}$,

$$0 = r_0 < r_1 < \cdots < r_N = \varepsilon/2$$

and

$$\varepsilon/2 = \tilde{r}_0 < \tilde{r}_1 < \cdots < \tilde{r}_N = 3\varepsilon/2,$$

such that the measure of

$$B_{ij} = \{ r(\cos\theta, \sin\theta) \ : \ \theta_{i-1} < \theta < \theta_i, \ r_{j-1} < r < r_j \}$$

is constant, that is, $|B_{ij}| = |\Omega_0|/N^2$, and the measure of

$$A_{ij} = \{ r(\cos\theta, \sin\theta) \ : \ \theta_{i-1} < \theta < \theta_i, \ \tilde{r}_{j-1} < r < \tilde{r}_j \}$$

is also constant, that is, $|A_{ij}| = |\Omega_1 \setminus \Omega_0|/N^2$. In this way we have partitioned Ω_0 and $\Omega_1 \setminus \Omega_0$ as a disjoint family of N^2 sets such that

$$\left| \Omega_0 \setminus \bigcup_{i,j=1}^{N} B_{ij} \right| = 0, \qquad \left| (\Omega_1 \setminus \Omega_0) \setminus \bigcup_{i,j=1}^{N} A_{ij} \right| = 0.$$

By construction, if we take

$$x_{ij} = r_j(\cos\theta_{i-1}, \sin\theta_{i-1}) \in B_{ij}, \quad \tilde{x}_{ij} = \tilde{r}_{j-1}(\cos\theta_{i-1}, \sin\theta_{i-1}) \in A_{ij},$$

then $|x_{ij} - \tilde{x}_{ij}| \le \varepsilon$ for all $i, j = 1, \dots, N$.

Given a continuous function $v \in K_\varepsilon$, by uniform continuity of v, for $\delta > 0$, there exists $\rho > 0$ such that

$$|v(x) - v(y)| \le \frac{\delta}{2} \qquad \text{if } |x - y| \le \rho.$$

Hence, if we take N big enough such that diameter$(B_{ij}) \le \rho$ and diameter$(A_{ij}) \le \rho$, we have

$$\left| \int_{\Omega_0} v(x) - \sum_{i,j=1}^{N} v(x_{ij})|B_{ij}| \right| \le \frac{\delta|\Omega_0|}{2}$$

and

$$\left| \int_{\Omega_1 \setminus \Omega_0} v(x) - \sum_{i,j=1}^{N} v(\tilde{x}_{ij})|A_{ij}| \right| \le \frac{\delta|\Omega_1 \setminus \Omega_0|}{2}.$$

Since $v \in K_\varepsilon$ and $|x_{ij} - \tilde{x}_{ij}| \le \varepsilon$, $|v(x_{ij}) - v(\tilde{x}_{ij})| \le \varepsilon$,

$$\left| \frac{1}{|\Omega_0|} \int_{\Omega_0} v - \frac{1}{|\Omega_1 \setminus \Omega_0|} \int_{\Omega_1 \setminus \Omega_0} v \right|$$

$$\le \left| \frac{1}{|\Omega_0|} \sum_{i,j=1}^{N} v(x_{ij})|B_{ij}| - \frac{1}{|\Omega_1 \setminus \Omega_0|} \sum_{i,j=1}^{N} v(\tilde{x}_{ij})|A_{ij}| \right| + \delta$$

$$= \left| \frac{1}{N^2} \sum_{i,j=1}^{N} v(x_{ij}) - \frac{1}{N^2} \sum_{i,j=1}^{N} v(\tilde{x}_{ij}) \right| + \delta$$

$$\le \varepsilon + \delta.$$

Therefore, since $\delta > 0$ is arbitrary, (8.30) is obtained.

Again the explicit formula (8.29) can be easily generalized to the case where the source depends on t in the form

$$f(x, t) = \varphi(t)\chi_{\Omega_0}(x).$$

8.1.5.2. *An estimate of the support of u_t.* Taking a source $f \ge 0$ supported in a set A, let us see where the material is added (places where u_t is positive). Compute a set that we will call $\Omega^*(t)$ as follows. Let

$$\Omega_0(t) = A,$$

and define inductively

$$\Omega_1(t) = \left\{ x \in \mathbb{R}^N \setminus \Omega_0(t) \ : \ \exists y \in \Omega_0(t) \text{ with } |x - y| < \varepsilon \text{ and } u(y, t) - u(x, t) = \varepsilon \right\}$$

and

$$\Omega_j(t) = \left\{ \begin{array}{c} x \in \mathbb{R}^N \setminus \Omega_{j-1}(t) \ : \ \exists y \in \Omega_{j-1}(t) \text{ with } |x - y| < \varepsilon \\ \text{and } u(y, t) - u(x, t) = \varepsilon \end{array} \right\}.$$

With these sets $\Omega_i(t)$ (observe that there exists a finite number of such sets, since $u(x, t)$ is bounded) let

$$\Omega^*(t) = \bigcup_i \Omega_i(t).$$

We have that

$$u_t(x, t) = 0, \qquad \text{for } x \notin \Omega^*(t).$$

EXAMPLE 8.9. Finally, note that a description analogous to those in the above examples can be given for an initial condition that is of the form

$$u_0(x) = \sum_{i=-K}^{K} a_i \chi_{[i\varepsilon, (i+1)\varepsilon]}(x),$$

where

$$|a_i - a_{i\pm 1}| \le \varepsilon, \qquad a_{-K} = a_K = 0$$

(this last condition is needed just to imply that $u_0 \in K_\varepsilon$), together with the sum of a finite number of delta functions placed at points $x_l = l\varepsilon$, or a finite sum of functions of time times the characteristic functions of some intervals of the form $[l\varepsilon, (l+1)\varepsilon]$, as the source term.

For example, consider a source placed in just one interval, $f(x,t) = \chi_{[0,\varepsilon]}(x)$. Initially, $u(x,0) = l\varepsilon$ for $x \in [0,\varepsilon]$. Let us take for $w_1(x)$ the regular and symmetric pyramid centered at $[0,\varepsilon]$ of height $(l+1)\varepsilon$ (and base of length $(2l-1)\varepsilon$). With this pyramid and the initial condition, consider the set

$$\Lambda_1 = \{j \in \mathbb{Z} : w_1(x) > u(x,0) \text{ for } x \in (j\varepsilon, (j+1)\varepsilon)\}.$$

This set contains the indices of the intervals in which the sand is being added in the first stage. During this first stage, $u(x,t)$ is given by

$$u(x,t) = u(x,0) + \frac{t}{\mathrm{Card}(\Lambda_1)} \sum_{j \in \Lambda_1} \chi_{[j\varepsilon,(j+1)\varepsilon]}(x),$$

for $t \in [0,t_1]$, where $t_1 = \mathrm{Card}(\Lambda_1)\varepsilon$ is the first time at which u is of size $(l+1)\varepsilon$ in the interval $[0,\varepsilon]$.

From now on the evolution follows the same scheme. In fact,

$$u(x,t) = u(x,t_i) + \frac{t - t_i}{\mathrm{Card}(\Lambda_i)} \sum_{j \in \Lambda_i} \chi_{[j\varepsilon,(j+1)\varepsilon]}(x),$$

for

$$t \in [t_i, t_{i+1}], \qquad t_{i+1} - t_i = \mathrm{Card}(\Lambda_i)\varepsilon.$$

From the pyramid w_i of height $(l+i)\varepsilon$, we obtain

$$\Lambda_i = \{j : w_i(x) > u(x,t_i) \text{ for } x \in (j\varepsilon, (j+1)\varepsilon)\}.$$

Remark that eventually the pyramid w_k is bigger than the initial condition, and from this time on the evolution is the same as described for $u_0 = 0$ in the first example.

In case we have two sources, the pyramids w_i, $\tilde{w}_i$ corresponding to the two sources eventually intersect. In the interval where the intersection takes place, u_t is given by the greater of the two possible speeds (that correspond to the different sources). If the two possible speeds are the same, this interval has to be computed as corresponding to both sources simultaneously.

8.1.5.3. *Recovering the sandpile model.* Note that any initial condition w_0 with $|\nabla w_0| \le 1$ can be approximated by an u_0 such as the one described above. Hence we can obtain an explicit solution of the nonlocal model that approximates the solutions constructed in [**23**].

8.1.5.4. *Compact support of the solutions.* Observe also that when the source f and the initial condition u_0 are compactly supported and bounded, then the solution is compactly supported and bounded for all positive times. This property has to be contrasted with the fact that solutions of the nonlocal p-Laplacian $P_p^J(u_0, f)$ are not compactly supported even if u_0 is.

8.1.5.5. *Formation of pyramids.* Let $N = 2$, J having support the square $\{(x_1, x_2) : \max\{|x_1|, |x_2|\} \le 1\}$, $u_0 = 0$ and δ_a as source. Then the evolution of the corresponding generalized problem similar to (8.16) will form the pyramid given in Figure 3.

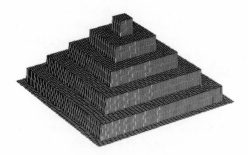

FIGURE 3. A pyramid

Observe that in this case

$$G_\infty^\varepsilon(u) = \begin{cases} 0 & \text{if } u \in L^2(\mathbb{R}^2), \ |u(x) - u(y)| \leq \varepsilon, \ \text{for } |x - y|_\infty \leq \varepsilon, \\ +\infty & \text{otherwise,} \end{cases}$$

where $|x|_\infty = \max\{|x_1|, |x_2|\}$.

Note that the evolution depends in a nontrivial way on the geometry of the support of J. In fact, compare the previous examples (in which supp(J) is a ball in the Euclidean norm) and the present one (where the support is a square).

8.1.6. A mass transport interpretation. We can also give an interpretation of the limit problem (8.4) in terms of Monge-Kantorovich mass transport theory as in [**100**], [**103**] (see [**151**] for a general introduction to mass transportation problems). To this end consider the distance

$$d(x, y) = \begin{cases} 0 & \text{if } x = y, \\ [\![|x - y|]\!] + 1 & \text{if } x \neq y, \end{cases}$$

where $[\![r]\!]$ is defined for $r > 0$ by

$$[\![r]\!] := n \quad \text{if } n < r \leq n + 1, \ n = 0, 1, 2, \dots.$$

Note that this function d measures distances with jumps of length one. Then, given two measures (that for simplicity we will take absolutely continuous with respect to Lebesgue measure in $\mathbb{R}^N$) f_+, f_- in $\mathbb{R}^N$, and supposing the overall condition of mass balance

$$\int_{\mathbb{R}^N} f_+ \, dx = \int_{\mathbb{R}^N} f_- \, dy,$$

the Monge problem associated to the distance d is given by

$$\text{minimize} \int d(x, s(x)) \, f_+(x) dx$$

among the set of maps s that transport f_+ into f_-, which means that

$$\int_{\mathbb{R}^N} h(s(x)) f_+(x) \, dx = \int_{\mathbb{R}^N} h(y) f_-(y) \, dy$$

for each continuous function $h : \mathbb{R}^N \to \mathbb{R}$. The dual formulation of this minimization problem, introduced by Kantorovich (see [99] or [151]), is given by

$$(8.31) \qquad \max_{u,v \in \mathcal{L}} \int_{\mathbb{R}^N} u(x) f_+(x) dx + \int_{\mathbb{R}^N} v(y) f_-(y) dy,$$

where the set $\mathcal{L}$ is defined as

$$\mathcal{L} := \left\{ u, v \in L^1(\mathbb{R}^N) \ : \ u(x) + v(y) \leq d(x,y) \text{ for almost all } x, y \in \mathbb{R}^N \right\}.$$

Since d is a lower semicontinuous metric on $\mathbb{R}^N$, by the Kantorovich-Rubinstein theorem (see Theorem 1.14 in [151]), we can rewrite the dual problem (8.31) as

$$(8.32) \qquad \max_{u \in K_\infty} \int_{\mathbb{R}^N} u(x)(f_+(x) - f_-(x)) dx,$$

where the set K_∞ is given by

$$K_\infty := \left\{ u \in L^2(\mathbb{R}^N) \ : \ |u(x) - u(y)| \leq 1 \text{ for } |x - y| \leq 1 \right\}.$$

We are assuming that $\mathrm{supp}(J) = \overline{B}(0,1)$ (otherwise we have to redefine the distance d accordingly).

With these definitions and notation we have the following result.

THEOREM 8.10. *The solution $u_\infty(\cdot, t)$ of the limit problem (8.4) is a solution of the dual problem*

$$\max_{u \in K_\infty} \int_{\mathbb{R}^N} u(x)(f_+(x) - f_-(x)) dx$$

when the involved measures are the source term $f_+ = f(x,t)$ and the time derivative of the solution $f_- = u_t(x,t)$.

PROOF. It is easy to obtain that the solution $u_\infty(\cdot, t)$ of the limit problem (8.4) is a solution to the dual problem (8.32) when the involved measures are the source $f(x,t)$ and the time derivative of the solution $u_{\infty,t}(x,t)$. In fact, we have

$$G_\infty^J(v) \geq G_\infty^J(u_\infty) + \langle f - (u_\infty)_t, \ v - u_\infty \rangle \quad \text{for all } v \in L^2(\mathbb{R}^N),$$

which is equivalent to

$$u_\infty(\cdot, t) \in K_\infty$$

and

$$(8.33) \qquad 0 \geq \int_{\mathbb{R}^N} (f(x,t) - (u_\infty)_t(x,t))(v(x) - u_\infty(x,t)) \, dx$$

for every $v \in K_\infty$. Now, we just observe that (8.33) is

$$\int_{\mathbb{R}^N} (f(x,t) - (u_\infty)_t(x,t)) u_\infty(x,t) \, dx \geq \int_{\mathbb{R}^N} (f(x,t) - (u_\infty)_t(x,t)) v(x) \, dx.$$

Therefore, $u_\infty(\cdot, t)$ is a solution of the dual mass transport problem. $\square$

Consequently, we conclude that the mass of sand added by the source $f(\cdot, t)$ is transported (via $u(\cdot, t)$ as the transport potential) to $(u_\infty)_t(\cdot, t)$ at each time t. This mass transport interpretation of the problem can be clearly observed looking at the previous concrete examples.

8.1.7. Neumann boundary conditions. Analogous results to the previous ones are also valid when we consider the Neumann problem in a bounded convex domain Ω, that is, when all the involved integrals are taken in Ω.

Let Ω be a convex domain in $\mathbb{R}^N$ and consider the evolution problem

$$(8.34) \quad \begin{cases} u_t(x,t) = \displaystyle\int_\Omega J(x-y)|u(y,t) - u(x,t)|^{p-2}(u(y,t) - u(x,t))dy + f(x,t), \\ u(x,0) = u_0(x), \qquad\qquad\qquad\qquad\qquad\qquad\quad x \in \Omega,\, t > 0. \end{cases}$$

Associated to this problem is the energy functional

$$G_p^{J,\Omega}(u) = \frac{1}{2p}\int_\Omega \int_\Omega J(x-y)|u(y) - u(x)|^p \, dy \, dx.$$

Consider the rescaled problems and the corresponding limit problems

$$(8.35) \quad \begin{cases} f(\cdot,t) - u_t(\cdot,t) \in \partial G_\infty^{\varepsilon,\Omega}(u(\cdot,t)) & \text{a.e. } t \in (0,T), \\ u(x,0) = u_0(x) & \text{in } \Omega, \end{cases}$$

with associated functionals

$$G_\infty^{\varepsilon,\Omega}(u) = \begin{cases} 0 & \text{if } |u(x) - u(y)| \le \varepsilon,\ \text{for } |x-y| \le \varepsilon;\ \ x,y \in \Omega, \\ +\infty & \text{otherwise.} \end{cases}$$

The limit problem for the local p-Laplacians is

$$(8.36) \quad \begin{cases} f(\cdot,t) - (v_\infty)_t(\cdot,t) \in \partial F_\infty^\Omega(v_\infty(\cdot,t)) & \text{a.e. } t \in (0,T), \\ v_\infty(x,0) = g(x) & \text{in } \Omega, \end{cases}$$

where the functional F_∞^Ω is defined in $L^2(\Omega)$ by

$$F_\infty^\Omega(v) = \begin{cases} 0 & \text{if } |\nabla v| \le 1, \\ +\infty & \text{otherwise.} \end{cases}$$

In these limit problems we assume that the material is confined in a domain Ω; thus we are looking at models for sandpiles inside a container (see [**103**] for a local model).

Working as in the previous sections we can prove the following result.

THEOREM 8.11. *Let Ω be a convex domain in $\mathbb{R}^N$.*

(1) *Let $T > 0$, $u_0 \in L^\infty(\Omega)$ such that $|u_0(x) - u_0(y)| \le 1$ for $x - y \in \Omega \cap \operatorname{supp}(J)$. Take $f \in L^2(0,T;L^\infty(\Omega))$ and let u_p be the unique solution of (8.34). Then, if u_∞ is the unique solution of (8.35) with $\varepsilon = 1$,*

$$\lim_{p\to\infty}\sup_{t\in[0,T]} \|u_p(\cdot,t) - u_\infty(\cdot,t)\|_{L^2(\Omega)} = 0.$$

(2) *Let $T > 0$, $u_0 \in W^{1,\infty}(\Omega)$ such that $|\nabla u_0| \le 1$, take $f \in L^2(0,T;L^2(\Omega))$ and consider $u_{\infty,\varepsilon}$, the unique solution of (8.35). Then, if v_∞ is the unique solution of (8.36), we have*

$$\lim_{\varepsilon\to 0}\sup_{t\in[0,T]} \|u_{\infty,\varepsilon}(\cdot,t) - v_\infty(\cdot,t)\|_{L^2(\Omega)} = 0.$$

EXAMPLE 8.12. In this case, let us also compute an explicit solution to the limit problem (8.35) (to simplify we have taken $\varepsilon = 1$ in this example). Consider a recipient $\Omega = (0, l)$ with l an integer greater than 1, $u_0 = 0$ and a source given by $f(x, t) = \mathcal{X}_{[0,1]}(x)$. Then the solution is given by

$$u(x, t) = t\mathcal{X}_{[0,1]}(x)$$

for times $t \in [0, 1]$. For $t \in [1, 3]$ we get

$$u(x, t) = \begin{cases} 1 + \dfrac{t-1}{2}, & x \in [0, 1), \\[2mm] \dfrac{t-1}{2}, & x \in [1, 2), \\[2mm] 0, & x \notin [0, 2). \end{cases}$$

In general we have, until the recipient is full, for any $k = 1, \ldots, l$ and for $t \in [t_{k-1}, t_k)$,

$$u(x, t) = \begin{cases} k - 1 + \dfrac{t - t_{k-1}}{k}, & x \in [0, 1), \\[2mm] k - 2 + \dfrac{t - t_{k-1}}{k}, & x \in [1, 2), \\[2mm] \ldots \\[2mm] \dfrac{t - t_{k-1}}{k}, & x \in [k-1, k), \\[2mm] 0, & x \notin [0, k). \end{cases}$$

Here $t_k = t_{k-1} + k$ is the first time when the solution reaches the level k, that is, $u(t_k, 0) = k$.

For times even greater, $t \geq t_l$, the solution turns out to be

$$u(x, t) = \begin{cases} l + \dfrac{t - t_l}{l}, & x \in [0, 1), \\[2mm] l - 1 + \dfrac{t - t_l}{l}, & x \in [1, 2), \\[2mm] \ldots \\[2mm] 1 + \dfrac{t - t_l}{l}, & x \in [l-1, l). \end{cases}$$

Hence, when the recipient is full, the solution grows with speed $1/l$ uniformly in $(0, l)$.

8.2. A nonlocal version of the Prigozhin model for sandpiles

In this section we introduce a nonlocal version of the Prigozhin model for sandpiles. We first recall the Prigozhin model for sandpiles. Next, we introduce the nonlocal version of the Prigozhin model as the limit as $p \to \infty$ in the nonlocal p-Laplacian Dirichlet problem. By rescaling, we also prove the convergence of the nonlocal model to the Prigozhin model for sandpiles. Finally, we present concrete examples of explicit solutions that illustrate the general convergence result.

8.2.1. The Prigozhin model for sandpiles. Suppose sand is poured out onto a rigid surface, $y = u_0(x)$, given in a bounded open subset Ω of $\mathbb{R}^2$ with Lipschitz boundary $\partial\Omega$. If the support boundary is open and we assume that the angle of stability is equal to $\frac{\pi}{4}$, a model for pile surface evolution was proposed by L. Prigozhin [138] [139] as

$$(8.37) \qquad \begin{cases} \partial_t u + \operatorname{div} \mathbf{q} = f, \\[2mm] u|_{t=0} = u_0, \\[2mm] u|_{\partial\Omega} = u_0|_{\partial\Omega}, \end{cases}$$

where $u(x,t)$ is the unknown pile surface, $f(x,t) \geq 0$ is the given source density and $\mathbf{q}(x,t)$ is the unknown horizontal projection of the flux of sand pouring down the pile surface. If the support has no slopes steeper than the sand angle of repose, $\|\nabla u_0\|_\infty \leq 1$, Prigozhin ([139]; see also [30], [94], [95], [116] and the references therein) proposed to take $\mathbf{q} = -m\nabla u$, where $m \geq 0$, the Lagrange multiplier related to the constraint $\|\nabla u\|_\infty \leq 1$, satisfies $m(\|\nabla u\|^2 - 1) = 0$, and reformulated this model as the following variational inequality:

$$\begin{cases} f(\cdot,t) - u_t(\cdot,t) \in \partial I_{K(u_0)}(u(\cdot,t)), \quad \text{a.e. } t \in (0,T), \\[2mm] u(x,0) = u_0(x), \end{cases}$$

where

$$K(u_0) := \left\{ v \in W^{1,\infty}(\Omega) \ : \ \|\nabla v\|_\infty \leq 1, \ v|_{\partial\Omega} = u_0|_{\partial\Omega} \right\}.$$

Our aim now is to approximate the Prigozhin model for sandpile by a nonlocal model obtained as the limit as $p \to +\infty$ of the nonlocal p-Laplacian problem with Dirichlet boundary condition studied in Chapter 6.

8.2.2. Limit as $p \to +\infty$ in the nonlocal p-Laplacian Dirichlet problem. In Section 6.2, we have obtained existence and uniqueness of solutions of the nonlocal p-Laplacian Dirichlet problem

$$(8.38) \qquad \begin{cases} u_t(x,t) = \displaystyle\int_\Omega J(x-y)|u(y,t) - u(x,t)|^{p-2}(u(y,t) - u(x,t))dy + f(x,t), \\[2mm] \qquad\qquad\qquad\qquad\qquad\qquad\qquad\qquad\qquad\qquad\qquad x \in \Omega,\ t > 0, \\[2mm] u(x,t) = \psi(x), \quad x \in \Omega_J \setminus \overline{\Omega},\ t > 0, \\[2mm] u(x,0) = u_0(x), \quad x \in \Omega. \end{cases}$$

This problem is associated to the energy functional

$$G_{p,\psi}^J(u) = \frac{1}{2p} \int_\Omega \int_\Omega J(x-y)|u(y) - u(x)|^p \, dy \, dx$$
$$+ \frac{1}{p} \int_\Omega \int_{\Omega_J \setminus \overline{\Omega}} J(x-y)|\psi(y) - u(x)|^p \, dy \, dx.$$

Using a formal computation, and taking the limit as $p \to +\infty$, we arrive at the functional

$$G_{\infty,\psi}^J(u) = \begin{cases} 0 & \text{if } |u(x) - u(y)| \leq 1 \text{ for } x, y \in \overline{\Omega}, \\ & \text{and } |\psi(y) - u(x)| \leq 1 \text{ for } x \in \Omega, y \in \Omega_J \setminus \overline{\Omega}, \\ & \text{with } x - y \in \operatorname{supp}(J), \\ +\infty & \text{otherwise.} \end{cases}$$

Hence, if we define

$$K_{\infty,\psi}^J := \left\{ u \in L^2(\Omega) \ : \ \begin{array}{l} |u(x) - u(y)| \leq 1, \ x, y \in \Omega, \\ \text{and } |\psi(y) - u(x)| \leq 1 \text{ for } x \in \Omega, y \in \Omega_J \setminus \overline{\Omega}, \\ \text{with } x - y \in \operatorname{supp}(J) \end{array} \right\},$$

we have that the functional $G_{\infty,\psi}^J$ is given by the indicator function of $K_{\infty,\psi}^J$, that is, $G_{\infty,\psi}^J = I_{K_{\infty,\psi}^J}$. Then the *nonlocal limit problem* can be written as

$$(8.39) \qquad \begin{cases} f(\cdot,t) - u_t(\cdot,t) \in \partial I_{K_{\infty,\psi}^J}(u(\cdot,t)), & \text{a.e. } t \in (0,T), \\ u(x,0) = u_0(x). \end{cases}$$

In order to prove the next result we proceed in a similar way as in the proof of Theorem 8.1.

THEOREM 8.13. *Let $\psi \in L^\infty(\Omega_J \setminus \overline{\Omega})$ such that $K_{\infty,\psi}^J \neq \emptyset$. Let $T > 0$, $f \in L^2(0,T;L^\infty(\Omega))$, $u_0 \in L^\infty(\Omega)$ such that $u_0 \in K_{\infty,\psi}^J$, and let u_p be the unique solution of (8.38), $p \geq 2$. Then, if u_∞ is the unique solution of (8.39),*

$$\lim_{p \to \infty} \sup_{t \in [0,T]} \|u_p(\cdot,t) - u_\infty(\cdot,t)\|_{L^2(\Omega)} = 0.$$

PROOF. Let $T > 0$. By Theorem A.38, to prove the result it is enough to show that the functionals $G_{p,\psi}^J(u)$ converge to $G_{\infty,\psi}^J(u)$ as $p \to +\infty$, in the sense of Mosco.

First, let us check that

$$(8.40) \qquad \operatorname{Epi}(G_{\infty,\psi}^J) \subset \operatorname*{s-lim\,inf}_{p \to +\infty} \operatorname{Epi}(G_{p,\psi}^J).$$

To this end let $(u, \lambda) \in \operatorname{Epi}(G_{\infty,\psi}^J)$. We can assume that $u \in K_{\infty,\psi}^J$ and $\lambda \geq 0$ (as $G_{\infty,\psi}^J(u) = 0$). Now take

$$v_p = u \qquad \text{and} \qquad \lambda_p = G_{p,\psi}^J(u) + \lambda.$$

Then, as $\lambda \geq 0$, we have $(v_p, \lambda_p) \in \operatorname{Epi}(G_{p,\psi}^J)$. Obviously,

$$v_p = u \to u \quad \text{in } L^2(\Omega),$$

and, as $u \in K_{\infty,\psi}^J$,

$$G_{p,\psi}^J(u) \;=\; \frac{1}{2p} \int_\Omega \int_\Omega J(x-y)|u(y)-u(x)|^p \, dy \, dx$$

$$+ \frac{1}{p} \int_\Omega \int_{\Omega_J \setminus \overline{\Omega}} J(x-y)|\psi(y)-u(x)|^p \, dy \, dx$$

$$\leq \frac{1}{2p} \int_\Omega \int_\Omega J(x-y) \, dy \, dx + \frac{1}{p} \int_\Omega \int_{\Omega_J \setminus \overline{\Omega}} J(x-y) \, dy \, dx;$$

and consequently

$$\lambda_p \to \lambda \quad \text{as } p \to +\infty.$$

Therefore, we get (8.40).

Finally, let us prove that

$$\text{w-}\limsup_{p \to +\infty} \text{Epi}(G_{p,\psi}^J) \subset \text{Epi}(G_{\infty,\psi}^J).$$

Consider a sequence $(u_{p_j}, \lambda_{p_j}) \in \text{Epi}(G_{p_j,\psi}^J)$, that is, $G_{p_j,\psi}^J(u_{p_j}) \leq \lambda_{p_j}$, with

$$u_{p_j} \rightharpoonup u \quad \text{and} \quad \lambda_{p_j} \to \lambda.$$

Since $0 \leq G_{p_j,\psi}^J(u_{p_j}) \leq \lambda_{p_j} \to \lambda$, then $0 \leq \lambda$. On the other hand, there exists a constant $C > 0$ such that

$$(p_j C)^{1/p_j} \;\geq\; \left(p_j G_{p,\psi}^J(u_{p_j})\right)^{1/p_j}$$

$$= \left(\frac{1}{2} \int_\Omega \int_\Omega J(x-y)|u_{p_j}(y)-u_{p_j}(x)|^{p_j} \, dy \, dx \right.$$

$$\left. + \int_\Omega \int_{\Omega_J \setminus \overline{\Omega}} J(x-y)|\psi(y)-u_{p_j}(x)|^{p_j} \, dy \, dx \right)^{1/p_j}.$$

Then, by the above inequality, applying Hölder's inequality, we get

$$\left(\int_\Omega \int_\Omega J(x-y) \left|u_{p_j}(y)-u_{p_j}(x)\right|^q \, dy \, dx \right)^{1/q}$$

$$\leq \left(\int_\Omega \int_\Omega J(x-y) \, dy \, dx \right)^{(p_j-q)/p_j q}$$

$$\times \left(\int_\Omega \int_\Omega J(x-y) \left|u_{p_j}(y)-u_{p_j}(x)\right|^{p_j} \, dy \, dx \right)^{1/p_j}$$

$$\leq \left(\int_\Omega \int_\Omega J(x-y) \, dy \, dx \right)^{(p_j-q)/p_j q} (C p_j)^{1/p_j}.$$

Hence, we can extract a subsequence, if necessary, and let $p_j \to +\infty$, obtaining

$$\left(\int_\Omega \int_\Omega J(x-y) \left|u(y)-u(x)\right|^q \, dy \, dx \right)^{1/q} \leq \left(\int_\Omega \int_\Omega J(x-y) \, dy \, dx \right)^{1/q}.$$

Now, just letting $q \to +\infty$, we get

$$|u(x) - u(y)| \leq 1 \quad \text{a.e. } (x,y) \in \Omega \times \Omega, \; x-y \in \text{supp}(J).$$

With a similar argument we deduce that

$$|u(x) - \psi(y)| \leq 1 \qquad \text{a.e. } x \in \Omega, \ y \in \Omega_J \setminus \overline{\Omega}, \text{ with } x - y \in \text{supp}(J).$$

Hence, we conclude that $u \in K_{\infty,\psi}^J$. This ends the proof. $\qquad\square$

8.2.3. Convergence to the Prigozhin model. Suppose that $\text{supp}(J) = \overline{B}(0,1)$. We also assume now that Ω is convex and ψ satisfies $\|\nabla\psi\|_\infty \leq 1$. For $\varepsilon > 0$, we rescale the functional $G_{\infty,\psi}^J$ as follows:

$$G_{\infty,\psi}^\varepsilon(u) = \begin{cases} 0 & \text{if } |u(x) - u(y)| \leq \varepsilon \text{ for } x, y \in \Omega, \\ & \text{and } |\psi(y) - u(x)| \leq \varepsilon \text{ for } x \in \Omega, y \in \Omega_J \setminus \overline{\Omega}, \\ & \text{with } |x - y| \leq \varepsilon, \\ +\infty & \text{otherwise.} \end{cases}$$

In other words, $G_{\infty,\psi}^\varepsilon = I_{K_{\infty,\psi}^\varepsilon}$, where

$$K_{\infty,\psi}^\varepsilon := \left\{ u \in L^2(\Omega) \ : \ \begin{array}{l} |u(x) - u(y)| \leq \varepsilon, \ x, y \in \Omega, \\ \text{and } |\psi(y) - u(x)| \leq \varepsilon \text{ for } x \in \Omega, y \in \Omega_J \setminus \overline{\Omega}, \\ \text{with } |x - y| \leq \varepsilon \end{array} \right\}.$$

Consider the gradient flow associated to the functional $G_{\infty,\psi}^\varepsilon$ with source

$$(8.41) \qquad \begin{cases} f(\cdot,t) - u_t(\cdot,t) \in \partial I_{K_{\infty,\psi}^\varepsilon}(u(\cdot,t)), & \text{a.e. } t \in (0,T), \\ u(x,0) = u_0(x) & \text{in } \Omega, \end{cases}$$

and the problem

$$(8.42) \qquad \begin{cases} f(\cdot,t) - (v_\infty)_t(\cdot,t) \in \partial I_{K_\psi}(v_\infty(\cdot,t)), & \text{a.e. } t \in (0,T), \\ v_\infty(x,0) = u_0(x) & \text{in } \Omega, \end{cases}$$

where

$$K_\psi := \left\{ u \in W^{1,\infty}(\Omega) \ : \ \|\nabla u\|_\infty \leq 1, \ u|_{\partial\Omega} = \psi|_{\partial\Omega} \right\}.$$

Observe that if $u \in K_\psi$, $\|\nabla u\|_\infty \leq 1$. Then, since $\|\nabla\psi\|_\infty \leq 1$ and Ω is convex, we have $|u(x) - u(y)| \leq |x - y|$ and $|u(x) - \psi(y)| \leq |x - y|$, from which it follows that $u \in K_{\infty,\psi}^\varepsilon$, that is, $K_\psi \subset K_{\infty,\psi}^\varepsilon$.

With all these definitions and notation, we can pass to the limit as $\varepsilon \to 0$ for the sandpile model.

THEOREM 8.14. *Suppose Ω is a convex bounded domain in $\mathbb{R}^N$. Let $T > 0$, $f \in L^2(0,T;L^2(\Omega))$, $\psi \in W^{1,\infty}(\Omega_J \setminus \overline{\Omega})$ such that $\|\nabla\psi\|_\infty \leq 1$, $u_0 \in W^{1,\infty}(\Omega)$ such that $\|\nabla u_0\|_\infty \leq 1$ and $u_0|_{\partial\Omega} = \psi|_{\partial\Omega}$ (this means that $u_0 \in K_\psi$), and consider $u_{\infty,\varepsilon}$, the unique solution of (8.41). Then, if v_∞ is the unique solution of (8.42), we have*

$$\lim_{\varepsilon\to 0} \sup_{t\in[0,T]} \|u_{\infty,\varepsilon}(\cdot,t) - v_\infty(\cdot,t)\|_{L^2(\Omega)} = 0.$$

PROOF. Since $u_0 \in K_\psi$, $u_0 \in K_{\infty,\psi}^\varepsilon$ for all $\varepsilon > 0$. Again we are using that $\|\nabla\psi\|_\infty \leq 1$. Consequently there exists $u_{\infty,\varepsilon}$, the unique solution of (8.41).

By Theorem A.38, as above, to prove the result it is enough to show that $I_{K_{\infty,\psi}^{\varepsilon}}$ converges to I_{K_ψ} in the sense of Mosco. Using that $\|\nabla\psi\|_\infty \leq 1$ it is easy to obtain that

$$(8.43) \qquad\qquad K_{\infty,\psi}^{\varepsilon_1} \subset K_{\infty,\psi}^{\varepsilon_2} \quad \text{if } \varepsilon_1 \leq \varepsilon_2.$$

Since $K_\psi \subset K_{\infty,\psi}^{\varepsilon}$ for all $\varepsilon > 0$, we have

$$K_\psi \subset \bigcap_{\varepsilon > 0} K_{\infty,\psi}^{\varepsilon}.$$

On the other hand, if

$$u \in \bigcap_{\varepsilon > 0} K_{\infty,\psi}^{\varepsilon},$$

we get

$$|u(y) - u(x)| \leq |y - x|, \quad \text{a.e. } x, y \in \Omega,$$

and moreover

$$|u(y) - \psi(x)| \leq |y - x|, \quad \text{a.e. } x \in \Omega_J \setminus \overline{\Omega},\ y \in \Omega,$$

from which it follows that $u \in K_\psi$. Therefore, we obtain

$$(8.44) \qquad\qquad K_\psi = \bigcap_{\varepsilon > 0} K_{\infty,\psi}^{\varepsilon}.$$

Note that

$$(8.45) \qquad \text{Epi}(I_{K_\psi}) = K_\psi \times [0, \infty[, \quad \text{Epi}(I_{K_{\infty,\psi}^{\varepsilon}}) = K_{\infty,\psi}^{\varepsilon} \times [0, \infty[\quad \forall \varepsilon > 0.$$

By (8.44) and (8.45),

$$(8.46) \qquad \text{Epi}(I_{K_\psi}) \subset \text{s-}\liminf_{\varepsilon \to 0} \text{Epi}(I_{K_{\infty,\psi}^{\varepsilon}}).$$

Given $(u, \lambda) \in \text{w-}\limsup_{\varepsilon \to 0} \text{Epi}(I_{K_{\infty,\psi}^{\varepsilon}})$ there exists $(u_{\varepsilon_k}, \lambda_k) \in K_{\varepsilon_k,\psi} \times [0, \infty[$ such that $\varepsilon_k \to 0$ and

$$u_{\varepsilon_k} \rightharpoonup u \quad \text{in } L^2(\Omega), \quad \lambda_k \to \lambda \quad \text{in } \mathbb{R}.$$

By (8.43), given $\varepsilon > 0$, there exists k_0 such that $u_{\varepsilon_k} \in K_{\infty,\psi}^{\varepsilon}$ for all $k \geq k_0$. Then, since $K_{\infty,\psi}^{\varepsilon}$ is a closed convex set, we get $u \in K_{\infty,\psi}^{\varepsilon}$, and, by (8.44), $u \in K_0$. Consequently,

$$(8.47) \qquad\qquad \text{w-}\limsup_{n \to \infty} \text{Epi}(I_{K_{\infty,\psi}^{\varepsilon}}) \subset \text{Epi}(I_{K_\psi}).$$

Finally, by (8.46), (8.47) and having in mind (A.19), we obtain that $I_{K_{\infty,\psi}^{\varepsilon}}$ converges to I_{K_ψ} in the sense of Mosco. $\qquad\qquad\qquad\qquad\qquad \square$

In the above theorem we are approximating the sandpile model described in Subsection 8.2.1 by a nonlocal model. In this nonlocal approximation a configuration of sand is stable when its height u satisfies $|u(x) - u(y)| \leq \varepsilon$ if $|x - y| \leq \varepsilon$.

8.2.4. Explicit solutions. Our goal now is to present some explicit examples that illustrate the behaviour of the solutions when $p = +\infty$.

REMARK 8.15. There is a natural upper bound (and of course also a natural lower bound) for the solutions with boundary datum ψ outside Ω (regardless of the source term f). Indeed, given a bounded domain $\Omega \subset \mathbb{R}^N$, let us define inductively

$$\Omega_1 = \left\{ x \in \Omega \;:\; |x - y| < 1 \text{ for some } y \in \Omega_J \setminus \overline{\Omega} \right\}$$

and, for $j \geq 2$,

$$\Omega_j = \left\{ x \in \Omega \setminus \bigcup_{i=1}^{j-1} \Omega_i \;:\; |x - y| < 1 \text{ for some } y \in \Omega_{j-1} \right\}.$$

Then, since $u(\cdot, t) \in K^J_{\infty, \psi}$, we must have

$$u(x, t) \leq \psi(y) + 1 \quad \text{if } |x - y| \leq 1, \; x \in \Omega_1, \; y \in \Omega_J \setminus \overline{\Omega},$$

and for any $j \geq 2$,

$$u(x, t) \leq u(y, t) + 1 \quad \text{if } |x - y| \leq 1, \; x \in \Omega_j, \; y \in \Omega_{j-1} \setminus \Omega_j.$$

Therefore we have an upper bound for $u(x, t)$ in the whole Ω,

$$u(x, t) \leq \Psi_1(x),$$

where Ψ_1 is defined by the inductive formulas

$$\Psi_1(x) = \max \left\{ \psi(y) + 1 \;:\; y \in \Omega_J \setminus \overline{\Omega}, \; |x - y| \leq 1 \right\} \text{ for } x \in \Omega_1,$$

and

$$\Psi_1(x) = \max \left\{ \Psi_1(y) + 1 \;:\; y \in \Omega_{j-1}, \; |x - y| \leq 1 \right\} \text{ for } x \in \Omega_j \text{ if } j \geq 2.$$

Analogously, we can obtain a lower bound for $u(x, t)$,

$$u(x, t) \geq \Phi_1(x),$$

where Φ_1 is defined by the inductive formulas

$$\Phi_1(x) = \min \left\{ \psi(y) - 1 \;:\; y \in \Omega_J \setminus \overline{\Omega}, \; |x - y| \leq 1 \right\} \text{ for } x \in \Omega_1,$$

and

$$\Phi_1(x) = \min \left\{ \Phi_1(y) - 1 \;:\; y \in \Omega_{j-1}, \; |x - y| \leq 1 \right\} \text{ for } x \in \Omega_j \text{ if } j \geq 2.$$

With this remark in mind we give some explicit examples of solutions of

$$(8.48) \quad \begin{cases} f(\cdot, t) - u_t(\cdot, t) \in \partial G^J_{\infty, \psi}(u(\cdot, t)), & \text{a.e. } t \in {]0, T[}, \\[2mm] u(x, 0) = u_0(x) & \text{in } \Omega, \end{cases}$$

where

$$G^J_{\infty, \psi}(u) = \begin{cases} 0 & \text{if } u \in L^2(\Omega), \; |u(x) - u(y)| \leq 1 \text{ for } x, y \in \Omega, \; |x - y| \leq 1, \\[2mm] & \text{and } |u(x) - \psi(y)| \leq 1 \text{ for } x \in \Omega, \; y \in \Omega_J \setminus \overline{\Omega}, \; |x - y| \leq 1, \\[2mm] +\infty & \text{otherwise.} \end{cases}$$

In order to verify that a function $u(x, t)$ is a solution to (8.48) we need to check that

$$(8.49) \quad G^J_{\infty, \psi}(v) \geq G^J_{\infty, \psi}(u) + \langle f - u_t, \; v - u \rangle \quad \text{for all } v \in L^2(\Omega).$$

To this end we can assume that $v \in K^J_{\infty,\psi}$ (otherwise $G^J_{\infty,\psi}(v) = +\infty$ and then (8.49) becomes trivial). Therefore, we need to check that

$$(8.50) \qquad u(\cdot,t) \in K^J_{\infty,\psi}$$

and, by (8.49), that

$$(8.51) \qquad \int_\Omega (f(x,t) - u_t(x,t))(v(x) - u(x,t))\, dx \le 0$$

for every $v \in K^J_{\infty,\psi}$.

EXAMPLE 8.16. Consider a nonnegative source f and take as initial condition the upper bound defined in the previous remark, $u_0(x) = \Psi_1(x)$. Then the solution to (8.48) is given by

$$u(x,t) \equiv \Psi_1(x)$$

for every $t > 0$. Indeed, $\Psi_1(x) \in K^J_{\infty,\psi}$ and for every $v \in K^J_{\infty,\psi}$ we have that $v(x) \le \Psi_1(x)$ and therefore

$$\int_\Omega (f(x,t) - u_t(x,t))(v(x) - u(x,t))\, dx = \int_\Omega f(x,t)(v(x) - \Psi_1(x))\, dx \le 0$$

as was to be proved.

In general, given a nonnegative source f supported in $D \subset \Omega$, any initial condition $u_0 \in K^J_{\infty,\psi}$ that satisfies $u_0(x) = \Psi_1(x)$ in D produces a stationary solution $u(x,t) \equiv u_0(x)$.

Analogously, it can be shown that $u(x,t) \equiv \Phi_1(x)$ when $u_0(x) = \Phi_1(x)$ and $f(x,t) \le 0$.

EXAMPLE 8.17. Now, let us assume that we are in an interval $\Omega = (-L,L)$, $\psi = 0$, $\varepsilon = L/n$, $n \in \mathbb{N}$, $u_0 = 0$, which belongs to $K_{\varepsilon,0}$, and the source f is an approximation of a delta function,

$$f(x,t) = \frac{1}{\eta}\chi_{[-\frac{\eta}{2},\frac{\eta}{2}]}(x), \quad 0 < \eta \le 2\varepsilon.$$

Using the same ideas as in Example 8.5, it is easy to verify the following general formula that describes the solution of the corresponding rescaled problem (8.48) for every $t \ge 0$. For any given integer $l \ge 0$ we have

$$u(x,t) = \begin{cases} l\varepsilon + k_l(t - t_l), & x \in [-\frac{\eta}{2},\frac{\eta}{2}], \\[2mm] (l-1)\varepsilon + k_l(t - t_l), & x \in [-\frac{\eta}{2}-\varepsilon, \frac{\eta}{2}+\varepsilon] \setminus [-\frac{\eta}{2},\frac{\eta}{2}], \\[2mm] \dots \\[2mm] k_l(t - t_l), & x \in [-\frac{\eta}{2}-l\varepsilon, \frac{\eta}{2}+l\varepsilon] \setminus [-\frac{\eta}{2}-(l-1)\varepsilon, \frac{\eta}{2}+(l-1)\varepsilon], \\[2mm] 0, & x \notin [-\frac{\eta}{2}-l\varepsilon, \frac{\eta}{2}+l\varepsilon], \end{cases}$$

for $t \in [t_l, t_{l+1})$, where

$$k_l = \frac{1}{2l\varepsilon + \eta} \quad \text{and} \quad t_{l+1} = t_l + \frac{\varepsilon}{k_l}, \quad t_0 = 0.$$

This general formula is valid until the time at which the solution satisfies $u(x,t) = \Psi_\varepsilon(x)$ for $x \in [-\frac{\eta}{2}, \frac{\eta}{2}]$ (the support of f), that is, until $T = t_{l^*+1}$, where

$$l^* \text{ is the first } l \text{ such that } l\varepsilon + k_l(t_{l+1} - t_l) = \Psi_\varepsilon(0)$$

and

$$\Psi_\varepsilon \text{ is the natural upper bound defined in Remark 8.15}$$

for the corresponding rescaled kernel. Observe that for this l^*, $\frac{\eta}{2} + l^*\varepsilon \leq L$. From that time on the solution is stationary, that is, $u(x,t) = u(x,T)$ for all $t > T$.

From the above formula, taking limits as $\eta \to 0$, we get that the expected solution when the source is δ_0 is given by

$$(8.52) \quad u(x,t) = \begin{cases} (l-1)\varepsilon + k_l(t - t_l), & x \in [-\varepsilon, \varepsilon], \\[2mm] (l-2)\varepsilon + k_l(t - t_l), & x \in [-2\varepsilon, 2\varepsilon] \setminus [-\varepsilon, \varepsilon], \\[2mm] \dots \\[2mm] k_l(t - t_l), & x \in [-l\varepsilon, l\varepsilon] \setminus [-(l-1)\varepsilon, (l-1)\varepsilon], \\[2mm] 0, & x \notin [-l\varepsilon, l\varepsilon], \end{cases}$$

for any integer $l \geq 1$ and $t \in [t_l, t_{l+1})$, where $k_l = \frac{1}{2l\varepsilon}$, $t_{l+1} = t_l + \frac{\varepsilon}{k_l}$, $t_1 = 0$, until $T = t_{l^*+1}$, where

$$l^* \text{ is the first } l \text{ such that } l\varepsilon + k_l(t_{l+1} - t_l) = \Psi_\varepsilon(0).$$

From that time on the solution is stationary, that is, $u(x,t) = u(x,T)$ for all $t > T$.

Note that, since the space of functions $K^\varepsilon_{\infty,\psi}$ is not contained in $C(\mathbb{R})$, formula (8.51) with $f = \delta_0$ does not make sense. Hence the function $u(x,t)$ described by (8.52) must be understood as a *generalized solution* obtained as a limit of solutions of approximating problems.

Note that the function $u(x,T)$ is a "regular and symmetric pyramid" composed by squares of side ε which is one step below the upper profile Ψ_ε.

8.2.4.1. *Recovering the local sandpile model as $\varepsilon \to 0$.* Now, to recover the local sandpile model, take the limit as $\varepsilon \to 0$ in the previous example to get that $u(x,t) \to v(x,t)$, where

$$v(x,t) = (l - |x|)^+ \qquad \text{for } t = l^2,$$

until the time at which $t = L^2$, and from that time on the solution is stationary.

A similar argument shows that, for any $a \in (0, L)$, the *generalized solution* when the source is δ_a converges as $\varepsilon \to 0$ to $v(x,t)$, where

$$v(x,t) = (l - |x - a|)^+ \qquad \text{for } t = l^2,$$

until the time at which $t = (L-a)^2$, and from that time on the solution is stationary.

These concrete examples illustrate the general convergence result given in Theorem 8.14.

Bibliographical notes

The references [16] and [17] are the sources for this chapter.

The nonlocal version of the Aronsson-Evans-Wu model for sandpiles given in Section 8.1 has been characterized in [118], where, moreover, the connection with the stochastic process introduced in [102] is shown (see also [117]).

In Subsection 8.1.6 we introduced a Monge-Kantorovich mass transportation problem when the cost function is given by a discrete distance $d(x, y)$ that counts the number of steps needed to go from x to y. A more detailed study of this mass transportation problem, where the limit as p goes to $+\infty$ of the nonlocal p-Laplacian plays a crucial role, can be found in [119].

Nonlinear semigroups

A.1. Introduction

In this Appendix we outline some of the main points of the theory of nonlinear semigroups and evolution equations governed by accretive operators. We refer the reader to [27], [43], [46], [56], [87], [88] and [89].

The linear part of this theory started in the 1930s with the works of E. Hille, Y. Yosida and R. Phillips on semigroups of linear operator in Banach spaces. One of the first ideas came from a paper by G. Peano of 1887, where he wrote the system of differential equations

$$\begin{cases} \dfrac{du_1}{dt} = a_{11}u_1 + \cdots + a_{1n}u_n + f_1(t), \\ \vdots \\ \dfrac{du_n}{dt} = a_{n1}u_1 + \cdots + a_{nn}u_n + f_n(t), \end{cases}$$

in a matrix form as

$$\mathbf{u}'(t) = A\mathbf{u}(t) + \mathbf{f}(t),$$

where $\mathbf{u}(t) = (u_1(t), u_2(t), \ldots, u_n(t))$, $\mathbf{f}(t) = (f_1(t), f_2(t), \ldots, f_n(t))$ and $A = (a_{ij})$, and solved it by means of the explicit formula

$$\mathbf{u}(t) = e^{tA}\mathbf{u}(0) + \int_0^t e^{(t-s)A}\mathbf{f}(s) \, ds,$$

where

$$e^{tA} = \sum_{k=0}^{\infty} \frac{1}{k!} t^k A^k.$$

Thus, he transformed a complicated problem in one dimension to a formally simpler problem in higher dimension. That is the essence of the Nonlinear Semigroup Theory. Now, since we want to apply this abstract theory to solving nonlocal evolution equations, we must work in infinite dimension. So our main object will be the study of evolution problems of the form

(A.1)
$$\begin{cases} u'(t) + Au(t) = f(t) \quad \text{on } (0, T), \\ u(0) = u_0, \end{cases}$$

where X is a Banach space, $f : (0, T) \to X$ and $A : D(A) \to X$ is an operator.

Let us give one example about how to write a PDE problem as a problem of the form (A.1).

EXAMPLE A.1. Let Ω be a bounded domain in $\mathbb{R}^N$ with smooth boundary $\partial\Omega$. Consider the classical initial-boundary problem for the heat equation, that is, the problem

$$(A.2) \qquad \begin{cases} \dfrac{\partial w}{\partial t}(x,t) = \Delta w(x,t) & \text{in } \Omega \times (0,\infty), \\[2mm] w(x,t) = 0 & \text{on } \partial\Omega \times (0,\infty), \\[2mm] w(x,0) = f(x) & \text{in } \Omega. \end{cases}$$

Write $u(t) = w(\cdot,t)$, regarded as a function of x, and take X to be a space of functions on Ω, for example, $X = L^p(\Omega)$ for some $p \geq 1$ or $X = C(\overline{\Omega})$. Suppose we are in this last case. Let A be the operator with domain

$$D(A) := \big\{ v \in C(\overline{\Omega}) \; : \; \Delta v \in C(\overline{\Omega}) \text{ and } v(x) = 0 \; \forall x \in \partial\Omega \big\}$$

and defined by $Av := -\Delta v$, for $v \in D(A)$. Then we can write the problem (A.2) in the form (A.1). Note that the boundary condition of (A.2) is absorbed into the domain of the operator A and into the requirement that $u(t) \in D(A)$ for all $t \geq 0$.

A.2. Abstract Cauchy problems

From now on, X will be a real Banach space with norm denoted by $\| \; \|$ and dual X^*.

We will use multivalued nonlinear operators not only because they permit us to obtain a coherent theory but also because it is often necessary in applications. So let us recall some notation and basic facts concerning multivalued operators.

A mapping $A : X \to 2^X$ from X into 2^X (the collection of subsets of X) will be called an *operator* in X. For $x \in X$, Ax denotes the value of A at x, $D(A) := \{x \in X \; : \; Ax \neq \emptyset\}$ will be called the *effective domain* of A, and $\mathrm{R}(A) := \bigcup\{Ax \; : \; x \in D(A)\}$ its *range*.

If A is an operator in X, it determines the subset

$$G(A) = \{(x,y) \in X \times X \; : \; y \in Ax\},$$

called the graph of A; conversely, a subset G of $X \times X$ determines a unique operator A whose graph is G; the operator A is given by $Ax := \{y \; : \; (x,y) \in G\}$. Whenever it is convenient we will identify an operator with its graph.

Given two operators A and B in X and $\alpha \in \mathbb{R}$, we define new operators $A+B$, αA and A^{-1} according to

$$(A+B)x := Ax + Bx,$$
$$(\alpha A)x := \alpha(Ax),$$
$$A^{-1}x := \{y \in X \; : \; x \in Ay\}.$$

The *closure* of the operator A, denoted by $\overline{A}$, is defined to be the closure of the graph of A in $X \times X$, that is,

$$y \in \overline{A}x \text{ iff } \exists \, y_n \in Ax_n \; : \; x_n \to x, \; y_n \to y.$$

Before proceeding we fix some notation. By $L^1(a,b;X)$ we denote the vector space of all Bochner integrable functions $f : [a,b] \to X$ with respect to Lebesgue measure (i.e., of all strong measurable functions f such that $\int_a^b \|f(t)\| \, dt < +\infty$).

If I is an interval in $\mathbb{R}$, $L^1_{\text{loc}}(I; X)$ is the space of those functions $f : I \to X$ which are Bochner integrable on compact subintervals of I. As in the case of real functions, if $f \in L^1(a, b; X)$, then for almost all $t \in (a, b)$ one has

$$\text{(A.3)} \qquad \lim_{h \downarrow 0} \frac{1}{h} \int_{t-h}^{t+h} \|f(s) - f(t)\| \, ds = 0.$$

If (A.3) holds, t is called a *Lebesgue point* of f.

The space $W^{1,1}(a, b; X)$ consists of those functions f which have the form

$$\text{(A.4)} \qquad f(t) = f(0) + \int_0^t h(s) \, ds$$

for some $h \in L^1(a, b; X)$. It is well known that $W^{1,1}(a, b; X)$ consists of exactly those absolutely continuous functions $f : [a, b] \to X$ which are differentiable a.e. on $[a, b]$, and if (A.4) holds, then $f'(t) = h(t)$ a.e.

In a general Banach space X, the absolute continuity of a function $f : [a, b] \to X$ does not imply the existence of $f'(t)$ almost everywhere. When this happens, it is said that the Banach space X has the *Radon-Nikodym property*. For instance, every reflexive Banach space has the Radon-Nikodym property. However, there are important Banach spaces such as $L^1(\Omega)$, $L^\infty(\Omega)$ or $C(\overline{\Omega})$ without the Radon-Nikodym property.

As we mentioned before, our aim is to study evolution problems of the form

$$\text{(A.5)} \qquad \begin{cases} u'(t) + Au(t) \ni f(t) & \text{on } t \in (0, T), \\ u(0) = x, \end{cases}$$

where $f : (0, T) \to X$ and A is an operator in X. A problem of the form (A.5) is called an *abstract Cauchy problem*, and it will be denoted by $(\text{CP})_{x,f}$. In the homogeneous case, that is, for $f = 0$, we will write $(\text{CP})_x$ instead $(\text{CP})_{x,0}$.

In principle, one natural notion of solution for $(\text{CP})_{x,f}$ is the classical one, that is, a function u satisfying

$$\begin{cases} u \in C([0, T]; X) \cap C^1((0, T); X), \\ u'(t) + Au(t) \ni f(t) & \forall t \in (0, T), \\ u(0) = x. \end{cases}$$

In fact, this is a common notion of solution in the classical theory of ordinary differential equation (i.e., for $X = \mathbb{R}^N$) when A and f are continuous. But as soon as discontinuities arise, the notion of classical solution turns out to be too restrictive as may be illustrated by the following example.

EXAMPLE A.2. Let $X = \mathbb{R}$, $f = 0$, $x = 1$ and let A be the Heaviside function

$$A(r) = \begin{cases} 1 & \text{if } r > 0, \\ 0 & \text{if } r \le 0. \end{cases}$$

Then $(CP)_{x,f}$ becomes

(A.6)
$$\begin{cases} u'(t) = -1 & \text{if } u(t) > 0, \\ u'(t) = 0 & \text{if } u(t) \leq 0, \\ u(0) = 1. \end{cases}$$

The solution of problem (A.6) is given by

$$u(t) = \begin{cases} 1 - t & \text{if } 0 \leq t \leq 1, \\ 0 & \text{if } t \geq 1. \end{cases}$$

But u is not a classical solution since it is not differentiable at $t = 1$.

This example motivates the following weaker notion of solution for $(CP)_{x,f}$.

DEFINITION A.3. A function u is called a *strong solution* of $(CP)_{x,f}$ if

$$\begin{cases} u \in C([0,T];X) \cap W^{1,1}_{\text{loc}}((0,T);X), \\ u' + Au(t) \ni f(t) \quad \text{a.e. } t \in (0,T), \\ u(0) = x. \end{cases}$$

Clearly, the previous example is covered by this notion of solution. However, it is still not sufficient in general, as the following simple example due to G. Webb in [**152**] shows.

EXAMPLE A.4. Consider the problem

(A.7)
$$\begin{cases} w_t - w_x + w^+ = 0 & \text{on } \mathbb{R} \times [0, +\infty), \\ w(x,0) = u_0(x), & x \in \mathbb{R}. \end{cases}$$

We are interested in solving (A.7) in the space $X = C_0(\mathbb{R})$. To this end, we define the operator A in X by $Au := -u' + u^+$ with domain

$$D(A) := \left\{ u \in C^1(\mathbb{R}) \ : \ u, u' \in C_0(\mathbb{R}) \right\}.$$

We rewrite the problem (A.7) as an evolution problem in X:

(A.8)
$$\begin{cases} u'(t) + Au(t) = 0 & \text{in } [0, +\infty), \\ u(0) = u_0. \end{cases}$$

Observe that this is a semilinear problem with $A = A_0 + F$, where $A_0 u = -u'$ and $F(u) = u^+$. Then, since $-A_0$ is the infinitesimal generator of a C_0-semigroup $(S(t))_{t \geq 0}$ in X and F is Lipschitz continuous, it is well known that for every $u_0 \in X$ there is a unique solution of (A.8) given by the classical Duhamel formula

$$u(t) = S(t)u_0 - \int_0^t S(t-s)F(u(s)) \, ds \qquad \forall t \geq 0.$$

Nevertheless, u need not be a strong solution of problem (A.8), even if $u_0 \in D(A)$. In fact: Let $u_0 \in X$ such that there exists $x_0 \in \mathbb{R}$ satisfying: $u_0(x) > 0$ if $x > x_0$

and $u_0(x) < 0$ if $x < x_0$. Then, using the classical method of characteristics, it is not difficult to see that the solution of (A.7) is given by

$$w(x,t) = \begin{cases} e^{-t}u_0(x+t) & \text{if } x+t > x_0, \\ \\ u_0(x+t) & \text{if } x+t \leq x_0. \end{cases}$$

From this it follows that if $u_0'(x_0) \neq 0$, this solution is not a strong solution.

Consequently, we need to introduce a more general concept of solution for $(CP)_{x,f}$. The more adequate notion of solution for $(CP)_{x,f}$ in general Banach spaces is the concept of mild solution, introduced by M. G. Crandall and T. M. Liggett in [**89**] and Ph. Bénilan in [**43**], which is studied in the next section.

A.3. Mild solutions

Let A be an operator in X and $f \in L^1(a,b;X)$. Roughly speaking a mild solution of the problem

(A.9) $$u' + Au \ni f \qquad \text{on } [a,b]$$

is a continuous function $u \in C([a,b];X)$ which is the uniform limit of solutions of time-discretized problems, given by the implicit Euler scheme of the form

$$\frac{v(t_i) - v(t_{i-1})}{t_i - t_{i-1}} + Av(t_i) \ni f_i,$$

where f_i are approximations of f as $|t_i - t_{i-1}| \to 0$. So the underlying idea of the notion of mild solution is simple, and from the point of view of numerical analysis, even classical. Formally, the definition is as follows.

DEFINITION A.5. Let $\varepsilon > 0$. An ε-*discretization* of $u' + Au \ni f$ on $[a,b]$ consists of a partition $t_0 < t_1 < \cdots < t_N$ and a finite sequence $f_1, f_2, \ldots, f_N$ of elements of X such that

$$a \leq t_0 < t_1 < \cdots < t_N \leq b, \quad \text{with}$$

$$t_i - t_{i-1} \leq \varepsilon, \ i = 1, \ldots, N, \quad t_0 - a \leq \varepsilon \ \text{ and } \ b - t_N \leq \varepsilon.$$

and

$$\sum_{i=1}^{N} \int_{t_{i-1}}^{t_i} \|f(s) - f_i\| \, ds \leq \varepsilon.$$

We will denote this discretization by $D_A(t_0, \ldots, t_N; f_1, \ldots, f_N)$.

A *solution of the discretization* $D_A(t_0, \ldots, t_N; f_1, \ldots, f_N)$ is a piecewise constant function $v : [t_0, t_N] \to X$ whose values $v(t_0) = v_0$, $v(t) = v_i$ for $t \in]t_{i-1}, t_i]$, $i = 1, \ldots, N$ satisfy

(A.10) $$\frac{v_i - v_{i-1}}{t_i - t_{i-1}} + Av_i \ni f_i, \quad i = 1, \ldots, N.$$

A *mild solution* of $u' + Au \ni f$ on $[a,b]$ is a continuous function $u \in C([a,b];X)$ such that, for each $\varepsilon > 0$ there is an ε-discretization $D_A(t_0, \ldots, t_N; f_1, \ldots, f_N)$ of $u' + Au \ni f$ on $[a,b]$ which has a solution v satisfying

$$\|u(t) - v(t)\| \leq \varepsilon \quad \text{for } t_0 \leq t \leq t_N.$$

It is easy to see that if u is a mild solution of $u' + Au \ni f$ on $[a, b]$ and $[c, d] \subset [a, b]$, then $u_{|[c,d]}$ is a mild solution of $u' + Au \ni f$ on $[c, d]$. Therefore, the following definition is consistent.

DEFINITION A.6. Let I an interval of $\mathbb{R}$, and $f \in L^1_{\mathrm{loc}}(I; X)$. A *mild solution* of $u' + Au \ni f$ on I is a function $u \in C(I; X)$ whose restriction to each compact subinterval $[a, b]$ of I is a mild solution of $u' + Au \ni f$ on $[a, b]$.

In the next result we will see that mild solutions generalize the concept of the strong solutions.

THEOREM A.7. *Let $f \in L^1_{\mathrm{loc}}(I; X)$ and u be a strong solution of $u' + Au \ni f$ on I. Then u is a mild solution of $u' + Au \ni f$ on I.*

The heart of the proof of the above theorem is the following result concerning the approximation of Bochner integrals by Riemann sums in a strong sense.

LEMMA A.8. *Let Y be a Banach space, $g \in L^1(a, b; Y)$ and let K be a subset of $[a, b]$ such that $[a, b] \setminus K$ has measure zero. Then, given $\delta > 0$, there is a partition $a = t_0 < t_1 < \cdots < t_N \leq b$ satisfying:*

$$t_i \in K \text{ and } t_i \text{ is a Lebesgue point of } g \text{ for all } i = 1, \ldots, N,$$

$$b - t_N < \delta \text{ and } t_i - t_{i-1} < \delta, \quad i = 1, \ldots, N,$$

and

$$\sum_{i=1}^{N} \int_{t_{i-1}}^{t_i} \|g(t) - g(t_i)\| \, dt < \delta.$$

The converse of Theorem A.7 is false; mild solutions need not be strong solutions. One counterexample is given by the equation of Example A.4.

The next result collects some of the properties of mild solutions.

THEOREM A.9. *Let A be an operator in X and $f \in L^1_{\mathrm{loc}}(I; X)$. Then*

(i) *If u is a mild solution of $u' + Au \ni f$ on I, then $u(t) \in \overline{D(A)}$ for all $t \in I$.*

(ii) *Let I_1, I_2 be subintervals of I with $I \subset \overline{I_1 \cup I_2}$. If $u \in C(I; X)$ is a mild solution of $u' + Au \ni f$ on I_1 and on I_2, then u is a mild solution of $u' + Au \ni f$ on I.*

(iii) *Let $\overline{A}$ be the closure of the operator A. Then u is a mild solution of $u' + Au \ni f$ on I if and only if u is a mild solution of $u' + \overline{A}u \ni f$ on I.*

(iv) *Let $\{u_n\} \subset C(I; X)$, $\{f_n\} \subset L^1_{\mathrm{loc}}(I; X)$ and let u_n be a mild solution of $u'_n + Au_n \ni f_n$ on I. Suppose $u \in C(I; X)$, $f \in L^1_{\mathrm{loc}}(I; X)$ and for each compact subinterval $[a, b]$ of I,*

$$\lim_{n \to \infty} \left(\int_a^b \|f_n(t) - f(t)\| \, dt + \sup_{a \leq t \leq b} \|u_n(t) - u(t)\| \right) = 0;$$

then u is a mild solution of $u' + Au \ni f$ on I.

DEFINITION A.10. Let D be a subset of X. A family of mappings $S(t) : D \to D$, $(t \geq 0)$ satisfying

$$S(t+s)x = S(t)S(s)x \quad \text{for all } t, s \geq 0, \ x \in D,$$

$$\lim_{t \to 0} S(t)x = x \quad \text{for } x \in D,$$

is called a *strongly continuous semigroup* on D.

One may now associate to every operator A in X a strongly continuous semigroup $(S^A(t))_{t \geq 0}$ by the following definition:

$$D\left(S^A\right) := \big\{ x \in X \ : \ \text{there exists a unique mild solution } u_x \text{ of}$$

$$u' + Au \ni 0 \text{ on } (0, +\infty) \text{ with } u_x(0) = x \big\}.$$

For $t \geq 0$ and $x \in D(S^A)$, we set

$$S^A(t)x := u_x(t).$$

It is an immediate consequence of the properties of mild solutions that, in fact, $\left(S^A(t)\right)_{t \geq 0}$ is a strongly continuous semigroup on $D(S^A)$.

In the linear case, that is, if $S(t) \in \mathcal{L}(X)$, the strongly continuous semigroups are called C_0-semigroups. In this situation, each C_0-semigroup $(S(t))_{t \geq 0}$ has its associated *infinitesimal generator B* defined by

$$Bx := \lim_{t \to 0} \frac{S(t)x - x}{t} \quad \text{for } \ x \in D(B)$$

and

$$D(B) := \left\{ x \in X \ : \ \exists \lim_{t \to 0} \frac{S(t)x - x}{t} \right\}.$$

In the linear case it is well known that $-A$ is the infinitesimal generator of a C_0-semigroup $(S(t))_{t \geq 0}$ of bounded linear operators on X if and only if A is linear, closed and $D(S^A) = X$, and then $S^A(t) = S(t)$ for all $t \geq 0$.

This motivates the development of a nonlinear semigroup theory analogous to the classical linear one. We will see that in the nonlinear case the situation is very different from the linear one, and has more difficulties.

A.4. Accretive operators

Now we are going to introduce the class of operators for which we could obtain existence and uniqueness results of mild solutions.

The existence of mild solutions requires, as we pointed out before, the existence of solutions of discretized equations of the form

$$\frac{x_i - x_{i-1}}{t_i - t_{i-1}} + Ax_i \ni f_i, \qquad i = 1, \ldots, N$$

or equivalently

(A.11) $$x_i + (t_i - t_{i-1})Ax_i \ni (t_i - t_{i-1})f_i + x_{i-1}, \qquad i = 1, \ldots, N.$$

Then, to solve (A.11) we need the inverse of the operator $(I + \lambda A)$ to be a single-valued operator. Operators satisfying this property are the following.

DEFINITION A.11. An operator A in X is *accretive* if

$$\|x - \hat{x}\| \le \|x - \hat{x} + \lambda(y - \hat{y})\| \quad \text{whenever} \ \lambda > 0 \ \text{and} \ (x, y), (\hat{x}, \hat{y}) \in A.$$

Note that A is accretive if and only if for $\lambda > 0$ and $z \in X$, $x + \lambda y = z$ has at most one solution $(x, y) \in A$ and the relations $x + \lambda y = z$, $(x, y) \in A$, $\hat{x} + \lambda \hat{y} = \hat{z}$, $(\hat{x}, \hat{y}) \in A$ imply

$$\|x - \hat{x}\| = \left\|(I + \lambda A)^{-1} z - (I + \lambda A)^{-1} \hat{z}\right\| \le \|z - \hat{z}\|.$$

Therefore, we have

A is accretive if and only if $(I + \lambda A)^{-1}$ is a single-valued nonexpansive map for $\lambda \ge 0$.

In case A is accretive, we denote $J_\lambda^A = (I + \lambda A)^{-1}$ and we call J_λ^A the *resolvent* of A. Note that $D(J_\lambda^A) = \mathrm{R}(I + \lambda A)$.

It is easy to see that if β is an operator in $\mathbb{R}$, then β is accretive if and only if $(y - \hat{y})(x - \hat{x}) \ge 0$ for all $(x, y), (\hat{x}, \hat{y}) \in \beta$. Thus, if β is single-valued, then β is accretive if and only if β is nondecreasing. The following operators are examples of accretive operators in $\mathbb{R}$:

$$\mathrm{sgn}_0(r) = \begin{cases} -1 & \text{if } r < 0, \\ 0 & \text{if } r = 0, \\ 1 & \text{if } r > 0, \end{cases} \quad \text{and} \quad \mathrm{sgn}(r) = \begin{cases} -1 & \text{if } r < 0, \\ [-1, 1] & \text{if } r = 0, \\ 1 & \text{if } r > 0. \end{cases}$$

In order to verify accretivity of a given operator, it is useful to take into account alternative characterizations of this property. To do that we need to introduce the bracket and the duality map.

For each $\lambda \ne 0$ define $[\cdot, \cdot]_\lambda : X \times X \to \mathbb{R}$ by

$$[x, y]_\lambda := \frac{\|x + \lambda y\| - \|x\|}{\lambda}.$$

For fixed $(x, y) \in X \times X$, $\lambda \mapsto [x, y]_\lambda$ is nondecreasing for $\lambda > 0$. Indeed, if $\lambda \ge \mu > 0$ then

$$\|x + \mu y\| = \left\|\left(1 - \frac{\mu}{\lambda}\right) x + \frac{\mu}{\lambda}(x + \lambda y)\right\| \le \left(1 - \frac{\mu}{\lambda}\right) \|x\| + \frac{\mu}{\lambda} \|x + \lambda y\|,$$

from which it follows that $[x, y]_\mu \le [x, y]_\lambda$. Therefore for every $(x, y) \in X \times X$ we can define

$$[x, y] := \lim_{\lambda \downarrow 0} [x, y]_\lambda = \inf_{\lambda > 0} [x, y]_\lambda.$$

The number $[x, y]$ is the right-hand derivative of the norm of x in the direction y. In the next proposition we collect some of the useful properties of the bracket $[\cdot, \cdot]$.

PROPOSITION A.12. *If $x, y, z \in X$ and $\alpha, \beta \in \mathbb{R}$, then*

(i) $[\cdot, \cdot] : X \times X \to \mathbb{R}$ *is upper semicontinuous.*
(ii) $[\alpha x, \beta y] = |\beta|[x, y]$ *if $\alpha \cdot \beta > 0$.*
(iii) $[x, \alpha x + y] = \alpha \|x\| + [x, y]$.
(iv) $[x, y] \ge 0$ *if and only if $\|x + \lambda y\| \ge \|x\|$ for $\lambda \ge 0$.*
(v) $|[x, y]| \le \|y\|$ *and* $[0, y] = \|y\|$.
(vi) $[x, y] \ge -[x, -y]$.
(vii) $[x, y + z] \le [x, y] + [x, z]$.

(viii) *Let* $u :]a, b[\to \mathbb{R}$ *and* $t_0 \in]a, b[$, *such that* u *is differentiable at* t_0; *then* $t \mapsto$ $\|u(t)\|$ *is differentiable at* t_0 *if and only if* $[u(t_0), u'(t_0)] = -[u(t_0), -u'(t_0)]$. *In this case*

$$\frac{d}{dt}\|u(t)\|_{|t=t_0} = [u(t_0), u'(t_0)].$$

As a consequence of (iv) of the above proposition we obtain the following characterization of accretive operators.

COROLLARY A.13. *An operator A in X is accretive if and only if*

$$[x - \hat{x}, y - \hat{y}] \geq 0$$

whenever $(x, y), (\hat{x}, \hat{y}) \in A$.

In some concrete Banach spaces the bracket $[\cdot, \cdot]$ can be computed explicitly. We give some examples.

EXAMPLE A.14. Suppose $(H, (\ |\ ))$ is a Hilbert space. Then for $x, y \in H$,

$$\left(\|x + \lambda y\| - \|x\|\right)\left(\|x + \lambda y\| + \|x\|\right) = \|x + \lambda y\|^2 - \|x\|^2 = 2\lambda(x|y) + \lambda^2\|y\|^2.$$

Dividing this equality by λ yields

$$\left(\|x + \lambda y\| + \|x\|\right)[x, y]_\lambda = 2(x|y) + \lambda\|y\|^2,$$

so we find

$$\|x\|[x, y] = (x|y).$$

Then, by Corollary A.13, it follows that an operator A in H is accretive if and only if

(A.12) $$(x - \hat{x}|y - \hat{y}) \geq 0 \quad \text{for all} \quad (x, y), (\hat{x}, \hat{y}) \in A.$$

An operator in a Hilbert space satisfying (A.12) is called *monotone,* and therefore in Hilbert spaces monotone and accretive operators coincide.

EXAMPLE A.15. Let $X = L^p(\Omega)$, where $1 < p < \infty$. By the convexity of the map $t \mapsto |t|^p$, and applying the dominated convergence theorem, it is easy to see that

$$[f, g] = \|f\|_p^{1-p}\int_\Omega g|f|^{p-1}\,\mathrm{sgn}_0(f).$$

In the case $p = 1$, i.e., for $X = L^1(\Omega)$, we have

$$[f, g] = \int_\Omega g\,\mathrm{sgn}_0(f) + \int_{\{f=0\}}|g|.$$

The formulas for the bracket given in the above examples are very useful for proving that some concrete operator is accretive. Another useful tool to study the accretivity of concrete operators is the *duality map* $\mathcal{J} : X \to 2^{X^*}$, defined as

$$\mathcal{J}(x) := \{x^* \in X^* \ : \ \|x^*\| \leq 1, \ \langle x, x^* \rangle = \|x\|\}.$$

By the Hanh-Banach theorem, we have $\mathcal{J}(x) \neq \emptyset$ for every $x \in X$.

Given $x^* \in \mathcal{J}(x)$, since $\|x^*\| \leq 1$, we have

$$|\langle x^*, x + \lambda y \rangle| \leq \|x + \lambda y\|$$

and
$$\langle x^*, y \rangle = \frac{1}{\lambda} \left(\langle x^*, x + \lambda y \rangle - \|x\| \right) \leq [x, y]_\lambda.$$
Hence
$$\langle x^*, y \rangle \leq [x, y] \qquad \forall\, x^* \in \mathcal{J}(x).$$
On the other hand, if $V = \mathrm{LIN}\{x, y\}$ and we define $\xi^* \in V^*$ by
$$\langle \xi^*, \alpha x + \beta y \rangle := \alpha \|x\| + \beta [x, y],$$
then, by the Hanh-Banach theorem, there exists $x^* \in X^*$ such that $x^*|_V = \xi^*$, so
$$\langle x^*, x \rangle = -\|x\| \quad \text{and} \quad \langle x^*, y \rangle = [x, y].$$
Moreover, it is not difficult to see that $\|x^*\| \leq 1$, therefore $x^* \in \mathcal{J}(x)$. Consequently, we have the following result.

PROPOSITION A.16. *For $x, y \in X$*
$$[x, y] = \max_{x^* \in \mathcal{J}(x)} \langle x^*, y \rangle.$$

As a consequence of the above proposition and Corollary A.13, we have the following characterization of accretive operators.

COROLLARY A.17. *An operator A in X is accretive if and only if, whenever $(x, y), (\hat{x}, \hat{y}) \in A$, there exists $x^* \in \mathcal{J}(x - \hat{x})$ such that*
$$\langle x^*, y - \hat{y} \rangle \geq 0.$$

EXAMPLE A.18. Let $X = L^p(\Omega)$ where $1 < p < \infty$; then by Hölder's inequality we have
$$\mathcal{J}(f) = \mathrm{sgn}_0(f)|f|^{p-1}\|f\|_p^{1-p}.$$

In $L^1(\Omega)$, we have
$$\mathcal{J}(f) = \mathrm{sgn}(f) = \left\{ g \in L^\infty(\Omega) \; : \; |g| \leq 1, \;\; gf = |f| \;\; \text{a.e.} \right\}.$$

Given $w \in \mathbb{R}$, we define:
$$\mathcal{A}(w) := \{ A \subset X \times X \; : \; A + wI \text{ is accretive} \}.$$

PROPOSITION A.19. *Let A be an operator in X. The following statements are equivalent:*

   (i) *$A \in \mathcal{A}(w)$.*
   (ii) *$(1 - \lambda w)\|x - \hat{x}\| \leq \|x - \hat{x} + \lambda(y - \hat{y})\| \quad \forall \lambda < 0, \; (x, y), (\hat{x}, \hat{y}) \in A$.*
   (iii) *$[x - \hat{x}, y - \hat{y}] + w\|x - \hat{x}\| \geq 0$.*
   (iv) *For $\lambda > 0$, $\lambda w < 1$, $J_\lambda^A = (I + \lambda A)^{-1}$ is Lipschitz continuous with Lipschitz constant $(1 - \lambda w)^{-1}$.*
   (v) *For $(x, y), (\hat{x}, \hat{y}) \in A$, there exists $x^* \in \mathcal{J}(x - \hat{x})$ such that*
$$\langle x^*, y - \hat{y} \rangle + w\|x - \hat{x}\| \geq 0.$$

We have that accretivity implies uniqueness of strong solutions. More precisely, we have the following result.

THEOREM A.20. *Let $f, \hat{f} \in L^1(0, T; X)$, $A \in \mathcal{A}(w)$ and let $u, \hat{u}$ be strong solutions of $u' + Au \ni f$, $\hat{u}' + A\hat{u} \ni \hat{f}$, respectively, on $[0, T]$. Then*

$$\|u(t) - \hat{u}(t)\| \le e^{wt}\|u(0) - \hat{u}(0)\| + \int_0^t e^{w(t-s)} \left[u(s) - \hat{u}(s), f(s) - \hat{f}(s) \right] ds$$

$$\le e^{wt}\|u(0) - \hat{u}(0)\| + \int_0^t e^{w(t-s)} \|f(s) - \hat{f}(s)\| ds$$

for $t \in [0, T]$.

In particular, the strong solutions of $(\mathrm{CP})_{x,f}$ are unique.

PROOF. For simplicity, we assume that $w = 0$, i.e., A is accretive. Since u and $\hat{u}$ are differentiable a.e. in $]0, T[$, by (viii) of Proposition A.12, we have

$$\frac{d}{dt}\|u(t) - \hat{u}(t)\| = -[u(t) - \hat{u}(t), \hat{u}'(t) - u'(t)]$$

$$= -\left[u(t) - \hat{u}(t), (f(t) - u'(t)) - (\hat{f}(t) - \hat{u}'(t)) + (\hat{f}(t) - f(t)) \right]$$

for almost all $t \in (0, T)$.

Moreover, for almost all $t \in (0, T)$, we have that $(u(t), f(t) - u'(t)) \in A$ and $(\hat{u}(t), \hat{f}(t) - \hat{u}'(t)) \in A$. Then, by Corollary A.13 and (vi), (vii) of Proposition A.12, we get

$$\left[u(t) - \hat{u}(t), (f(t) - u'(t)) - (\hat{f}(t) - \hat{u}'(t)) + (\hat{f}(t) - f(t)) \right]$$

$$\ge \left[u(t) - \hat{u}(t), (f(t) - u'(t)) - (\hat{f}(t) - \hat{u}'(t)) \right] - \left[u(t) - \hat{u}(t), f(t) - \hat{f}(t) \right]$$

$$\ge -\left[u(t) - \hat{u}(t), f(t) - \hat{f}(t) \right].$$

Hence

$$\frac{d}{dt}\|u(t) - \hat{u}(t)\| \le \left[u(t) - \hat{u}(t), f(t) - \hat{f}(t) \right].$$

From this, applying Gronwall's inequality we obtain

$$\|u(t) - \hat{u}(t)\| \le \|u(0) - \hat{u}(0)\| + \int_0^t \left[u(s) - \hat{u}(s), f(s) - \hat{f}(s) \right] ds$$

$$\le \|u(0) - \hat{u}(0)\| + \int_0^t \|f(s) - \hat{f}(s)\| ds. \qquad \square$$

We have seen that the accretivity of the operator A implies uniqueness of the solution x_i of the discretized equation

$$\frac{x_i - x_{i-1}}{t_i - t_{i-1}} + Ax_i \ni f_i, \qquad i = 1, \dots, N,$$

which, if they exist, are given by

$$x_i = J^A_{(t_i - t_{i-1})} \left((t_i - t_{i-1})f_i + x_{i-1} \right), \qquad i = 1, \dots, N.$$

This formula indicates that apart from accretivity one should expect a range condition (i.e., a condition on $\mathrm{R}(I + \lambda A) = D(J_\lambda^A)$) to hold in order to get the existence of solution as well. This motivates the following definition.

DEFINITION A.21. An operator A is called *m-accretive* in X if and only if A is accretive and $\mathrm{R}(I + \lambda A) = X$ for all $\lambda > 0$.

Applying the Banach fix point theorem it is not hard to see that if A is accretive, then A is m-accretive if there exists $\lambda > 0$ such that $\mathrm{R}(I + \lambda A) = X$.

It is easy to see that each m-accretive operator A in X is *maximal accretive* in the sense that every accretive extension of A coincides with A. In general, the converse is not true, but it is true in Hilbert spaces due to the following classical result of G. Minty [132]:

THEOREM A.22 (Minty's Theorem). *Let H be a Hilbert space and A an accretive operator in H. Then A is m-accretive if and only if A is maximal monotone.*

One of the most important examples of maximal monotone operators in Hilbert spaces comes from optimization theory. This is the example of subdifferentials of convex functions, which we introduce next.

Let $(H, (\mid))$ be a Hilbert space and $\varphi : H \to (-\infty, +\infty]$. We denote

$$D(\varphi) = \{x \in H \; : \; \varphi(x) \neq +\infty\} \quad \text{(effective domain)}.$$

We say that φ is *proper* if $D(\varphi) \neq \emptyset$ and that φ is *convex* if

$$\varphi(\alpha x + (1 - \alpha)y) \leq \alpha\varphi(x) + (1 - \alpha)\varphi(y)$$

for all $\alpha \in [0, 1]$ and $x, y \in H$.

Some of the properties of φ are reflected in its epigraph defined by

$$\mathrm{Epi}(\varphi) := \{(x, r) \in H \times \mathbb{R} \; : \; r \geq \varphi(x)\}.$$

For instance, φ is convex if and only if $\mathrm{Epi}(\varphi)$ is a convex subset of $H \times \mathbb{R}$; and φ is lower semicontinuous if and only if $\mathrm{Epi}(\varphi)$ is closed.

The *subdifferential* $\partial\varphi$ of φ is the operator defined by

$$w \in \partial\varphi(z) \iff \varphi(x) \geq \varphi(z) + (w|x - z) \quad \forall x \in H.$$

Observe that $0 \in \partial\varphi(z)$ if and only if $\varphi(x) \geq \varphi(z)$ for all $x \in H$ if and only if $\varphi(z) = \min_{x \in D(\varphi)} \varphi(x)$. Therefore, we have that $0 \in \partial\varphi(z)$ is the *Euler equation* of the variational problem

$$\varphi(z) = \min_{x \in D(\varphi)} \varphi(x).$$

If $(z, w), (\hat{z}, \hat{w}) \in \partial\varphi$, then $\varphi(z) \geq \varphi(\hat{z}) + (\hat{w}|z - \hat{z})$ and $\varphi(\hat{z}) \geq \varphi(z) + (\hat{w}|\hat{z} - z)$. Adding this inequalities we get

$$(w - \hat{w}|z - \hat{z}) \geq 0.$$

Thus, $\partial\varphi$ is a monotone operator. Now, if φ is convex, lower semicontinuous and proper, it can be proved that $\partial\varphi$ is maximal monotone and $\overline{D(\partial\varphi)} = \overline{D(\varphi)}$ (see [56], [27]).

Given a closed convex subset K of H, the *indicator function* of K is defined by

$$I_K(u) = \begin{cases} 0 & \text{if } u \in K, \\ +\infty & \text{if } u \notin K. \end{cases}$$

Then it is easy to see that the subdifferential is characterized as follows:

$$v \in \partial I_K(u) \iff u \in K \text{ and } (v, w - u) \leq 0 \quad \forall w \in K.$$

As we mentioned in the linear case, the existence and uniqueness of mild solutions is equivalent to the fact that $-A$ is the infinitesimal generator of a C_0-semigroup. Now, there are classical results connecting this fact with the m-accretivity of the operator A, for example:

THEOREM A.23 (Lumer-Phillips Theorem). *$-A$ is the infinitesimal generator of a C_0-semigroup $(S(t))_{t \geq 0}$ of linear contractions on X if and only if A is linear, m-accretive and $\overline{D(A)} = X$. Moreover, in this case*

$$S(t)x = \lim_{n \to \infty} \left(I + \frac{t}{n} A \right)^{-n} x.$$

A first extension to the nonlinear case of this type of results has been given by Y. Komura in [**128**].

THEOREM A.24 (Komura Theorem). (i) *Let A be a maximal monotone operator in the Hilbert space H. Then $\overline{D(A)}$ is a closed convex subset of H and $D(S^A) = \overline{D(A)}$.*

(ii) *Given some closed convex set $C \subset H$ and a strongly continuous semigroup of contractions $(S(t))_{t \geq 0}$ on C, there exists a unique maximal monotone operator A in H such that $\overline{D(A)} = C$ and $S^A(t) = S(t)$ for all $t \geq 0$.*

This result has been extended to some Banach spaces with good geometrical properties, but it turns out to be false for general Banach spaces. The good extension to nonlinear operators for general Banach spaces was done by Crandall-Liggett [**89**] and Ph. Bénilan [**43**] in the early 1970s. In the next section we give the outline of this theory.

A.5. Existence and uniqueness theorem

Suppose A is an operator in X and $f \in L^1(0, T; X)$. Consider the abstract Cauchy problem

$$(\text{CP})_{x_0, f} \begin{cases} u'(t) + Au(t) \ni f(t) & \text{on } t \in (0, T), \\ u(0) = x. \end{cases}$$

DEFINITION A.25. An ε-*approximate solution* of $(\text{CP})_{x_0, f}$ is a solution v of an ε-discretization $D_A(0 = t_0, \ldots, t_N, f_1, \ldots, f_N)$ of $u' + Au \ni f$ on $[0, T]$ with $\|v(0) - x_0\| < \varepsilon$.

It follows from this definition that u is a mild solution of $(\text{CP})_{x_0, f}$ on $[0, T]$ if and only if $u \in C([0, T]; X)$ and for each $\varepsilon > 0$ there is an ε-approximate solution v of $(\text{CP})_{x_0, f}$ such that $\|u(t) - v(t)\| < \varepsilon$ on the domain of v.

DEFINITION A.26. Suppose that for each $\varepsilon > 0$ there are ε-approximate solutions of $(\mathrm{CP})_{x_0,f}$ on $[0,T]$. We say that the ε-*approximate solutions converge* on $[0,T]$ as $\varepsilon \downarrow 0$ to $u \in C([0,T];X)$ if there exists a function $\psi : [0,+\infty[\to [0,+\infty[$ with $\lim_{\varepsilon\downarrow 0} \psi(\varepsilon) = 0$ such that $\|u(t) - v(t)\| \leq \psi(\varepsilon)$ whenever $\varepsilon > 0$, v is an ε-approximate solution of $(\mathrm{CP})_{x_0,f}$ on $[0,T]$ and t is in the domain of v.

THEOREM A.27. *Suppose that* $A \in \mathcal{A}(w)$, $f \in L^1(0,T;X)$ *and* $x_0 \in \overline{D(A)}$. *If the problem* $(\mathrm{CP})_{x_0,f}$ *has an* ε-*approximate solution on* $[0,T]$ *for every* $\varepsilon > 0$, *then it has a unique mild solution on* $[0,T]$ *to which the* ε-*approximate solutions of* $(\mathrm{CP})_{x_0,f}$ *converge as* $\varepsilon \downarrow 0$.

This theorem was proved by Ph. Bénilan in his Thesis [**43**] as an extension of Crandall-Liggett's theorem (which corresponds to $f = 0$). We also have the following result.

THEOREM A.28. *Let* A *be an accretive operator in* X *and let* u *be a mild solution of* $u' + Au \ni 0$ *on* $[0,T]$. *Then:*

(i) *If* v *is an* ε-*approximate solution of* $u' + Au \ni 0$ *on* $[0,T]$ *with* $[0,s]$ *in its domain,* $0 \leq t \leq T$, *and* $(x,y) \in A$, *then*

$$\|u(t) - v(s)\| \leq 2\|u(0) - x\| + \|y\| |t - s|, \qquad 0 \leq s,t \leq T.$$

(ii) *If* $\hat{u}$ *is a mild solution of* $\hat{u}' + A\hat{u} \ni 0$ *on* $[0,T]$, *then*

$$\|u(t) - \hat{u}(t)\| \leq \|u(0) - \hat{u}(0)\|, \qquad 0 \leq t \leq T.$$

Theorem A.27 tells us that, for accretive operators, to have existence and uniqueness of mild solutions is enough for the existence of ε-approximate solutions for each $\varepsilon > 0$. Now we have seen that this is the case for m-accretive operators; consequently, we have the following result.

THEOREM A.29. *Let* A *be an operator in* X, $f \in L^1(0,T;X)$ *and* $x_0 \in \overline{D(A)}$. *If* $A + wI$ *is* m-*accretive, then the problem*

$$u' + Au \ni f \quad \text{on} \quad [0,T], \quad u(0) = x_0$$

has a unique mild solution u *on* $[0,T]$.

Recall that

$$D\left(S^A\right) := \left\{x \in X \; : \; \text{there exists a unique mild solution } u_x \text{ of}\right.$$
$$\left. u' + Au \ni 0 \text{ on } (0,+\infty) \text{ with } u_x(0) = x\right\}.$$

and for $t \geq 0$ and $x \in D(S^A)$, $S^A(t)x := u_x(t)$. From now on, we denote $S^A(t)$ by e^{-tA}, and we call $(e^{-tA})_{t\geq 0}$ the *semigroup generated* by $-A$.

As a consequence of Theorem A.28, if A is accretive, then $(e^{-tA})_{t\geq 0}$ is a contraction semigroup, i.e.,

$$\|e^{-tA}x - e^{-tA}\hat{x}\| \leq \|x - \hat{x}\| \qquad \forall x, \hat{x} \in D(S^A), \; \forall t \geq 0.$$

Moreover, by the properties of mild solutions, it is easy to see that $D(S^A)$ is closed and, by Theorem A.28, the map $(t,x) \mapsto e^{-tA}x$ is continuous in $[0,+\infty) \times D(S^A)$.

As a consequence of Theorem A.29 we have that if A is m-accretive in X, then $D(S^A) = \overline{D(A)}$ and $(e^{-tA})_{t\geq 0}$ is a contraction semigroup in $\overline{D(A)}$.

We now see that in the homogeneous case we can debilitate the m-accretivity of the operator and get an explicit representation of the mild solution. Suppose for the moment that A is m-accretive. Let $\lambda > 0$ and let v be a solution of the discretization $D_A(0, \lambda, 2\lambda, \ldots, N\lambda; 0, \ldots, 0)$ satisfying $v(0) = x_0$. Due to the fact that the discretization has a constant step size λ, the difference equation for v is equivalent to

(A.13)
$$\begin{cases} v(t) = x_0 \quad \text{for} \ -\lambda < t \le 0, \\[2mm] \dfrac{v(t) - v(t - \lambda)}{\lambda} + Av(t) \ni 0 \quad \text{for } 0 < t \le N\lambda. \end{cases}$$

Moreover, $v(k\lambda) = J_\lambda^A v((k-1)\lambda)$ or, iterating,

$$v(k\lambda) = (J_\lambda^A)^k v(0) = (J_\lambda^A)^k x_0.$$

Then in order to solve (A.13) we only need that $\overline{D(A)} \subset D(J_\lambda^A)$ for $\lambda > 0$ and of course the accretivity of the operator A.

DEFINITION A.30. An accretive operator A satisfies the *range condition* if $\overline{D(A)} \subset \mathrm{R}(I + \lambda A)$ for all $\lambda > 0$.

THEOREM A.31 (Crandall-Liggett Theorem). *If A is accretive and satisfies the range condition, then $-A$ generates a semigroup of contractions $(e^{-tA})_{t \ge 0}$ on $\overline{D(A)}$ and:*

(i) *For $x_0 \in \overline{D(A)}$ and $0 \le t < \infty$,*

$$\lim_{\lambda \downarrow 0, k\lambda \to t} (J_\lambda^A)^k x_0 = e^{-tA} x_0$$

holds uniformly for t on compact subintervals of $[0, \infty)$.

(ii) *If $x_0 \in \overline{D(A)}$, $t > 0$ and $n \in \mathbb{N}$, then*

$$\left\| (J_{t/n}^A)^n x_0 - e^{-tA} x_0 \right\| \le \frac{t}{\sqrt{n}} \|y\| + 2\|x_0 - x\|$$

for every $(x, y) \in A$.

From either (i) or (ii) of the last theorem we deduce

(A.14) $$e^{-tA} x = \lim_{n \to \infty} \left(I + \frac{t}{n} A \right)^{-n} x \qquad \text{for } x \in \overline{D(A)}.$$

This representation of the semigroup $(e^{-tA})_{t \ge 0}$ is called the *exponential formula* by analogy with the formula $\lim_{n \to \infty} (1 + \frac{t}{n} a)^{-n} = e^{-ta}$ for $a \in \mathbb{C}$.

Observe the analogy of (A.14) with the exponential formula given by the Lumer-Phillips theorem for the linear case. Now, there are strong differences between the linear and nonlinear cases. For instance, in the linear case, $-A$ is the infinitesimal generator of the C_0-semigroup $(e^{-tA})_{t \ge 0}$, and in the nonlinear case there are examples of operators A satisfying the assumptions of the Crandall-Liggett theorem, such that the domain of the infinitesimal generator of the semigroup $(e^{-tA})_{t \ge 0}$ is empty (see [**89**]):

We now give an example of how to apply the Crandall-Liggett theorem.

EXAMPLE A.32. Consider the nonlinear partial differential equation

(A.15)
$$\begin{cases} u_t(x,t) = \Delta\varphi(u(x,t)) & \text{in } \Omega \times (0,\infty), \\ \varphi(u(x,t)) = 0, & \text{on } \partial\Omega \times (0,\infty), \\ u(x,0) = u_0(x) & \text{in } \Omega, \end{cases}$$

where $\varphi : \mathbb{R} \to \mathbb{R}$ is a nondecreasing function and Ω is a smooth domain in $\mathbb{R}^N$. This equation is called the *Filtration Equation* and different choices of φ correspond to equations that appear in applications. For instance, if $\varphi(r) = |r|^m \mathrm{sgn}_0(r)$, we have, for $m > 1$, the *Porous Medium Equation*, an equation that appears in the study of a gas flow through a porous medium; moreover, this equation also appears in models for population dynamic ([111]). The case $0 < m < 1$ occurs in the theory of plasma, and in this case the equation is called the *Fast Diffusion Equation*.

To simplify the discussion we will assume that $\varphi \in C(\mathbb{R}) \cap C^1(\mathbb{R} \setminus \{0\})$, $\varphi(0) = 0$ and $\varphi'(s) > 0$ for $s \neq 0$.

Associated to the problem (A.15) is the operator A in $L^1(\Omega)$ defined by

$$D(A) := \left\{ u \in L^1(\Omega) \ : \ \varphi(u) \in W_0^{1,1}(\Omega), \ \Delta\varphi(u) \in L^1(\Omega) \right\},$$

$$Au := -\Delta\varphi(u) \quad \text{for} \ \ u \in D(A).$$

We rewrite problem (A.15) as the abstract Cauchy problem,

(A.16)
$$\begin{cases} u'(t) + Au(t) = 0, & t \in (0,+\infty), \\ u(0) = u_0. \end{cases}$$

Since $\left\{ u \in L^1(\Omega) \ : \ \varphi(u) \in D(\Delta) \right\} \subset D(A)$, where

$$D(\Delta) = \left\{ v \in W_0^{1,1}(\Omega) \ : \ \Delta v \in L^1(\Omega) \right\},$$

we have $\overline{D(A)} = L^1(\Omega)$. Therefore, if we prove that A is m-accretive in $L^1(\Omega)$, for each $u_0 \in L^1(\Omega)$, $e^{-tA}u_0$ solves problem (A.15) in the mild sense, i.e., $e^{-tA}u_0$ is the unique mild solution of (A.16). Let us show that A is m-accretive in $L^1(\Omega)$. To see the accretivity of A we need to show that

(A.17) $\quad 0 \leq [u - \hat{u}, Au - A\hat{u}] = \displaystyle\int_\Omega (Au - A\hat{u})\mathrm{sgn}_0(u - \hat{u}) + \int_{\{u=\hat{u}\}} |Au - A\hat{u}|.$

To achieve this goal, choose $p_n \in C^1(\mathbb{R})$ with the properties: $p_n(0) = 0$, $|p_n(s)| \leq 1$, $p_n'(s) \geq 0$, $\lim_{n\to\infty} p_n(s) = \mathrm{sgn}_0(s)$ for all $s \in \mathbb{R}$. For example, consider

$$p_n(s) = \frac{ns}{n|s| + 1}, \quad s \in \mathbb{R}.$$

Applying Green's formula we have

$$\int_\Omega (Au - A\hat{u})p_n\left(\varphi(u) - \varphi(\hat{u})\right) = -\int_\Omega \Delta\left(\varphi(u) - \varphi(\hat{u})\right)p_n\left(\varphi(u) - \varphi(\hat{u})\right)$$

$$= \int_\Omega \nabla\left(\varphi(u) - \varphi(\hat{u})\right) \cdot \nabla\left(p_n\left(\varphi(u) - \varphi(\hat{u})\right)\right)$$

$$= \int_\Omega p_n'\left(\varphi(u) - \varphi(\hat{u})\right)\left|\nabla\left(\varphi(u) - \varphi(\hat{u})\right)\right|^2 \geq 0.$$

Then, letting $n \to +\infty$, we obtain

$$\int_\Omega (Au - A\hat{u})\mathrm{sgn}_0\left(\varphi(u) - \varphi(\hat{u})\right) \geq 0.$$

Now, since φ is increasing,

$$\mathrm{sgn}_0\left(\varphi(u) - \varphi(\hat{u})\right) = \mathrm{sgn}_0(u - \hat{u}).$$

Hence, we get

$$\int_\Omega (Au - A\hat{u})\mathrm{sgn}_0(u - \hat{u}) \geq 0$$

and consequently, (A.17) holds.

It remains to prove that for each $f \in L^1(\Omega)$ there exists a (unique) $u \in D(A)$, such that

(A.18) $$u - \Delta\varphi(u) = f.$$

The proof of (A.18) is more complicated than the proof of the accretivity and is a consequence of a result due to H. Brezis and W. Strauss [**60**].

A.6. Regularity of the mild solution

As we have already pointed out, mild solutions may not satisfy any additional regularity properties; in general, they cannot be interpreted as a solution of the Cauchy problem in a pointwise sense, and they are not strong solutions.

Nevertheless, a question arises naturally whether under certain additional assumptions one may obtain more regularity of mild solutions. This will be done now. We emphasize, before this, that even in applications one does not want to be limited to strong solutions, since there are important partial differential equations which simply do not have strong solutions.

A basic fact is the following consistence between the accretivity of A and the differentiability of mild solutions of $u' + Au \ni f$.

THEOREM A.33. *Let A be an accretive operator in X, $f \in L^1(0, T; X)$ and let u be a mild solution of $u' + Au \ni f$ on $[0, T]$. If u has a right derivative $\frac{d^+u}{dt}(\tau)$ at $\tau \in]0, T[$ and*

$$\lim_{h\downarrow 0} \frac{1}{h}\int_\tau^{\tau+h} \|f(t) - f(\tau)\|\, dt = 0,$$

that is, τ is a right Lebesgue point of f, then the operator $\hat{A}$ given by

$$\hat{A}x = Ax \quad \text{for } x \neq u(\tau),$$

$$\hat{A}u(\tau) = Au(\tau) \cup \left\{ f(\tau) - \frac{d^+u}{dt}(\tau) \right\},$$

is accretive.

Since every m-accretive operator is maximal accretive, as a consequence of the above theorem we have the following result.

COROLLARY A.34. *Suppose A is an m-accretive operator in X, $f \in L^1(0, T; X)$ and u is a mild solution of $u' + Au \ni f$ on $[0, T]$. Then*

(i) *If u is differentiable at $t \in (0, T)$ and t is a right Lebesgue point of f, then*

$$u'(t) + Au(t) \ni f(t).$$

(ii) *If $u \in W^{1,1}(0, T; X)$, then u is a strong solution of $u' + Au \ni f$ on $[0, T]$.*

Then the problem is: *When does a mild solution belong to $W^{1,1}(0, T; X)$?*

We denote by $BV(0, T; X)$ the subspace of functions in $L^1(0, T; X)$ which are of *bounded variation*, i.e., $f \in BV(0, T; X)$ if $f \in L^1(0, T; X)$ and

$$\text{Var}(f, T) := \limsup_{h \downarrow 0} \int_0^{T-h} \frac{\|f(\tau + h) - f(\tau)\|}{h} \, d\tau < +\infty.$$

The principal conditions guaranteeing that a mild solution is in $W^{1,1}(0, T; X)$ are given by the following result.

PROPOSITION A.35. *Let A be an accretive operator in X, $f \in BV(0, T; X)$ and $x \in D(A)$. If u is a mild solution of $(CP)_{x,f}$ on $[0, T]$, then u is locally Lipschitz continuous on $[0, T]$. Moreover, if X has the Radon-Nikodym property, then $u \in W^{1,1}(0, T; X)$ and consequently u is a strong solution of $(CP)_{x,f}$ on $[0, T]$.*

In the case that the operator is the subdifferential of a convex lower semicontinuous function in a Hilbert space, we have good regularity. More precisely, we have the following result.

THEOREM A.36. *Let H be a Hilbert space and $\varphi : H \to (-\infty, +\infty]$ a proper, convex and lower semicontinuous function such that $\text{Min } \varphi = 0$, and let $K := \{v \in H : \varphi(v) = 0\}$. Suppose $f \in L^2(0, T; H)$ and $u_0 \in \overline{D(\partial\varphi)}$; then the mild solution $u(t)$ of*

$$\begin{cases} u' + \partial\varphi(u) \ni f & \text{on } [0, T], \\ u(0) = u_0, \end{cases}$$

is a strong solution and we have the following estimates:

$$\|u'(t)\|_{L^2(\delta, T; H)} \leq \|f\|_{L^2(0, T; H)} + \frac{1}{\sqrt{2\delta}} \int_0^\delta \|f(t)\| \, dt + \frac{1}{\sqrt{2\delta}} \text{dist}\,(u_0, K)$$

for $0 < \delta < T$, and

$$\left(\int_0^T \|u'(t)\|^2 t \, dt \right)^{\frac{1}{2}} \leq \left(\int_0^T \|f(t)\|^2 t \, dt \right)^{\frac{1}{2}} + \frac{1}{\sqrt{2}} \int_0^T \|f(t)\|^2 \, dt + \frac{1}{\sqrt{2}} \text{dist}\,(u_0, K).$$

Moreover, for almost all $t \in [0, T]$, we have

$$\frac{d}{dt}\varphi(u(t)) = (h|u'(t)) \quad \forall h \in \partial\varphi(u(t)).$$

In the homogeneous case, i.e., $f = 0$, we have

$$\|u'(t)\|_{L^\infty(\delta,T;H)} \leq \frac{1}{\delta}\|u_0\| \quad \text{for } 0 < \delta < T.$$

A.7. Convergence of operators

One important property of the mild solutions is the dependence on the operator that was discovered by H. Brezis and A. Pazy [59] and that says that in order to obtain convergence of mild solutions of a sequence of abstract Cauchy problems, it is enough to know the convergence of the resolvent. In other words, if for the corresponding elliptic problems we have convergence, then we have convergence of the solutions of the parabolic problems. More precisely, the result states the following.

THEOREM A.37 (Brezis-Pazy Theorem). *Let A_n be m-accretive in X, $x_n \in \overline{D(A_n)}$ and $f_n \in L^1(0,T;X)$ for $n = 1, 2, \ldots, \infty$. Let u_n be the mild solution of*

$$u'_n + A_n u_n \ni f_n \quad in \; [0,T], \quad u_n(0) = x_n.$$

If $f_n \to f_\infty$ in $L^1(0,T;X)$ and $x_n \to x_\infty$ as $n \to \infty$ and

$$\lim_{n\to\infty} (I + \lambda A_n)^{-1} z = (I + \lambda A_\infty)^{-1} z,$$

for some $\lambda > 0$ and all $z \in D$, with D dense in X, then

$$\lim_{n\to\infty} u_n(t) = u_\infty(t) \quad uniformly \; on \; [0,T].$$

In case that the operators are the subdifferentials of convex lower semicontinuous functionals in Hilbert spaces, to prove the convergence of the resolvent it is enough to show the convergence of the functionals in the following sense introduced by U. Mosco in [134] (see [25]). Suppose X is a metric space and $A_n \subset X$. We define

$$\liminf_{n\to\infty} A_n = \{x \in X \; : \; \exists x_n \in A_n, \; x_n \to x\}$$

and

$$\limsup_{n\to\infty} A_n = \{x \in X \; : \; \exists x_{n_k} \in A_{n_k}, \; x_{n_k} \to x\}.$$

In case X is a normed space, we denote by s-lim inf and w-lim sup the above limits associated respectively to the strong and to the weak topology of X.

Given a sequence $\Psi_n, \Psi : H \to (-\infty, +\infty]$ of convex lower semicontinuous functionals, we say that Ψ_n converges to Ψ in the sense of Mosco if

(A.19) $\text{w-}\limsup_{n\to\infty} \text{Epi}(\Psi_n) \subset \text{Epi}(\Psi) \subset \text{s-}\liminf_{n\to\infty} \text{Epi}(\Psi_n).$

It is easy to see that (A.19) is equivalent to the following two conditions:

(1) $\forall u \in D(\Psi) \; \exists u_n \in D(\Psi_n) \; : \; u_n \to u$ and $\Psi(u) \geq \limsup_{n\to\infty} \Psi_n(u_n)$;

(2) for every subsequence n_k, as $u_k \rightharpoonup u$, we have $\Psi(u) \leq \liminf_k \Psi_{n_k}(u_k)$.

As a consequence of Theorem A.37 and using the results in [25] we can state the following result.

THEOREM A.38. *Let* $\Psi_n, \Psi : H \to (-\infty, +\infty]$ *be convex lower semicontinuous functionals. Then the following statements are equivalent:*

(i) Ψ_n *converges to* Ψ *in the sense of Mosco.*

(ii) $(I + \lambda \partial \Psi_n)^{-1} u \to (I + \lambda \partial \Psi)^{-1} u, \quad \forall \lambda > 0, \ u \in H.$

Moreover, any of these two conditions (i) *or* (ii) *imply that*

(iii) *for every* $u_0 \in \overline{D(\partial \Psi)}$ *and* $u_{0,n} \in \overline{D(\partial \Psi_n)}$ *such that* $u_{0,n} \to u_0$, *and every* $f_n, f \in L^2(0, T; H)$ *with* $f_n \to f$, *if* $u_n(t)$, $u(t)$ *are the strong solutions of the abstract Cauchy problems*

$$\begin{cases} u_n'(t) + \partial \Psi_n(u_n(t)) \ni f_n, & a.e. \ t \in (0, T), \\ u_n(0) = u_{0,n}, \end{cases}$$

and

$$\begin{cases} u'(t) + \partial \Psi(u(t)) \ni f, & a.e. \ t \in (0, T), \\ u(0) = u_0, \end{cases}$$

respectively, then

$$u_n \to u \quad in \ C([0, T]; H).$$

A.8. Completely accretive operators

Many nonlinear semigroups that appear in the applications are also order-preserving and contractions in every L^p. Ph. Bénilan and M. G Crandall [45] introduced a class of operators, named completely accretive, for which the semigroup generated by the Crandall-Liggett exponential formula enjoys these properties. In this section we outline some of the main points given in [45].

Let $(\Omega, \mathcal{B}, \mu)$ be a σ-finite measure space and let $M(\Omega)$ denote the space of measurable functions from Ω into $\mathbb{R}$. We denote by $L(\Omega)$ the space

$$L(\Omega) := L^1(\Omega) + L^\infty(\Omega)$$

$$= \left\{ u \in M(\Omega) \ : \ \int_\Omega (|u| - k)^+ < \infty \ \text{for some} \ k > 0 \right\};$$

$L(\Omega)$ is exactly the subset of $M(\Omega)$ on which the functional

$$\|u\|_{1+\infty} := \inf \{ \|f\|_1 + \|g\|_\infty \ : \ f, g \in M(\Omega), \ f + g = u \}$$

is finite, and $L(\Omega)$ equipped with $\| \cdot \|_{1+\infty}$ is a Banach space.

Let

$$L_0(\Omega) := \{ u \in L(\Omega) \ : \ \mu(\{|u| > k\}) < \infty \ \text{for any} \ k > 0 \}$$

$$= \left\{ u \in M(\Omega) \ : \ \int_\Omega (|u| - k)^+ < \infty \ \text{for any} \ k > 0 \right\}.$$

$L_0(\Omega)$ is a closed subspace of $L(\Omega)$; in fact, it is the closure in $L(\Omega)$ of the linear span of the set of characteristic functions of sets of finite measure. Hereafter, $L_0(\Omega)$ carries the norm $\| \|_{1+\infty}$; it is then a Banach space. With the natural pairing $\langle u, v \rangle = \int_\Omega uv$, the dual space of $L_0(\Omega)$ is isometrically isomorphic to

$$L^{1 \cap \infty}(\Omega) := L^1(\Omega) \cap L^\infty(\Omega),$$

when the norm in $L^{1 \cap \infty}(\Omega)$ is given by

$$\|u\|_{1 \cap \infty} := \max\{\|u\|_1, \|u\|_\infty\}.$$

For $u, v \in M(\Omega)$, we write

(A.20) $\qquad u \ll v \quad$ if and only if $\displaystyle\int_\Omega j(u)dx \leq \int_\Omega j(v)dx$

for all $j \in J_0$, where

(A.21) $\qquad J_0 = \{j : \mathbb{R} \to [0, \infty], \text{ convex, l.s.c., } j(0) = 0\}$

(l.s.c. is an abbreviation for lower semicontinuous).

DEFINITION A.39. A functional $N : M(\Omega) \to (-\infty, +\infty]$ is called *normal* if $N(u) \leq N(v)$ whenever $u \ll v$.

A map $S : D(S) \subset M(\Omega) \to M(\Omega)$ is a *complete contraction* if it is an N-contraction for every normal functional N, i.e., if

$$N(Su - Sv) \leq N(u - v) \quad \text{for } u, v \in D(S).$$

A Banach space $(X, \|\ \|_X)$, with $X \subset M(\Omega)$, is a *normal Banach space* if it has the following property:

(A.22) $\qquad u \in X, v \in M(\Omega), v \ll u \Rightarrow v \in X \text{ and } \|v\|_X \leq \|u\|_X.$

Simple examples of normal Banach spaces are: $L^p(\Omega)$, $1 \leq p \leq \infty$, and $L(\Omega)$, $L_0(\Omega)$, $L^{1 \cap \infty}(\Omega)$.

PROPOSITION A.40. *Let $S : D(S) \subset M(\Omega) \to M(\Omega)$ and assume that*

(A.23) $\qquad u, v \in D(S)$ and $k \geq 0 \Rightarrow u \wedge (v + k)$ or $v \vee (u - k) \in D(S).$

Then S is a complete contraction if and only if it is order-preserving and a contraction for $\|\ \|_1$ and $\|\ \|_\infty$.

DEFINITION A.41. Let A be an operator in $M(\Omega)$. We say that A is *completely accretive* if

(A.24) $\qquad u - \hat{u} \ll u - \hat{u} + \lambda(v - \hat{v}) \quad$ for all $\lambda > 0$ and all $(u, v), (\hat{u}, \hat{v}) \in A.$

In other words, A is completely accretive if

(A.25) $\qquad N(u - \hat{u}) \leq N(u - \hat{u} + \lambda(v - \hat{v}))$

for all $\lambda > 0$, all $(u, v), (\hat{u}, \hat{v}) \in A$ and every normal functional N in $M(\Omega)$.

The definition of completely accretive operators does not refer explicitly to topologies or norms. However, if A is completely accretive in $M(\Omega)$ and $A \subset X \times X$, where X is a subspace of $M(\Omega)$ whose norm is given by a normal function, then A is accretive in X. Choices for X might be $L^p(\Omega)$, $1 \leq p \leq \infty$.

Let

$$P_0 = \{q \in C^\infty(\mathbb{R}) : 0 \leq q' \leq 1, \text{ supp}(q') \text{ is compact and } 0 \notin \text{supp}(q)\}.$$

The following result, which is a generalization of a result due to H. Brezis and W. Strauss [60], provides a very useful characterization of the complete accretivity.

PROPOSITION A.42. *Let* $u \in L_0(\Omega)$, $v \in L(\Omega)$. *Then*

$$u \ll u + \lambda v, \quad \forall \lambda > 0 \iff \int_\Omega q(u)v \geq 0, \quad \forall q \in P_0.$$

Observe that $L^p(\Omega) \subset L_0(\Omega)$ for any $1 \leq p < \infty$. If $\mu(\Omega) < \infty$, then $L_0(\Omega) = L(\Omega) = L^1(\Omega)$. Consequently, from the above proposition we obtain the following characterization.

COROLLARY A.43. *If* $A \subseteq L^p(\Omega) \times L^p(\Omega)$, $1 \leq p < \infty$, *then* A *is completely accretive if and only if*

$$\int_\Omega q(u - \hat{u})(v - \hat{v}) \geq 0 \quad \text{for any } q \in P_0, (u, v), (\hat{u}, \hat{v}) \in A.$$

PROPOSITION A.44. *Let* $u \in L_0(\Omega)$. *Then*

(i) $\{v \in M(\Omega) : v \ll u\}$ *is a weakly sequentially compact subset of* $L_0(\Omega)$.
(ii) *Let* $(X, \|\cdot\|_X)$ *be a normal Banach space satisfying* $X \subset L_0(\Omega)$ *and having the property*

(A.26) $u_n \ll u \in X$, $n = 1, 2, \ldots$, *and* $u_n \to u$ *a.e.* $\Rightarrow \|u_n - u\|_X \to 0$.

  If $\{u_n\}$ *is a sequence satisfying* $u_n \ll u \in X$ *for* $n = 1, 2, \ldots$, *and* $u_n \rightharpoonup u$ *weakly in* $L_0(\Omega)$, *then* $\|u_n - u\|_X \to 0$.

REMARK A.45. The assumption (A.26) is satisfied for $X = L^p(\Omega)$, $1 \leq p < +\infty$.

DEFINITION A.46. Let X be a linear subspace of $M(\Omega)$. An operator A in X is *m-completely accretive* in X if A is completely accretive and $\mathrm{R}(I + \lambda A) = X$ for $\lambda > 0$

REMARK A.47. The above definition does not require X to be a Banach space and so does not require A to be *m*-accretive in any Banach space. However, if A is completely accretive, then it is accretive in $L(\Omega)$, and if A is *m*-completely accretive in a subspace X of $L(\Omega)$, then the closure $\overline{A}$ of A in $L(\Omega)$ is completely accretive and *m*-accretive in the closure $\overline{X}$ of X in $L(\Omega)$. We also note that if A is completely accretive in a subspace X of $M(\Omega)$ and $\mathrm{R}(I + \lambda A) = X$ for some $\lambda > 0$, the only completely accretive operator B in X which extends A is A.

PROPOSITION A.48. *Let* X *be a normal Banach space,* $X \subset L_0(\Omega)$, *and let* A *be a completely accretive operator in* X. *Suppose there exists* $\lambda > 0$ *for which* $\mathrm{R}(I + \lambda A)$ *is dense in* $L_0(\Omega)$. *Then the operator* $A^X := \overline{A} \cap (X \times X)$ *is the unique m-completely accretive extension of* A *in* X.

DEFINITION A.49. Let A be an operator in $L_0(\Omega)$. Then A° is the restriction of A defined by

$$v \in A^\circ u \iff v \in Au \text{ and } v \ll w \quad \forall w \in Au.$$

In case X is a normal Banach space and A is *m*-completely accretive in X, by Crandall-Liggett's Theorem, A generates a contraction semigroup in X given by the exponential formula

$$e^{-tA}u_0 = \|\cdot\|_X\text{-}\lim_{n \to \infty} \left(I + \frac{t}{n}A\right)^{-n} u_0 \quad \text{for any } u_0 \in \overline{D(A)}^X.$$

Now, since $\overline{A}$ is m-completely accretive in $\overline{X}$ endowed with the norm of $L(\Omega)$, we may also consider the semigroup $e^{-t\overline{A}}$ on $\overline{D(A)}$. We have the following relation between these two semigroups.

PROPOSITION A.50. *Let X be a normal Banach space and A an m-completely accretive operator in X. Then we have*

(i) *e^{-tA} is a complete contraction for $t \geq 0$.*

(ii) *e^{-tA} is the restriction of $e^{-t\overline{A}}$ to $\overline{D(A)}^X$ and $e^{-t\overline{A}}$ is the closure of e^{-tA} in $L(\Omega)$.*

(iii) *$e^{-t\overline{A}}\big(\overline{D(A)} \cap X\big) \subset \overline{D(A)} \cap X$.*

As a consequence of (iii) of the above proposition, if we denote by $S^A(t)$ the restriction of $e^{-t\overline{A}}$ to $\overline{D(A)} \cap X$, we see that $S^A(t)$ is given by the exponential formula

$$S^A(t)u = \| \cdot \|_{1+\infty}\text{-}\lim_{n\to\infty} \left(I + \frac{t}{n}A\right)^{-n} u \quad \text{for} \ \ u \in \overline{D(A)} \cap X.$$

THEOREM A.51. *Let X be a normal Banach space with $X \subset L_0(\Omega)$ and A an m-completely accretive operator in X. Then we have*

(i)

$$D(A) = \left\{ u \in \overline{D(A)} \cap X \ : \exists v \in X \ \text{s.t.} \ \frac{S^A(t)u - u}{t} \ll v \ \text{for small} \ t > 0 \right\}.$$

(ii) *$S^A(t)D(A) \subset D(A)$ for $t > 0$.*

(iii) *If $u \in D(A)$, then*

$$\frac{u - S^A(t)u}{t} \ll v \ \text{for} \ t > 0 \ \text{and} \ v \in Au$$

and

$$\| \cdot \|_{1+\infty}\text{-}\lim_{t\to 0} \frac{S^A(t)u - u}{t} = -A^\circ u.$$

COROLLARY A.52. *Suppose $\mu(\Omega) < \infty$. If $A \subseteq L^1(\Omega) \times L^1(\Omega)$ is an m-completely accretive operator in $L^1(\Omega)$, then for every $u_0 \in D(A)$, the mild solution $u(t) = e^{-tA}u_0$ of the problem*

(A.27)
$$\begin{cases} \dfrac{du}{dt} + Au \ni 0, \\[2mm] u(0) = u_0, \end{cases}$$

is a strong solution.

The following result is a variant of the regularizing effect of the homogeneous evolution equation obtained in [45] in the m-completely accretive case.

THEOREM A.53. *In addition to the hypothesis of Theorem A.51, assume that A is positively homogeneous of degree $0 < m \neq 1$, i.e., $A(\lambda u) = \lambda^m Au$ for $u \in D(A)$. Then for $u \in \overline{D(A)} \cap X$ and $t > 0$, we have $S^A(t)u \in D(A)$ and*

$$|A^\circ S^A(t)u| \leq 2\frac{|u|}{|m-1|t}.$$

Using similar heuristics to that in the above result, the scaling argument applied in [100] by Evans, Feldman and Gariepy to study the collapsing of the initial condition phenomena for sandpiles was extended by Bénilan, Evans and Gariepy [47] to cover general nonlinear evolution equations governed by homogeneous accretive operators. Let us recall this result.

Suppose that for each $n \in \mathbb{N}$, A_n is an m-accretive operator defined on X homogeneous of degree m_n. Define

$$C := \{x \in X \ : \ \text{there exists } (x_n, y_n) \in A_n \text{ with } x_n \to x, \ y_n \to 0\}$$

and let

$$X_0 := \overline{\bigcup_{\lambda > 0} \lambda C}.$$

We assume that the degrees of homogeneity tend to infinity,

(A.28) $$\lim_{n \to \infty} m_n = \infty,$$

and furthermore that the limit

(A.29) $$Px := \lim_{n \to \infty} (I + A_n)^{-1} x$$

exists in X, for all $x \in X_0$.

THEOREM A.54. *Under the assumptions* (A.28) *and* (A.29), *there exists a nonlinear operator* $Q : X_0 \to C$ *such that if* $x_n \in \overline{D(A_n)}$ *for* $n \in \mathbb{N}$ *and* $x_n \to x \in X_0$, *then*

$$e^{-tA_n} x_n \to Qx,$$

uniformly for t *in compact subsets of* $(0, \infty)$. *More precisely, we assert that*

   (i) *C is a closed subset, $\lambda C \subset C$ for each $\lambda \in [0, 1]$. The mapping P is a contraction of X_0 onto C, with $Px = x$ for each $x \in C$.*
   (ii) *The operator $A := P^{-1} - I$ is an accretive operator on X with $D(A) = C$, $X_0 \subset R(I + \lambda A)$ and $P = (I + \lambda A)^{-1}$ on X_0 for each $\lambda > 0$.*
   (iii) *Q is a contraction of X_0 onto C, and $Qx = x$ for $x \in C$. If $x \in \lambda C$ for some $\lambda > 1$, we have*

$$Qx = v(1),$$

   where v is the unique mild solution of the evolution problem

$$\begin{cases} v' + Av \ni \dfrac{v}{t} & on \ (\delta, \infty), \\ v(\delta) = \delta x, \end{cases}$$

   for $\delta := \lambda^{-1} \in (0, 1)$.

To finish we summarize some results about T-accretive operators.

DEFINITION A.55. Let X be a Banach lattice and $S : D(S) \subset X \to X$. We say that S is a T-*contraction* if

$$\|(Su - Sv)^+\| \leq \|(u - v)^+\| \quad \text{for } u, v \in D(S).$$

Let A be an operator in X. We say that A is T-*accretive* if

$$\|(u - \hat{u})^+\| \leq \|(u - \hat{u} + \lambda(v - \hat{v}))^+\| \quad \text{for } (u, v), (\hat{u}, \hat{v}) \in A \text{ and } \lambda > 0.$$

It is clear that a T-contraction is order-preserving; also if A is a T-accretive operator, then its resolvents $(I + \lambda A)^{-1}$ are single-valued and order-preserving. Indeed, A is T-accretive if and only if its resolvents are T-contractions. Contractions are not in general T-contractions and conversely. Actually, T-contractions are contractions if the norm satisfies

$$\|u^+\| \le \|v^+\| \text{ and } \|u^-\| \le \|v^-\| \implies \|u\| \le \|v\|$$

for $u, v \in X$. This is the case for the spaces $X = L^p(\Omega)$ for $1 \le p \le \infty$. Therefore, in $L^p(\Omega)$ every T-accretive operator is an accretive operator and also every completely accretive operator is a T-accretive operator. The mild solutions of the abstract Cauchy problems associated with T-accretive operators satisfy a contraction principle. More precisely, we have the following result.

THEOREM A.56. *Let X be a Banach lattice and A an m-accretive operator in X. Then, the following assertions are equivalent:*

(i) *A is T-accretive.*
(ii) *If $f, \hat{f} \in L^1(0, T; X)$, and u, $\hat{u}$ are mild solutions of $u' + Au \ni f$ and $\hat{u}' + A\hat{u} \ni \hat{f}$ on $[0, T]$, then, for $0 \le s \le t \le T$,*

$$\|(u(t) - \hat{u}(t))^+\| \le \|(u(s) - \hat{u}(s))^+\| + \int_s^t [u(\tau) - \hat{u}(\tau), f(\tau) - \hat{f}(\tau)]_+\| \, d\tau,$$

where

$$[u, v]_+ := \lim_{\lambda \downarrow 0} \frac{\|(u + \lambda v)^+\| - \|u^+\|}{\lambda}.$$

Bibliography

[1] G. Alberti and G. Bellettini, *A nonlocal anisotropic model for phase transition. Part I: the Optimal Profile problem*. Math. Ann. **310** (1998), 527–560.

[2] G. Alberti and G. Bellettini, *A nonlocal anisopropic model for phase transition: asymptotic behaviour of rescaled*. European J. Appl. Math. **9** (1998), 261–284.

[3] N. Alibaud and C. Imbert, *Fractional semi-linear parabolic equations with unbounded data*. Trans. Amer. Math. Soc. **361** (2009), 2527–2566.

[4] L. Ambrosio, N. Fusco and D. Pallara, *Functions of Bounded Variation and Free Discontinuity Problems*. Oxford Mathematical Monographs, 2000.

[5] F. Andreu, C. Ballester, V. Caselles and J. M. Mazón, *Minimizing total variation flow*. Differential Integral Equations **14** (2001), 321–360.

[6] F. Andreu, C. Ballester, V. Caselles and J. M. Mazón, *The Dirichlet problem for the total variational flow*. J. Funct. Anal. **180** (2001), 347–403.

[7] F. Andreu, V. Caselles, and J.M. Mazón, *Parabolic Quasilinear Equations Minimizing Linear Growth Functionals*. Progress in Mathematics, vol. 223, Birkhäuser, 2004.

[8] F. Andreu, N. Igbida, J. M. Mazón and J. Toledo, *A degenerate elliptic-parabolic problem with nonlinear dynamical boundary conditions*. Interfaces Free Bound. **8** (2006), 447–479.

[9] F. Andreu, N. Igbida, J. M. Mazón and J. Toledo, *Obstacle problems for degenerate elliptic equations with nonlinear boundary conditions*. Math. Models Methods Appl. Sci. **18** (2008), 1–25.

[10] F. Andreu, N. Igbida, J. M. Mazón and J. Toledo, L^1 *existence and uniqueness results for quasi-linear elliptic equations with nonlinear boundary conditions*. Ann. Inst. H. Poincaré Anal. Non Linéaire **24** (2007), 61–89.

[11] F. Andreu, N. Igbida, J. M. Mazón and J. Toledo, *Renormalized solutions for degenerate elliptic-parabolic porblems with nonlinear dynamical boundary conditions and L^1-data*. J. Differential Equations **244** (2008), 2764–2803.

[12] F. Andreu, J. M. Mazón, S. Segura and J. Toledo, *Quasilinear elliptic and parabolic equations in L^1 with nonlinear boundary conditions*. Adv. Math. Sci. Appl. **7** (1997), 183–213.

[13] F. Andreu, J.M. Mazón, S. Segura and J. Toledo, *Existence and uniqueness for a degenerate parabolic equation with L^1-data*. Trans. Amer. Math. Soc. **351(1)** (1999), 285–306.

[14] F. Andreu, J. M. Mazón, J. D. Rossi and J. Toledo, *The Neumann problem for nonlocal nonlinear diffusion equations*. J. Evol. Equ. **8(1)** (2008), 189–215.

[15] F. Andreu, J. M. Mazón, J. D. Rossi and J. Toledo, *A nonlocal p-Laplacian evolution equation with Neumann boundary conditions*. J. Math. Pures Appl. (9) **90(2)** (2008), 201–227.

[16] F. Andreu, J. M. Mazón, J. D. Rossi and J. Toledo, *The limit as $p \to \infty$ in a nonlocal p-Laplacian evolution equation. A nonlocal approximation of a model for sandpiles*. Calc. Var. Partial Differential Equations **35** (2009), 279–316.

[17] F. Andreu, J. M. Mazón, J. D. Rossi and J. Toledo, *A nonlocal p-Laplacian evolution equation with non homogeneous Dirichlet boundary conditions*. SIAM J. Math. Anal. **40** (2009), 1815–1851.

[18] F. Andreu, J. M. Mazón, J. D. Rossi and J. Toledo. *Local and nonlocal weighted p-Laplacian evolution equations with Neumann boundary conditions*. To appear in Pub. Mat.

[19] F. Andreu, J.M. Mazón and J. Toledo, *Stabilization of solutions of the filtration equation with absorption and non-linear flux*. Nonlinear Differential Equations Appl. **2** (1995), 267–289.

[20] G. Anzellotti, *Pairings between measures and bounded functions and compensated compactness*. Ann. Mat. Pura Appl. (4) **135** (1983), 293–318.

[21] D. Applebaum, *Lévy Processes and Stochastic Calculus*. Cambridge Studies in Advanced Mathematics, vol. 93, 2004.

[22] I. S. Aranson and L. S. Tsimring, *Patterns and collective behavior in granular media: theoretical concepts*. Rev. Mod. Phy. **788** (2006), 641–692.

[23] G. Aronsson, L. C. Evans and Y. Wu. *Fast/slow diffusion and growing sandpiles*. J. Differential Equations, **131** (1996), 304–335.

[24] I. Athanasopoulos, L. Caffarelli and S. Salsa, *The structure of the free boundary for lower dimensional obstacle problems*. Amer. J. Math. **130** (2008), 485–498.

[25] H. Attouch, *Familles d'opérateurs maximaux monotones et mesurabilité*. Ann. Mat. Pura Appl. (4) **120** (1979), 35–111.

[26] C. Ballester, M. Bertalmio, V. Caselles, G. Sapiro and J. Verdera, *Filling-in by joint interpolation of vector fields and gray levels*. IEEE Trans. Image Process. **10(8)** (2001), 1200–1211.

[27] V. Barbu, *Nonlinear Semigroups and Differential Equations in Banach Spaces*. Noordhoff International Publisher, 1976.

[28] G. Barles, E. Chasseigne and C. Imbert, *On the Dirichlet problem for second-order elliptic integro-differential equations*. Indiana Univ. Math. J. **57** (2008), 213–246.

[29] G. Barles and C. Imbert, *Second-order elliptic integro-differential equations: Viscosity solutions theory revisited*. Ann. Inst. H. Poincaré Anal. Non Linéaire **25** (2008), 567–585.

[30] J. W. Barrett and L. Prigozhin, *Dual formulations in critical state problems*. Interfaces Free Bound. **8** (2006), 349–370.

[31] S. Barza, V. Burenkov, J. Pecaric and L.-E. Persson, *Sharp multidimensional multiplicative inequalities for weighted L_p spaces with homogeneous weights*. Math. Inequal. Appl. **1(1)** (1998), 53–67.

[32] P. Bates, *On some nonlocal evolution equations arising in materials science*. Nonlinear dynamics and evolution equations, 13–52, Fields Inst. Commun., 48, Amer. Math. Soc., Providence, RI, 2006.

[33] P. Bates and F. Chen, *Spectral analysis of traveling waves for nonlocal evolution equations*. SIAM J. Math. Anal. **38** (2006), 116–126.

[34] P. Bates and F. Chen, *Spectral analysis and multidimensional stability of traveling waves for nonlocal Allen-Cahn equation*. J. Math. Anal. Appl. **273** (2002), 45–57.

[35] P. Bates and F. Chen, *Periodic traveling waves for a nonlocal integro-differential model*. Electron. J. Differential Equations **1999 (26)** (1999), 1–19.

[36] P. Bates, X. Chen and A. Chmaj, *Heteroclinic solutions of a van der Waals model with indefinite nonlocal interactions*. Calc. Var. Partial Differential Equations **24** (2005), 261–281.

[37] P. Bates and A. Chmaj, *An integrodifferential model for phase transitions: stationary solutions in higher dimensions*. J. Statistical Phys. **95** (1999), 1119–1139.

[38] P. Bates and A. Chmaj, *A discrete convolution model for phase transitions*. Arch. Ration. Mech. Anal. **150** (1999), 281–305.

[39] P. Bates, P. Fife, X. Ren and X. Wang, *Travelling waves in a convolution model for phase transitions*. Arch. Ration. Mech. Anal. **138** (1997), 105–136.

[40] P. Bates, J. Han and G. Zhao, *On a nonlocal phase-field system*. Nonlinear Anal. **64** (2006), 2251–2278.

[41] P. Bates and C. Zhang, *Traveling pulses for the Klein-Gordon equation on a lattice or continuum with long-range interaction*. Discrete Contin. Dyn. Syst. **16** (2006), 235–252.

[42] P. Bates and G. Zhao, *Existence, uniqueness and stability of the stationary solution to a nonlocal evolution equation arising in population dispersal*. J. Math. Anal. Appl. **332** (2007), 428–440.

[43] Ph. Bénilan, Equations d'évolution dans un espace de Banach quelconque et applications. Thesis, Univ. Orsay, 1972.

[44] Ph. Bénilan, L. Boccardo, T. Gallouet, R. Gariepy, M. Pierre and J. L. Vázquez, *An L^1-theory of existence and uniqueness of solutions of nonlinear elliptic equations*. Ann. Sc. Norm. Super. Pisa Cl. Sci. (4) **XXII** (1995), 241–273.

[45] Ph. Bénilan and M. G. Crandall, *Completely accretive operators*. In *Semigroup Theory and Evolution Equations (Delft, 1989)*, volume 135 of *Lecture Notes in Pure and Appl. Math.*, pages 41–75, Dekker, New York, 1991.

[46] Ph. Bénilan, M. G. Crandall and A. Pazy. *Evolution Equations Governed by Accretive Operators*. Book to appear.

[47] Ph. Bénilan, L. C. Evans and R. F. Gariepy, *On some singular limits of homogeneous semigroups.* J. Evol. Equ. **3** (2003), 203–214.

[48] J. Bertoin, *Lévy processes.* Cambridge Tracts in Mathematics, vol. 121. Cambridge University Press, Cambridge, 1996.

[49] P. Biler, C. Imbert and G. Karch, *Fractal porous medium equation.* Preprint.

[50] I. H. Biswas, E. R. Jakobsen and K. H. Karlsen, *Error estimates for finite difference-quadrature schemes for fully nonlinear degenerate parabolic integro-PDEs.* J. Hyperbolic Differ. Equ. **5(1)** (2008), 187–219.

[51] M. Bodnar and J. J. L. Velázquez, *An integro-differential equation arising as a limit of individual cell-based models.* J. Differential Equations **222** (2006), 341–380.

[52] J. Bourgain, H. Brezis and P. Mironescu, *Another look at Sobolev spaces.* In: Menaldi, J. L. et al. (eds.), Optimal control and Partial Differential Equations. A volume in honour of A. Bensoussan's 60th birthday, pages 439–455, IOS Press, 2001.

[53] C. Brändle and E. Chasseigne, *Large deviations estimates for some non-local equations. Fast decaying kernels and explicit bounds.* Nonlinear Anal. **71** (2009), 5572–5586.

[54] P. Brenner, V. Thomée and L. B. Wahlbin, *Besov spaces and applications to difference methods for initial value problems,* Lecture Notes in Mathematics, vol. 434, Springer-Verlag, Berlin, 1975.

[55] H. Brezis, *Équations et inéquations non linéaires dans les espaces vectoriels en dualité.* Ann. Inst. Fourier (Grenoble) **18** (1968), 115–175.

[56] H. Brezis, *Opérateur Maximaux Monotones et Semi-groupes de Contractions dans les Espaces de Hilbert.* North-Holland, 1973.

[57] H. Brezis, *Analyse fonctionnelle. Théorie et applications.* Masson, 1978.

[58] H. Brezis, *Problèmes Unilatéraux.* J. Math. Pures Appl. (9) **51** (1992), 1–169.

[59] H. Brezis and A. Pazy, *Convergence and approximation of semigroups of nonlinear operators in Banach spaces.* J. Funct. Anal. **9** (1972), 63–74.

[60] H. Brezis and W. Strauss, *Semilinear elliptic equations in L^1.* J. Math. Soc. Japan **25** (1973), 565–590.

[61] A. Buades, B. Coll and J. M. Morel, *Neighborhood filters and PDE's.* Numer. Math. **150** (2006), 1–34.

[62] L. Caffarelli and L. Silvestre, *An extension problem related to the fractional Laplacian.* Comm. Partial Differential Equations **32** (2007), 1245–1260.

[63] L. Caffarelli and L. Silvestre, *Regularity theory for fully nonlinear integro-differential equations.* Comm. Pure Appl. Math. **62** (2009), 597–638.

[64] L. Caffarelli, S. Salsa and L. Silvestre, *Regularity estimates for the solution and the free boundary of the obstacle problem for the fractional Laplacian.* Invent. Math. **171** (2008), 425–461.

[65] C. Carrillo and P. Fife, *Spatial effects in discrete generation population models.* J. Math. Biol. **50(2)** (2005), 161–188.

[66] T. Cazenave and A. Haraux, *An Introduction to Semilinear Evolution Equations.* Transl. by Yvan Martel. Revised ed. Oxford Lecture Series in Mathematics and its Applications. 13. Oxford: Clarendon Press. xiv, 1998.

[67] E. Chasseigne, *The Dirichlet problem for some nonlocal diffusion equations.* Differential Integral Equations **20** (2007), 1389–1404.

[68] E. Chasseigne, M. Chaves and J. D. Rossi, *Asymptotic behaviour for nonlocal diffusion equations.* J. Math. Pures Appl. (9) **86** (2006), 271–291.

[69] X. Chen, *Existence, uniqueness and asymptotic stability of travelling waves in nonlocal evolution equations.* Adv. Differential Equations **2** (1997), 125–160.

[70] A. Chmaj, *Existence of traveling waves for the nonlocal Burgers equation.* Appl. Math. Lett. **20** (2007), 439–444.

[71] A. Chmaj and X. Ren, *Homoclinic solutions of an integral equation: existence and stability.* J. Differential Equations **155** (1999), 17–43.

[72] A. Chmaj and X. Ren. *Multiple layered solutions of the nonlocal bistable equation.* Phys. D **147** (2000), 135–154.

[73] A. Chmaj and X. Ren, *The nonlocal bistable equation: stationary solutions on a bounded interval.* Electronic J. Differential Equations **2002 (2)** (2002), 1–12.

[74] A. Cordoba and D. Cordoba, *A maximum principle applied to quasi-geostrophic equations.* Comm. Math. Phys. **249** (2004), 511–528.

[75] C. Cortázar, J. Coville, M. Elgueta and S. Martínez, *A non local inhomogeneous dispersal process*. J. Differential Equations **241** (2007), 332–358.

[76] C. Cortázar, M. Elgueta, S. Martínez and J. D. Rossi, *Random walks and the porous medium equation*. Rev. Unión Matemática Argentina **50** (2009), 149–155.

[77] C. Cortázar, M. Elgueta and J. D. Rossi, *A non-local diffusion equation whose solutions develop a free boundary*. Ann. Henri Poincaré **6(2)** (2005), 269–281.

[78] C. Cortázar, M. Elgueta and J. D. Rossi, *Nonlocal diffusion problems that approximate the heat equation with Dirichlet boundary conditions*. Israel J. Math. **170** (2009), 53–60.

[79] C. Cortázar, M. Elgueta, J. D. Rossi and N. Wolanski, *Boundary fluxes for non-local diffusion*. J. Differential Equations **234** (2007), 360–390.

[80] C. Cortázar, M. Elgueta, J. D. Rossi and N. Wolanski, *How to approximate the heat equation with Neumann boundary conditions by nonlocal diffusion problems*. Arch. Ration. Mech. Anal. **187(1)** (2008), 137–156.

[81] J. Coville, *On uniqueness and monotonicity of solutions of non-local reaction diffusion equations*. Ann. Mat. Pura Appl. (4) **185** (2006), 461–485.

[82] J. Coville, *Maximum principles, sliding techniques and applications to nonlocal equations*. Electron. J. Differential Equations **2007 (68)** (2007), 1–23.

[83] J. Coville, J. Dávila and S. Martínez, *Existence and uniqueness of solutions to a nonlocal equation with monostable nonlinearity*. SIAM J. Math. Anal. **39** (2008), 1693–1709.

[84] J. Coville, J. Dávila and S. Martínez, *Nonlocal anisotropic dispersal with monostable nonlinearity*. J. Differential Equations **244** (2008), 3080–3118.

[85] J. Coville and L. Dupaigne, *Propagation speed of travelling fronts in nonlocal reaction diffusion equations*. Nonlinear Anal. **60** (2005), 797–819.

[86] J. Coville and L. Dupaigne, *On a nonlocal equation arising in population dynamics*. Proc. Roy. Soc. Edinburgh Sect. A **137** (2007), 1–29.

[87] M. G. Crandall, *An introduction to evolution governed by accretive operators*. In Dynamical Systems (Proc. Internat. Sympos., Brown Univ., Providence, R.I., 1974), vol. I, pages 131–165. Academic Press, New York, 1976.

[88] M. G. Crandall, *Nonlinear Semigroups and Evolution Governed by Accretive Operators*. In Proc. of Sympos. in Pure Mathematics, Part I, vol. 45 (F. Browder ed.). A.M.S., Providence, 1986, pages 305–338.

[89] M. G. Crandall and T. M. Liggett, *Generation of Semigroups of Nonlinear Transformations on General Banach Spaces*. Amer. J. Math. **93** (1971), 265–298.

[90] J. Crank, *Free and Moving Boundary Problems*. North-Holland, Amsterdam, 1977.

[91] C. M. Dafermos, *Asymptotic Behavior of Solutions of Evolution Equations*. In *Nonlinear Evolution Equations* (M. G. Crandall, ed.) Academic Press, 1978, pp. 103–123.

[92] A. de Pablo, F. Quirós, A. Rodríguez and J. L. Vázquez, *A fractional porous medium equation*. Preprint.

[93] E. DiBenedetto and A. Friedman, *The ill-posed Hele-Shaw model and the Stefan problem for supercooler water*. Trans. Amer. Math. Soc. **282** (1984), 183–204.

[94] S. Dumont and N. Igbida, *On a dual formulation for the growing sandpile problem*. European J. Appl. Math. **20** (2009), 169–185.

[95] S. Dumont and N. Igbida, *On the collapsing sandpile problem*. Preprint.

[96] J. Duoandikoetxea and E. Zuazua, *Moments, masses de Dirac et decomposition de fonctions. (Moments, Dirac deltas and expansion of functions)*. C. R. Acad. Sci. Paris Sér. I Math. **315(6)** (1992), 693–698.

[97] C. M. Elliot and V. Janosky, *A variational inequality approach to the Hele-Shaw flow with a moving boundary*. Proc. Roy. Soc. Edinburgh Sect. A **88** (1981), 93–107.

[98] M. Escobedo and E. Zuazua, *Large time behavior for convection-diffusion equations in $\mathbb{R}^N$*. J. Funct. Anal. **100(1)** (1991), 119–161.

[99] L. C. Evans, *Partial differential equations and Monge-Kantorovich mass transfer*. Current Developments in Mathematics, 1997 (Cambridge, MA), pp. 65–126, Int. Press, Boston, MA, 1999.

[100] L. C. Evans, M. Feldman and R. F. Gariepy, *Fast/slow diffusion and collapsing sandpiles*. J. Differential Equations **137** (1997), 166–209.

[101] L. C. Evans and R. F. Gariepy, *Measure Theory and Fine Properties of Functions*, Studies in Advanced Math., CRC Press, 1992.

[102] L. C. Evans and Fr. Rezakhanlou, *A stochastic model for growing sandpiles and its continuum limit.* Comm. Math. Phys. **197** (1998), 325–345.

[103] M. Feldman, *Growth of a sandpile around an obstacle.* Monge Ampère Equation: Applications to Geometry and Optimization (Deerfield Beach, FL, 1997), pp. 55–78, Contemp. Math., 226, Amer. Math. Soc., Providence, RI, 1999.

[104] W. Feller, *An Introduction to the Probability Theory and its Applications* (vol. 2). Wiley, New York, 1971.

[105] P. Fife, *Travelling waves for a nonlocal doble-obstacle problem.* Euro. J. Applied Mathematics **8** (1997), 581–594.

[106] P. Fife, *Some nonclassical trends in parabolic and parabolic-like evolutions.* Trends in Nonlinear Analysis, pp. 153–191, Springer, Berlin, 2003.

[107] P. Fife and X. Wang, *A convolution model for interfacial motion: the generation and propagation of internal layers in higher space dimensions.* Adv. Differential Equations **3(1)** (1998), 85–110.

[108] N. Fournier and P. Laurençot, *Well-posedness of Smoluchowski's coagulation equation for a class of homogeneous kernels.* J. Funct. Anal. **233** (2006) 351–379.

[109] A. Friedman, *Partial Differential Equations of Parabolic Type.* Prentice-Hall, Englewood Cliffs, NJ, 1964.

[110] G. Gilboa and S. Osher, *Nonlocal linear image regularization and supervised segmentation.* UCLA CAM Report 06-47, (2006).

[111] M. E. Gurtin and R. C. MacCamy, *On the diffusion of biological populations.* Math. Biosci. **33** (1977), 35–49.

[112] J. Han, *The Cauchy problem and steady state solutions for a nonlocal Cahn-Hilliard equation.* Electronic J. Differential Equations **2004 (113)** (2004), 1–9.

[113] L. Hörmander, *The Analysis of Linear Partial Diferrential Operator I.* Springer-Verlag, 1983.

[114] V. Hutson and M. Grinfeld, *Non-local dispersal and bistability.* Euro. J. Applied Mathematics **17** (2006), 221–232.

[115] V. Hutson, S. Martínez, K. Mischaikow and G. T. Vickers, *The evolution of dispersal.* J. Math. Biol. **47** (2003), 483–517.

[116] N. Igbida, *A generalized collapsing sandpile model.* Arch. Math. **94** (2010), 193–200.

[117] N. Igbida, *Back on Stochastic Model for Sandpile.* Recent Developments in Nonlinear Analysis: Proceedings of the Conference in Mathematics and Mathematical Physics, Morocco, 2008. Edited by H. Ammari, A. Benkirane, and A. Touzani (2010).

[118] N. Igbida, *Partial integro-differential equations in granular matter and its connection with stochastic model.* Preprint.

[119] N. Igbida, J. M. Mazón, J. D. Rossi and J. Toledo, *A nonlocal Monge-Kantorovich problem.* Preprint.

[120] L. I. Ignat and J. D. Rossi, *A nonlocal convection-diffusion equation.* J. Funct. Anal. **251(2)** (2007), 399–437.

[121] L. I. Ignat and J. D. Rossi, *Refined asymptotic expansions for nonlocal diffusion equations.* J. Evol. Equ. **8** (2008), 617–629.

[122] L. I. Ignat and J. D. Rossi, *Asymptotic behaviour for a nonlocal diffusion equation on a lattice.* Z. Angew. Math. Phys. **59(5)** (2008), 918–925.

[123] C. Imbert, *A non-local regularization of first order Hamilton-Jacobi equations.* J. Differential Equations **211** (2005), 218–246.

[124] C. Imbert, *Level set approach for fractional mean curvature flows.* Interfaces Free Bound. **11** (2009), 153–176.

[125] H. Ishii and G. Nakamura, *A class of integral equations and approximation of p-Laplace equations.* Calc. Var. Partial Differential Equations **37** (2010), 485–522.

[126] E. R. Jakobsen and K. H. Karlsen, *Continuous dependence estimates for viscosity solutions of integro-PDEs.* J. Differential Equations **212** (2005), 278–318.

[127] D. Kinderlehrer and G. Stampacchia, *An Introduction to Variational Inequalities and their Applications.* Academic Press, 1980.

[128] Y. Komura, *Nonlinear semigroups in Hilbert spaces.* J. Math. Soc. Japan **19** (1967), 493–507.

[129] S. Kindermann, S. Osher and P. W. Jones, *Deblurring and denoising of images by nonlocal functionals.* Multiscale Model. Simul. **4** (2005), 1091–1115.

[130] N. S. Landkof, *Foundations of Modern Potential Theory*. New York: Springer- Verlag, 1972.

[131] C. Lederman and N. Wolanski, *Singular perturbation in a nonlocal diffusion problem*. Comm. Partial Differential Equations **31** (2006), 195–241.

[132] G. Minty, *Monotone (nonlinear) operators in Hilbert space*. Duke Math J. **29** (1962), 341–346.

[133] A. Mogilner and Leah Edelstein-Keshet, *A non-local model for a swarm*. J. Math. Biol. **38** (1999), 534–570

[134] U. Mosco, *Convergence of convex sets and solutions of variational inequalities*. Advances Math. **3** (1969), 510–585.

[135] Y. Nishiura, and I. Ohnishi, *Some mathematical aspects of the micro-phase separation in diblock copolymers*. Phys. D **84**, (1995), 31–39.

[136] T. Ohta, and K. Kawasaki, *Equilibrium Morphology of Block Copolymer Melts*. Macromolecules **19** (1986), 2621–2632.

[137] A. F. Pazoto and J. D. Rossi, *Asymptotic behavior for a semilinear nonlocal equation*. Asymptot. Anal. **52** (2007), 143–155.

[138] L. Prigozhin, *Sandpiles and river networks extended systems with nonlocal interactions*. Phys. Rev. E **49** (1994), 1161–1167.

[139] L. Prigozhin, *Variational models of sandpile growth*. Euro. J. Applied Mathematics **7** (1996), 225–236.

[140] J. D. Rossi and C. B. Schönlieb, *Nonlocal higher order evolution equations*. To appear in Appl. Anal.

[141] L. Rudin, S. Osher and E. Fatemi, *Nonlinear Total Variation based Noise Removal Algorithms*. Phys. D **60** (1992), 259–268.

[142] M. Schonbek. *Decay of solutions to parabolic conservation laws*. Comm. Partial Differential Equations **5(5)** (1980), 449–473.

[143] M. Schonbek. *Uniform decay rates for parabolic conservation laws*. Nonlinear Anal. **10(9)** (1986), 943–956.

[144] M. Schonbek, *The Fourier splitting method*. Advances in Geometric Analysis and Continuum Mechanics (Stanford, CA, 1993), Int. Press, Cambridge, MA, 1995, pp. 269–274.

[145] K. Schumacher, *Travelling-front solutions for integro-differential equations I*. J. Reine Angew. Math. **316** (1980), 54–70.

[146] L. Silvestre, *Holder estimates for solutions of integro-differential equations like the fractional laplace*. Indiana Univ. Math. J. **55(3)** (2006), 1155–1174.

[147] E. M. Stein, *Singular Integrals and Differentiabily Properties of Functions*. Princeton University Press, Princeton, 1970.

[148] E. M. Stein and G. Weiss, *Introduction to Fourier Analysis on Euclidean Spaces*. Princeton University Press, Princeton, 1971.

[149] C. Tomasi and R. Manduchi, *Bilateral filtering for gray and color images*. In Proceedings of the Sixth International Conference on Computer Vision, Bombay, India, 1998, pp. 839–846.

[150] J. L. Vázquez, *The porous medium equation. Mathematical theory*. Oxford Mathematical Monographs. The Clarendon Press, Oxford University Press, Oxford, 2007.

[151] C. Villani, *Topics in Optimal Transportation*. Graduate Studies in Mathematics. vol. 58, 2003.

[152] G. F. Webb, *Continuous nonlinear perturbations of linear accretive operators in Banach spaces*. J. Funct. Anal. **10** (1972), 191–203.

[153] L. P. Yaroslavsky, *Digital Picture Processing. An Introduction*. Springer, Berlin, 1985.

[154] L. P. Yaroslavsky and M. Eden, *Fundamentals of Digital Optics*. Birkhäuser, Boston, 1996.

[155] L. Zhang, *Existence, uniqueness and exponential stability of traveling wave solutions of some integral differential equations arising from neural networks*. J. Differential Equations **197** (2004), 162–196.

[156] W. P. Ziemer, *Weakly Differentiable Functions*. GTM 120, Springer-Verlag, 1989.

Index

Mengesha, T. 2017

U.T.K